COGALOIS THEORY

PURE AND APPLIED MATHEMATICS

A Program of Monographs, Textbooks, and Lecture Notes

MONOGRAPHS AND TEXTBOOKS IN
PURE AND APPLIED MATHEMATICS

1. *K. Yano,* Integral Formulas in Riemannian Geometry (1970)
2. *S. Kobayashi,* Hyperbolic Manifolds and Holomorphic Mappings (1970)
3. *V. S. Vladimirov,* Equations of Mathematical Physics (A. Jeffrey, ed.; A. Littlewood, trans.) (1970)
4. *B. N. Pshenichnyi,* Necessary Conditions for an Extremum (L. Neustadt, translation ed.; K. Makowski, trans.) (1971)
5. *L. Narici et al.,* Functional Analysis and Valuation Theory (1971)
6. *S. S. Passman,* Infinite Group Rings (1971)
7. *L. Dornhoff,* Group Representation Theory. Part A: Ordinary Representation Theory. Part B: Modular Representation Theory (1971, 1972)
8. *W. Boothby and G. L. Weiss, eds.,* Symmetric Spaces (1972)
9. *Y. Matsushima,* Differentiable Manifolds (E. T. Kobayashi, trans.) (1972)
10. *L. E. Ward, Jr.,* Topology (1972)
11. *A. Babakhanian,* Cohomological Methods in Group Theory (1972)
12. *R. Gilmer,* Multiplicative Ideal Theory (1972)
13. *J. Yeh,* Stochastic Processes and the Wiener Integral (1973)
14. *J. Barros-Neto,* Introduction to the Theory of Distributions (1973)
15. *R. Larsen,* Functional Analysis (1973)
16. *K. Yano and S. Ishihara,* Tangent and Cotangent Bundles (1973)
17. *C. Procesi,* Rings with Polynomial Identities (1973)
18. *R. Hermann,* Geometry, Physics, and Systems (1973)
19. *N. R. Wallach,* Harmonic Analysis on Homogeneous Spaces (1973)
20. *J. Dieudonné,* Introduction to the Theory of Formal Groups (1973)
21. *I. Vaisman,* Cohomology and Differential Forms (1973)
22. *B.-Y. Chen,* Geometry of Submanifolds (1973)
23. *M. Marcus,* Finite Dimensional Multilinear Algebra (in two parts) (1973, 1975)
24. *R. Larsen,* Banach Algebras (1973)
25. *R. O. Kujala and A. L. Vitter, eds.,* Value Distribution Theory: Part A; Part B: Deficit and Bezout Estimates by Wilhelm Stoll (1973)
26. *K. B. Stolarsky,* Algebraic Numbers and Diophantine Approximation (1974)
27. *A. R. Magid,* The Separable Galois Theory of Commutative Rings (1974)
28. *B. R. McDonald,* Finite Rings with Identity (1974)
29. *J. Satake,* Linear Algebra (S. Koh et al., trans.) (1975)
30. *J. S. Golan,* Localization of Noncommutative Rings (1975)
31. *G. Klambauer,* Mathematical Analysis (1975)
32. *M. K. Agoston,* Algebraic Topology (1976)
33. *K. R. Goodearl,* Ring Theory (1976)
34. *L. E. Mansfield,* Linear Algebra with Geometric Applications (1976)
35. *N. J. Pullman,* Matrix Theory and Its Applications (1976)
36. *B. R. McDonald,* Geometric Algebra Over Local Rings (1976)
37. *C. W. Groetsch,* Generalized Inverses of Linear Operators (1977)
38. *J. E. Kuczkowski and J. L. Gersting,* Abstract Algebra (1977)
39. *C. O. Christenson and W. L. Voxman,* Aspects of Topology (1977)
40. *M. Nagata,* Field Theory (1977)
41. *R. L. Long,* Algebraic Number Theory (1977)
42. *W. F. Pfeffer,* Integrals and Measures (1977)
43. *R. L. Wheeden and A. Zygmund,* Measure and Integral (1977)
44. *J. H. Curtiss,* Introduction to Functions of a Complex Variable (1978)
45. *K. Hrbacek and T. Jech,* Introduction to Set Theory (1978)
46. *W. S. Massey,* Homology and Cohomology Theory (1978)
47. *M. Marcus,* Introduction to Modern Algebra (1978)
48. *E. C. Young,* Vector and Tensor Analysis (1978)
49. *S. B. Nadler, Jr.,* Hyperspaces of Sets (1978)
50. *S. K. Segal,* Topics in Group Kings (1978)
51. *A. C. M. van Rooij,* Non-Archimedean Functional Analysis (1978)
52. *L. Corwin and R. Szczarba,* Calculus in Vector Spaces (1979)
53. *C. Sadosky,* Interpolation of Operators and Singular Integrals (1979)
54. *J. Cronin,* Differential Equations (1980)
55. *C. W. Groetsch,* Elements of Applicable Functional Analysis (1980)

56. *I. Vaisman,* Foundations of Three-Dimensional Euclidean Geometry (1980)
57. *H. I. Freedan,* Deterministic Mathematical Models in Population Ecology (1980)
58. *S. B. Chae,* Lebesgue Integration (1980)
59. *C. S. Rees et al.,* Theory and Applications of Fourier Analysis (1981)
60. *L. Nachbin,* Introduction to Functional Analysis (R. M. Aron, trans.) (1981)
61. *G. Orzech and M. Orzech,* Plane Algebraic Curves (1981)
62. *R. Johnsonbaugh and W. E. Pfaffenberger,* Foundations of Mathematical Analysis (1981)
63. *W. L. Voxman and R. H. Goetschel,* Advanced Calculus (1981)
64. *L. J. Corwin and R. H. Szczarba,* Multivariable Calculus (1982)
65. *V. I. Istrătescu,* Introduction to Linear Operator Theory (1981)
66. *R. D. Järvinen,* Finite and Infinite Dimensional Linear Spaces (1981)
67. *J. K. Beem and P. E. Ehrlich,* Global Lorentzian Geometry (1981)
68. *D. L. Armacost,* The Structure of Locally Compact Abelian Groups (1981)
69. *J. W. Brewer and M. K. Smith, eds.,* Emmy Noether: A Tribute (1981)
70. *K. H. Kim,* Boolean Matrix Theory and Applications (1982)
71. *T. W. Wieting,* The Mathematical Theory of Chromatic Plane Ornaments (1982)
72. *D. B. Gauld,* Differential Topology (1982)
73. *R. L. Faber,* Foundations of Euclidean and Non-Euclidean Geometry (1983)
74. *M. Carmeli,* Statistical Theory and Random Matrices (1983)
75. *J. H. Carruth et al.,* The Theory of Topological Semigroups (1983)
76. *R. L. Faber,* Differential Geometry and Relativity Theory (1983)
77. *S. Barnett,* Polynomials and Linear Control Systems (1983)
78. *G. Karpilovsky,* Commutative Group Algebras (1983)
79. *F. Van Oystaeyen and A. Verschoren,* Relative Invariants of Rings (1983)
80. *I. Vaisman,* A First Course in Differential Geometry (1984)
81. *G. W. Swan,* Applications of Optimal Control Theory in Biomedicine (1984)
82. *T. Petrie and J. D. Randall,* Transformation Groups on Manifolds (1984)
83. *K. Goebel and S. Reich,* Uniform Convexity, Hyperbolic Geometry, and Nonexpansive Mappings (1984)
84. *T. Albu and C. Năstăsescu,* Relative Finiteness in Module Theory (1984)
85. *K. Hrbacek and T. Jech,* Introduction to Set Theory: Second Edition (1984)
86. *F. Van Oystaeyen and A. Verschoren,* Relative Invariants of Rings (1984)
87. *B. R. McDonald,* Linear Algebra Over Commutative Rings (1984)
88. *M. Namba,* Geometry of Projective Algebraic Curves (1984)
89. *G. F. Webb,* Theory of Nonlinear Age-Dependent Population Dynamics (1985)
90. *M. R. Bremner et al.,* Tables of Dominant Weight Multiplicities for Representations of Simple Lie Algebras (1985)
91. *A. E. Fekete,* Real Linear Algebra (1985)
92. *S. B. Chae,* Holomorphy and Calculus in Normed Spaces (1985)
93. *A. J. Jerri,* Introduction to Integral Equations with Applications (1985)
94. *G. Karpilovsky,* Projective Representations of Finite Groups (1985)
95. *L. Narici and E. Beckenstein,* Topological Vector Spaces (1985)
96. *J. Weeks,* The Shape of Space (1985)
97. *P. R. Gribik and K. O. Kortanek,* Extremal Methods of Operations Research (1985)
98. *J.-A. Chao and W. A. Woyczynski, eds.,* Probability Theory and Harmonic Analysis (1986)
99. *G. D. Crown et al.,* Abstract Algebra (1986)
100. *J. H. Carruth et al.,* The Theory of Topological Semigroups, Volume 2 (1986)
101. *R. S. Doran and V. A. Belfi,* Characterizations of C*-Algebras (1986)
102. *M. W. Jeter,* Mathematical Programming (1986)
103. *M. Altman,* A Unified Theory of Nonlinear Operator and Evolution Equations with Applications (1986)
104. *A. Verschoren,* Relative Invariants of Sheaves (1987)
105. *R. A. Usmani,* Applied Linear Algebra (1987)
106. *P. Blass and J. Lang,* Zariski Surfaces and Differential Equations in Characteristic $p >$ 0 (1987)
107. *J. A. Reneke et al.,* Structured Hereditary Systems (1987)
108. *H. Busemann and B. B. Phadke,* Spaces with Distinguished Geodesics (1987)
109. *R. Harte,* Invertibility and Singularity for Bounded Linear Operators (1988)
110. *G. S. Ladde et al.,* Oscillation Theory of Differential Equations with Deviating Arguments (1987)
111. *L. Dudkin et al.,* Iterative Aggregation Theory (1987)
112. *T. Okubo,* Differential Geometry (1987)

113. *D. L. Stancl and M. L. Stancl*, Real Analysis with Point-Set Topology (1987)
114. *T. C. Gard*, Introduction to Stochastic Differential Equations (1988)
115. *S. S. Abhyankar*, Enumerative Combinatorics of Young Tableaux (1988)
116. *H. Strade and R. Farnsteiner*, Modular Lie Algebras and Their Representations (1988)
117. *J. A. Huckaba*, Commutative Rings with Zero Divisors (1988)
118. *W. D. Wallis*, Combinatorial Designs (1988)
119. *W. Więsław*, Topological Fields (1988)
120. *G. Karpilovsky*, Field Theory (1988)
121. *S. Caenepeel and F. Van Oystaeyen*, Brauer Groups and the Cohomology of Graded Rings (1989)
122. *W. Kozlowski*, Modular Function Spaces (1988)
123. *E. Lowen-Colebunders*, Function Classes of Cauchy Continuous Maps (1989)
124. *M. Pavel*, Fundamentals of Pattern Recognition (1989)
125. *V. Lakshmikantham et al.*, Stability Analysis of Nonlinear Systems (1989)
126. *R. Sivaramakrishnan*, The Classical Theory of Arithmetic Functions (1989)
127. *N. A. Watson*, Parabolic Equations on an Infinite Strip (1989)
128. *K. J. Hastings*, Introduction to the Mathematics of Operations Research (1989)
129. *B. Fine*, Algebraic Theory of the Bianchi Groups (1989)
130. *D. N. Dikranjan et al.*, Topological Groups (1989)
131. *J. C. Morgan II*, Point Set Theory (1990)
132. *P. Biler and A. Witkowski*, Problems in Mathematical Analysis (1990)
133. *H. J. Sussmann*, Nonlinear Controllability and Optimal Control (1990)
134. *J.-P. Florens et al.*, Elements of Bayesian Statistics (1990)
135. *N. Shell*, Topological Fields and Near Valuations (1990)
136. *B. F. Doolin and C. F. Martin*, Introduction to Differential Geometry for Engineers (1990)
137. *S. S. Holland, Jr.*, Applied Analysis by the Hilbert Space Method (1990)
138. *J. Okniński*, Semigroup Algebras (1990)
139. *K. Zhu*, Operator Theory in Function Spaces (1990)
140. *G. B. Price*, An Introduction to Multicomplex Spaces and Functions (1991)
141. *R. B. Darst*, Introduction to Linear Programming (1991)
142. *P. L. Sachdev*, Nonlinear Ordinary Differential Equations and Their Applications (1991)
143. *T. Husain*, Orthogonal Schauder Bases (1991)
144. *J. Foran*, Fundamentals of Real Analysis (1991)
145. *W. C. Brown*, Matrices and Vector Spaces (1991)
146. *M. M. Rao and Z. D. Ren*, Theory of Orlicz Spaces (1991)
147. *J. S. Golan and T. Head*, Modules and the Structures of Rings (1991)
148. *C. Small*, Arithmetic of Finite Fields (1991)
149. *K. Yang*, Complex Algebraic Geometry (1991)
150. *D. G. Hoffman et al.*, Coding Theory (1991)
151. *M. O. González*, Classical Complex Analysis (1992)
152. *M. O. González*, Complex Analysis (1992)
153. *L. W. Baggett*, Functional Analysis (1992)
154. *M. Sniedovich*, Dynamic Programming (1992)
155. *R. P. Agarwal*, Difference Equations and Inequalities (1992)
156. *C. Brezinski*, Biorthogonality and Its Applications to Numerical Analysis (1992)
157. *C. Swartz*, An Introduction to Functional Analysis (1992)
158. *S. B. Nadler, Jr.*, Continuum Theory (1992)
159. *M. A. Al-Gwaiz*, Theory of Distributions (1992)
160. *E. Perry*, Geometry: Axiomatic Developments with Problem Solving (1992)
161. *E. Castillo and M. R. Ruiz-Cobo*, Functional Equations and Modelling in Science and Engineering (1992)
162. *A. J. Jerri*, Integral and Discrete Transforms with Applications and Error Analysis (1992)
163. *A. Charlier et al.*, Tensors and the Clifford Algebra (1992)
164. *P. Biler and T. Nadzieja*, Problems and Examples in Differential Equations (1992)
165. *E. Hansen*, Global Optimization Using Interval Analysis (1992)
166. *S. Guerre-Delabrière*, Classical Sequences in Banach Spaces (1992)
167. *Y. C. Wong*, Introductory Theory of Topological Vector Spaces (1992)
168. *S. H. Kulkarni and B. V. Limaye*, Real Function Algebras (1992)
169. *W. C. Brown*, Matrices Over Commutative Rings (1993)
170. *J. Loustau and M. Dillon*, Linear Geometry with Computer Graphics (1993)
171. *W. V. Petryshyn*, Approximation-Solvability of Nonlinear Functional and Differential Equations (1993)

172. *E. C. Young*, Vector and Tensor Analysis: Second Edition (1993)
173. *T. A. Bick*, Elementary Boundary Value Problems (1993)
174. *M. Pavel*, Fundamentals of Pattern Recognition: Second Edition (1993)
175. *S. A. Albeverio et al.*, Noncommutative Distributions (1993)
176. *W. Fulks*, Complex Variables (1993)
177. *M. M. Rao*, Conditional Measures and Applications (1993)
178. *A. Janicki and A. Weron*, Simulation and Chaotic Behavior of α-Stable Stochastic Processes (1994)
179. *P. Neittaanmäki and D. Tiba*, Optimal Control of Nonlinear Parabolic Systems (1994)
180. *J. Cronin*, Differential Equations: Introduction and Qualitative Theory, Second Edition (1994)
181. *S. Heikkilä and V. Lakshmikantham*, Monotone Iterative Techniques for Discontinuous Nonlinear Differential Equations (1994)
182. *X. Mao*, Exponential Stability of Stochastic Differential Equations (1994)
183. *B. S. Thomson*, Symmetric Properties of Real Functions (1994)
184. *J. E. Rubio*, Optimization and Nonstandard Analysis (1994)
185. *J. L. Bueso et al.*, Compatibility, Stability, and Sheaves (1995)
186. *A. N. Michel and K. Wang*, Qualitative Theory of Dynamical Systems (1995)
187. *M. R. Darnel*, Theory of Lattice-Ordered Groups (1995)
188. *Z. Naniewicz and P. D. Panagiotopoulos*, Mathematical Theory of Hemivariational Inequalities and Applications (1995)
189. *L. J. Corwin and R. H. Szczarba*, Calculus in Vector Spaces: Second Edition (1995)
190. *L. H. Erbe et al.*, Oscillation Theory for Functional Differential Equations (1995)
191. *S. Agaian et al.*, Binary Polynomial Transforms and Nonlinear Digital Filters (1995)
192. *M. I. Gil'*, Norm Estimations for Operation-Valued Functions and Applications (1995)
193. *P. A. Grillet*, Semigroups: An Introduction to the Structure Theory (1995)
194. *S. Kichenassamy*, Nonlinear Wave Equations (1996)
195. *V. F. Krotov*, Global Methods in Optimal Control Theory (1996)
196. *K. I. Beidar et al.*, Rings with Generalized Identities (1996)
197. *V. I. Arnautov et al.*, Introduction to the Theory of Topological Rings and Modules (1996)
198. *G. Sierksma*, Linear and Integer Programming (1996)
199. *R. Lasser*, Introduction to Fourier Series (1996)
200. *V. Sima*, Algorithms for Linear-Quadratic Optimization (1996)
201. *D. Redmond*, Number Theory (1996)
202. *J. K. Beem et al.*, Global Lorentzian Geometry: Second Edition (1996)
203. *M. Fontana et al.*, Prüfer Domains (1997)
204. *H. Tanabe,* Functional Analytic Methods for Partial Differential Equations (1997)
205. *C. Q. Zhang*, Integer Flows and Cycle Covers of Graphs (1997)
206. *E. Spiegel and C. J. O'Donnell*, Incidence Algebras (1997)
207. *B. Jakubczyk and W. Respondek*, Geometry of Feedback and Optimal Control (1998)
208. *T. W. Haynes et al.*, Fundamentals of Domination in Graphs (1998)
209. *T. W. Haynes et al.*, Domination in Graphs: Advanced Topics (1998)
210. *L. A. D'Alotto et al.*, A Unified Signal Algebra Approach to Two-Dimensional Parallel Digital Signal Processing (1998)
211. *F. Halter-Koch*, Ideal Systems (1998)
212. *N. K. Govil et al.*, Approximation Theory (1998)
213. *R. Cross*, Multivalued Linear Operators (1998)
214. *A. A. Martynyuk*, Stability by Liapunov's Matrix Function Method with Applications (1998)
215. *A. Favini and A. Yagi*, Degenerate Differential Equations in Banach Spaces (1999)
216. *A. Illanes and S. Nadler, Jr.*, Hyperspaces: Fundamentals and Recent Advances (1999)
217. *G. Kato and D. Struppa*, Fundamentals of Algebraic Microlocal Analysis (1999)
218. *G. X.-Z. Yuan*, KKM Theory and Applications in Nonlinear Analysis (1999)
219. *D. Motreanu and N. H. Pavel*, Tangency, Flow Invariance for Differential Equations, and Optimization Problems (1999)
220. *K. Hrbacek and T. Jech*, Introduction to Set Theory, Third Edition (1999)
221. *G. E. Kolosov*, Optimal Design of Control Systems (1999)
222. *N. L. Johnson*, Subplane Covered Nets (2000)
223. *B. Fine and G. Rosenberger*, Algebraic Generalizations of Discrete Groups (1999)
224. *M. Väth*, Volterra and Integral Equations of Vector Functions (2000)
225. *S. S. Miller and P. T. Mocanu*, Differential Subordinations (2000)

226. *R. Li et al.*, Generalized Difference Methods for Differential Equations: Numerical Analysis of Finite Volume Methods (2000)
227. *H. Li and F. Van Oystaeyen*, A Primer of Algebraic Geometry (2000)
228. *R. P. Agarwal*, Difference Equations and Inequalities: Theory, Methods, and Applications, Second Edition (2000)
229. *A. B. Kharazishvili*, Strange Functions in Real Analysis (2000)
230. *J. M. Appell et al.*, Partial Integral Operators and Integro-Differential Equations (2000)
231. *A. I. Prilepko et al.*, Methods for Solving Inverse Problems in Mathematical Physics (2000)
232. *F. Van Oystaeyen,* Algebraic Geometry for Associative Algebras (2000)
233. *D. L. Jagerman*, Difference Equations with Applications to Queues (2000)
234. *D. R. Hankerson et al.*, Coding Theory and Cryptography: The Essentials, Second Edition, Revised and Expanded (2000)
235. *S. Dăscălescu et al.*, Hopf Algebras: An Introduction (2001)
236. *R. Hagen et al.*, C*-Algebras and Numerical Analysis (2001)
237. *Y. Talpaert*, Differential Geometry: With Applications to Mechanics and Physics (2001)
238. *R. H. Villarreal*, Monomial Algebras (2001)
239. *A. N. Michel et al.,* Qualitative Theory of Dynamical Systems: Second Edition (2001)
240. *A. A. Samarskii*, The Theory of Difference Schemes (2001)
241. *J. Knopfmacher and W.-B. Zhang*, Number Theory Arising from Finite Fields (2001)
242. *S. Leader*, The Kurzweil-Henstock Integral and Its Differentials (2001)
243. *M. Biliotti et al.,* Foundations of Translation Planes (2001)
244. *A. N. Kochubei*, Pseudo-Differential Equations and Stochastics over Non-Archimedean Fields (2001)
245. *G. Sierksma*, Linear and Integer Programming: Second Edition (2002)
246. *A. A. Martynyuk*, Qualitative Methods in Nonlinear Dynamics: Novel Approaches to Liapunov's Matrix Functions (2002)
247. *B. G. Pachpatte*, Inequalities for Finite Difference Equations (2002)
248. *A. N. Michel and D. Liu*, Qualitative Analysis and Synthesis of Recurrent Neural Networks (2002)
249. *J. R. Weeks,* The Shape of Space: Second Edition (2002)
250. *M. M. Rao and Z. D. Ren*, Applications of Orlicz Spaces (2002)
251. *V. Lakshmikantham and D. Trigiante*, Theory of Difference Equations: Numerical Methods and Applications, Second Edition (2002)
252. *T. Albu*, Cogalois Theory (2003)

Additional Volumes in Preparation

COGALOIS THEORY

TOMA ALBU
Atilim University
Ankara, Turkey, and
Bucharest University
Bucharest, Romania

MARCEL DEKKER, INC.   NEW YORK · BASEL

Library of Congress Cataloging-in-Publication Data
A catalog record for this book is available from the Library of Congress.

ISBN: 0-8247-0949-7

This book is printed on acid-free paper.

Headquarters
Marcel Dekker, Inc.
270 Madison Avenue, New York, NY 10016
tel: 212-696-9000; fax: 212-685-4540

Eastern Hemisphere Distribution
Marcel Dekker AG
Hutgasse 4, Postfach 812, CH-4001 Basel, Switzerland
tel: 41-61-260-6300; fax: 41-61-260-6333

World Wide Web
http://www.dekker.com

The publisher offers discounts on this book when ordered in bulk quantities. For more information, write to Special Sales/Professional Marketing at the headquarters address above.

Current printing (last digit):
10 9 8 7 6 5 4 3 2 1

PRINTED IN THE UNITED STATES OF AMERICA

To my beloved wife Marina

Preface

An interesting but difficult problem in Field Theory is to describe in a satisfactory manner the lattice $\underline{\text{Intermediate}}\,(E/F)$ of all intermediate fields of a given field extension E/F. If E/F is a finite Galois extension, then, by the Fundamental Theorem of Galois Theory, the lattice $\underline{\text{Intermediate}}\,(E/F)$ is anti-isomorphic to the lattice $\underline{\text{Subgroups}}(\text{Gal}(E/F))$ of all subgroups of the Galois group $\text{Gal}(E/F)$ of the extension E/F.

On the other hand, there exists a fairly large class of field extensions that are not necessarily Galois, but which enjoy a property that is dual to the previous one. Namely, these are the extensions E/F for which there exists a canonical lattice isomorphism (and *not* a lattice anti-isomorphism as in the Galois case) between $\underline{\text{Intermediate}}\,(E/F)$ and $\underline{\text{Subgroups}}\,(\Delta)$, where Δ is a certain group canonically associated with the extension E/F. We call the members of this class, *extensions with Δ-Cogalois correspondence*. Their prototype is the field extension $\mathbb{Q}(\sqrt[n_1]{a_1}\,,\ldots,\sqrt[n_r]{a_r})/\mathbb{Q}$, where $r, n_1,\ldots,n_r,a_1,\ldots,a_r$ are positive integers, and where $\sqrt[n_i]{a_i}$ is the positive real n_i-th root of a_i for each i, $1 \leq i \leq r$. For such an extension, the associated group Δ is the factor group $\mathbb{Q}^*\langle \sqrt[n_1]{a_1}\,,\ldots,\sqrt[n_r]{a_r}\rangle/\mathbb{Q}^*$.

The purpose of this monograph is to provide a systematic investigation of field extensions which possess a Cogalois correspondence. This topic, we named *Cogalois Theory*, is dual to the very classical one known as *Galois Theory* investigating field extensions possessing a Galois correspondence. As is well-known, Galois Theory has a finite part dealing with finite Galois extensions, as well as an infinite part dealing with infinite Galois extensions. Similarly, Cogalois Theory has a finite (resp. infinite) part, studying finite (resp. infinite) field extensions with a Cogalois correspondence.

The first part of this monograph is devoted to the study of finite field extensions with a Cogalois correspondence. The case of infinite extensions is discussed in the second part of the monograph.

An effort was made to keep the account as self-contained as possible. However, a certain knowledge of Field Theory and Group Theory, including the Fundamental Theorem of Galois Theory, on the level of a graduate course is needed for a good understanding of the most part of this volume. Thus, we assume from the reader some familiarity with the notions and basic facts on vector spaces, groups, rings, Galois extensions, as presented e.g., in the books of Bourbaki [**40**], Kaplansky [**73**], Karpilovsky [**76**], or Lang [**80**], although, whenever it was possible, the indispensable basic facts on these notions were included in the text.

The reader is also assumed to have familiarity with some Galois cohomology, including crossed homomorphisms, 1-coboundaries, the first group of cohomology, Hilbert's Theorem 90, etc., as presented e.g., in the books of Cassels and Fröhlich [**46**], Karpilovsky [**76**], or Serre [**96**].

Some basic facts on topological groups, projective limits, and Pontryagin duality for discrete or compact Abelian groups are needed only in Chapters 14 and 15, for which the reader can be referred to Hewitt and Ross [**72**], Karpilovsky [**76**], or Pontryagin [**87**].

Basics of Algebraic Number Theory are needed only in Section 9.3. They include algebraic number fields, algebraic integers, ideal class groups, Dedekind domains, class fields, etc., and may be found e.g., in the books of Hasse [**67**], Hecke [**71**], Neukirch [**85**], or Ribenboim [**89**].

Elements of graded rings and of Hopf algebras are required only in Chapter 10, and may be found e.g., in the books of Caenepeel [**42**], Dăscălescu, Năstăsescu, and Raianu [**54**], Năstăsescu and Van Oystaeyen [**84**], and Sweedler [**102**].

ACKNOWLEDGEMENTS

This work was started by the author during his stay at the University of California at Santa Barbara as a Visiting Professor in the Spring Quarter of 1999.

A large part of this work was done during his stay as a Humboldt Fellow at the University of Dortmund and Heinrich-Heine-University of Düsseldorf in July 2000 and April-June 2001. It was completed during his stay at the Atilim University, Ankara in the academic year 2001-2002.

The author thanks all these institutions for their hospitality and financial support.

He also gratefully acknowledges partial financial support from grant D-7 awarded by the CNCSIS, Romania.

For $\mathcal{A\!M\!S}$ - LaTeX assistance during the preparation of this monograph, the author is very indebted to Günter Krause and Robert Wisbauer.

The author would like to thank Şerban Basarab, Eberhard Becker, Sorin Dăscălescu, Laurenţiu Panaitopol, Şerban Raianu, Marcel Ţena, and Robert Wisbauer for helpful discussions on various subjects of this monograph. The author would also like to thank the anonymous referee for careful reading of the manuscript and valuable comments and suggestions which improved the presentation of the book, and especially for providing a completely different approach to the main result of Chapter 15 (Theorem 15.3.8), which is sketched at the end of that chapter.

Last, but not least, the author is very indebted to Ralph Svetic for his great stylistic help.

Toma Albu

Contents

COGALOIS THEORY

INTRODUCTION

Consider the following very general problem in Mathematics: *Describe in a satisfactory manner the collection of all subobjects of a given object of a category C.* In general, this is a difficult problem, but sometimes it can be reduced to describing the subobjects of an object in another more suitable category $\mathcal{D}$. For instance, let F be a given field and let C denote the category of all field extensions of F. Let E be an of object of C, i.e., a field extension E/F, and denote by $\mathcal{E}$ the set of all subfields of E containing F, in other words, the set of all subobjects of E in C. This set is, in general, a complicated-to-conceive, potentially infinite set of hard-to-describe-and-identify objects. However, when E/F is a finite Galois extension, then by the Fundamental Theorem of Galois Theory, there exists a canonical one-to-one order-reversing correspondence, or equivalently, a *lattice anti-isomorphism* between the lattice $\mathcal{E}$ and the lattice of all subgroups of a certain group Γ, namely the Galois group $\mathrm{Gal}\,(E/F)$, canonically associated with the extension E/F; we can say that such an E/F is an *extension with Γ-Galois correspondence.*

In this way, the lattice of all subobjects of an object $E \in C$, which has the additional property that it is a finite Galois extension of F, can be described by the lattice of all subobjects of the object Γ in the category of all finite groups; in principle, this category is more suitable than the category C of all field extensions of F, since the set of all subgroups of a finite group is a far more benign object. Thus, many questions concerning a field are best studied by transforming them into group theoretical questions in the group of automorphisms of the field. So, the following natural problem arises:

Problem 1. *Find large classes $\mathcal{F}$ of finite field extensions for which the lattice of all intermediate fields of every $E/F \in \mathcal{F}$ can be described in terms of the lattice of all subgroups of a certain group canonically associated with E/F.*

As we have already mentioned, finite Galois extensions are extensions as desired in Problem 1. On the other hand, there exist large classes of finite field extensions E/F which are not necessarily Galois, but are radical extensions of a special type, for which there exists a canonical one-to-one order-preserving correspondence, or equivalently, a *lattice isomorphism* (and not a lattice *anti-isomorphism* as in the Galois case) between the lattice $\mathcal{E}$ of all intermediate fields of the extension E/F and the lattice $\mathcal{D}$ of all subgroups of a certain group Δ canonically associated with the extension E/F. It seems therefore very natural to introduce the following definition: a field extension E/F is said to be an *extension with Δ-Cogalois correspondence* if there exists a canonical lattice isomorphism between the lattices $\mathcal{E}$ and $\mathcal{D}$ defined above.

The term *extension with Cogalois correspondence* was introduced by Albu and Nicolae in [**19**] to emphasize a situation which is, as shown above, somewhat dual to that appearing in Galois Theory.

To the best of our knowledge, the term of "Cogalois" appeared for the first time in the literature in 1986 in the fundamental paper of Greither and Harrison [**63**], where *Cogalois extensions* were introduced and investigated. Their pioneering work on Cogalois extensions was continued in 1991 by Barrera-Mora, Rzedowski-Calderón, and Villa-Salvador [**30**].

The prototype of extension with a Δ-Cogalois correspondence is any field extension

$$\mathbb{Q}(\sqrt[n_1]{a_1},\dots,\sqrt[n_r]{a_r})/\mathbb{Q},$$

where $r \in \mathbb{N}^* = \mathbb{N} \setminus \{0\}$, $n_1,\dots,n_r,a_1,\dots,a_r \in \mathbb{N}^*$, and for every i, $1 \leqslant i \leqslant r$, $\sqrt[n_i]{a_i}$ is the positive real n_i-th root of a_i. For such an extension, the associated group Δ is the factor group $\mathbb{Q}^*\langle\sqrt[n_1]{a_1},\dots,\sqrt[n_r]{a_r}\rangle/\mathbb{Q}^*$.

It seems surprising it was stated and proved explicitly only fairly late, in 1986, by Greither and Harrison [**63**] that such extensions are extensions with Δ-Cogalois correspondence. In particular, it follows that

$$[\mathbb{Q}(\sqrt[n_1]{a_1},\dots,\sqrt[n_r]{a_r}):\mathbb{Q}] = |\mathbb{Q}^*\langle\sqrt[n_1]{a_1},\dots,\sqrt[n_r]{a_r}\rangle/\mathbb{Q}^*|.$$

This equality was established for the first time in 1940, but only in the following particular case, by Besicovitch [**35**]: Let $r \in \mathbb{N}^*$, and let $p_1,\dots,p_r$ be different positive prime integers. If $b_1,\dots,b_r$ are positive integers not divisible by any of these primes, $a_1 = b_1p_1,\dots,a_r = b_rp_r$, and $n_1,\dots,n_r$ are arbitrary positive integers, then

$$[\mathbb{Q}(\sqrt[n_1]{a_1},\dots,\sqrt[n_r]{a_r}):\mathbb{Q}] = n_1\cdot\dots\cdot n_r.$$

As an immediate consequence of this equality one deduces that

$$[\mathbb{Q}(\sqrt[n]{p_1},\dots,\sqrt[n]{p_r}):\mathbb{Q}]=n^r$$

for any $n \in \mathbb{N}^*$.

Another proof of the last equality, using elementary Galois Theory, was given by Richards [90] (see also Gall's book [58], where the proof of Richards is reproduced). A result of the same nature was established by Ursell [106].

A generalization of Besicovitch's result to algebraic number fields satisfying certain conditions was established by Mordell [83]. Mordell's Theorem was further extended by Siegel [98] and Schinzel [91].

All these results deal with a particular case of the following:

Problem 2. *Let F be a field, let $\overline{F}$ be an algebraic closure of F, and let $x_1,\dots,x_r \in \overline{F}$ be elements of degree $n_1,\dots,n_r$ over F, respectively. When does the field $F(x_1,\dots,x_r)$ have the degree $n_1 \cdot \ldots \cdot n_r$ over F?*

A more general problem is the following:

Problem 3. *With the same notation and hypotheses as in Problem 2, when can we find an explicit formula to compute $[F(x_1,\dots,x_r):F]$?*

Partial answers to Problem 3 are given by a well-known result on classical finite Kummer extensions (see, e.g., Artin [26]), as well as by a result appearing in Kaplansky's book [74, Theorems 60], and by another one of similar nature due to Baker and Stark [28].

In his paper [77], Kneser answered Problem 3 for a large class of extensions that were called by Albu and Nicolae [19] *Kneser extensions*, honoring his crucial result.

It is well-known that for any finitely many elements $x_1,\dots,x_r$ in $\overline{F}$ which are separable over an infinite field F, there exist elements $c_1,\dots,c_r$ in F such that $\theta = \sum_{1\leqslant i\leqslant r} c_i x_i$ is a primitive element of the finite separable extension $F(x_1,\dots,x_r)/F$, i.e., $F(x_1,\dots,x_r) = F(\theta)$. The following natural problem arises:

Problem 4. *Let F be any field, and let $x_1,\dots,x_r$ in $\overline{F}$ be finitely many separable elements over F. When is $\sum_{1\leqslant i\leqslant r} x_i$ a primitive element of the finite separable extension $F(x_1,\dots,x_r)/F$?*

Partial answers to this problem are given e.g., by Albu [3], Kaplansky [74, Theorem 64], and Zhou [113]. These were extended by Albu and Nicolae in [20], where a general statement was proved for the large class of finite separable field extensions with Δ-Cogalois correspondence. In particular, from this general approach it follows very easily that

$$\mathbb{Q}(\sqrt[n_1]{a_1}, \ldots, \sqrt[n_r]{a_r}) = \mathbb{Q}(\sqrt[n_1]{a_1} + \ldots + \sqrt[n_r]{a_r}),$$

where $r \in \mathbb{N}^*$ and $n_1, \ldots, n_r, a_1, \ldots, a_r \in \mathbb{N}^*$.

An immediate consequence of this equality is the following result, which is folklore:

$$\sqrt[n_1]{a_1} + \ldots + \sqrt[n_r]{a_r} \in \mathbb{Q} \iff \sqrt[n_i]{a_i} \in \mathbb{Q} \text{ for all } 1 \leqslant i \leqslant r.$$

Besides the extension $\mathbb{Q}(\sqrt[n_1]{a_1}, \ldots, \sqrt[n_r]{a_r})/\mathbb{Q}$ considered above, Greither and Harrison produced in [63] two other large classes of field extensions with Δ-Cogalois correspondence, namely the *Cogalois extensions* and the *neat presentations*. As is known, the finite *classical Kummer* extensions have a privileged position: they are at the same time extensions with Galois and with Cogalois correspondences, and the two groups appearing in this setting are isomorphic.

Thus, the extensions of type $\mathbb{Q}(\sqrt[n_1]{a_1}, \ldots, \sqrt[n_r]{a_r})/\mathbb{Q}$, the Cogalois extensions, the neat presentations, and the finite classical Kummer extensions all are large classes of finite field extensions answering Problem 1.

A systematic investigation of separable finite radical extensions with Δ-Cogalois correspondence was initiated by Albu and Nicolae in [19], and continued in [20], [21], [22], and [23]. They introduced the general concept of *G-Cogalois extension*, which permitted a very simple, unifying, and even more general approach to the apparently different four classes of radical field extensions with a Cogalois correspondence presented above. These G-Cogalois extensions have many nice properties which answer in positive all the problems stated above.

The purpose of this monograph is to provide a systematic investigation of field extensions which possess a Cogalois correspondence. This subject can be named *Cogalois Theory*, and is dual to the very classical one known as *Galois Theory* investigating field extensions possessing a Galois correspondence. As is well-known, Galois Theory has a finite part dealing with finite Galois extensions, as well as an infinite part dealing with infinite Galois extensions. Similarly, Cogalois Theory has a finite (resp. infinite) part, studying finite (resp. infinite) field extensions with a Cogalois correspondence.

The investigation of finite G-Cogalois extensions is the body of the first part, comprising 10 chapters, of this monograph, while the investigation of infinite G-Cogalois extensions is performed in the second part, which contains 5 chapters. Now we will describe the contents of the first part of this volume.

Chapter 1 contains the necessary preliminaries, including the terminology and notation which will be used throughout the book, a short review of basic Field Theory, the Vahlen-Capelli Criterion dealing with the irreducibility over an arbitrary field of binomials $X^n - a$, and some general facts on bounded Abelian groups.

Chapter 2 introduces and investigates two basic concepts of Cogalois Theory, that of the *G-radical extension* and that of the *G-Kneser extension*. Roughly speaking, a radical extension is a field extension E/F such that E is obtained by adjoining to the base field F an arbitrary set of "radicals" over F, i.e., of elements $x \in E$ such that $x^n = a \in F$ for some $n \in \mathbb{N}^*$. Such an x is denoted by $\sqrt[n]{a}$ and is called an n-th radical of a. So, E/F is a radical extension when $E = F(R)$, where R is a set of radicals over F. Clearly, one can replace R by the subgroup $G = F^*\langle R \rangle$ of the multiplicative group E^* of E generated by F^* and R. Thus, any radical extension E/F has the form $E = F(G)$, where G is a subgroup of E^* containing F^*, with G/F^* a torsion group. Such an extension is called *G-radical*. A finite extension E/F is said to be *G-Kneser* when it is G-radical and $|G/F^*| = [E : F]$. These extensions were introduced by Albu and Nicolae in [19], honoring the nice criterion due to Kneser [77] evaluating the degrees of separable finite radical extensions. A complete proof of this criterion, which is a basic tool in this monograph, is provided in this chapter.

Chapter 3 studies Cogalois extensions, which were introduced by Greither and Harrison in [63]. Using the concept of Kneser extension, a *Cogalois extension* is nothing else than a field extension E/F which is $T(E/F)$-Kneser, where $T(E/F)$ is the subgroup of the multiplicative group E^* of the field E such that the factor group $T(E/F)/F^*$ is the torsion group of the factor group E^*/F^*. The group $T(E/F)/F^*$, called by Greither and Harrison the *Cogalois group* of the extension E/F, is denoted by $\mathrm{Cog}(E/F)$.

Using the Kneser Criterion we provide a very simple and short proof of the Greither-Harrison Criterion [63] characterizing Cogalois extensions. Let us mention that the original proof in [63] of this criterion involves cohomology of groups. Then, we calculate effectively the Cogalois group of quadratic extensions of $\mathbb{Q}$.

Chapter 4 contains the main results of this monograph. After presenting a very general discussion of Galois connections and Cogalois connections, we associate with any G-radical extension E/F, finite or not, a canonical Cogalois connection:

$$\mathcal{E} \xrightarrow[\psi]{\varphi} \mathcal{G},$$

where

$$\varphi : \mathcal{E} \longrightarrow \mathcal{G}, \quad \varphi(K) = K \cap G,$$
$$\psi : \mathcal{G} \longrightarrow \mathcal{E}, \quad \psi(H) = F(H),$$

$\mathcal{E}$ is the lattice $\underline{\text{Intermediate}}\,(E/F)$ of all intermediate fields of the extension E/F, and $\mathcal{G}$ is the lattice $\{\, H \mid F^* \leqslant H \leqslant G \,\}$.

Then, the basic notion of *strongly G-Kneser extension* is introduced: a finite field extension E/F is said to be *strongly G-Kneser* if it is a G-radical extension such that for any subfield K of E containing F the extension K/F is $K^* \cap G$-Kneser, or equivalently, the extension E/K is $K^* G$-Kneser. These are precisely the G-Kneser extensions for which the maps

$$\alpha : \underline{\text{Intermediate}}\,(E/F) \to \underline{\text{Subgroups}}\,(G/F^*), \quad \alpha(K) = (K \cap G)/F^*,$$

and

$$\beta : \underline{\text{Subgroups}}\,(G/F^*) \to \underline{\text{Intermediate}}\,(E/F), \quad \beta(H/F^*) = F(H),$$

are isomorphisms of lattices, inverse to one another; in other words, the G-Kneser extensions E/F with G/F^*-Cogalois correspondence.

In the theory of field extensions with G/F^*-Cogalois correspondence the most interesting are those which additionally are separable. They were called *G-Cogalois extensions* by Albu and Nicolae [**19**], and are completely characterized within the class of G-Kneser extensions by means of a very useful *n-Purity Criterion*, where n is the exponent of the finite group G/F^*. This allows to obtain in a simple and unified manner, and even in a more general setting, a series of results from the classical Kummer Theory, as well as results of Albu [**3**] concerning Kummer extensions with few roots of unity, of Barrera-Mora, Rzedowski-Calderón, and Villa-Salvador [**30**], and of Greither and Harrison [**63**] concerning Cogalois extensions and neat presentations.

It is shown that a separable G-Kneser extension E/F is G-Cogalois if and only if the group G/F^* has a prescribed structure. As a consequence, the uniqueness of the group G is deduced. This means that if the extension

E/F is simultaneously G-Cogalois and H-Cogalois, then necessarily $G =$ H. Consequently, it makes sense to define the *Kneser group* of a G-Cogalois extension as the group G/F^*, which will be denoted by $\text{Kne}\,(E/F)$.

Chapter 4 ends with a study of *almost G-Cogalois extensions*, introduced by Lam-Estrada, Barrera-Mora, and Villa-Salvador [**79**] under the name of *pseudo G-Cogalois extensions*.

Chapter 5 investigates field extensions which are simultaneously Galois and G-Cogalois. It is shown that whenever E/F is an Abelian G-Cogalois extension, then the Kneser group $\text{Kne}\,(E/F)$ and the Galois group $\text{Gal}\,(E/F)$ of E/F are isomorphic in a noncanonical way. Applications of this result to the elementary arithmetic of fields are also given.

In the next chapter we investigate finite Galois extensions which are radical, Kneser, or G-Cogalois, in terms of crossed homomorphisms. The basic result of this chapter is description of the Cogalois group $\text{Cog}(E/F)$ of any finite Galois extension E/F by means of crossed homomorphisms of the Galois group $\text{Gal}(E/F)$ with coefficients in the group $\mu(E)$ of all roots of unity in E.

Chapter 7 provides three main classes of G-Cogalois extensions, namely the *generalized Kummer extensions* including the finite *Kummer extensions with few roots of unity* as well as the finite *classical Kummer extensions*, the *Cogalois extensions*, and the *quasi-Kummer extensions*. In particular, a series of results by Albu [**3**] and Greither and Harrison [**63**], as well as the essential part of the classical Finite Kummer Theory are re-obtained in a unified manner, more easily, and in a more general setting. Note that the *neat presentations* introduced and investigated by Greither and Harrison [**63**] are very particular cases of quasi-Kummer extensions, namely they are precisely those quasi-Kummer extensions which are Galois extensions of fields of characteristic 0.

The prototype of finite Kummer extensions with few roots of unity is any subextension of $\mathbb{R}/\mathbb{Q}$ of the form $\mathbb{Q}(\sqrt[n_1]{a_1},\dots,\sqrt[n_r]{a_r})/\mathbb{Q}$, with $n_1,\dots,n_r,a_1,\dots,a_r$ positive integers. Notice that these extensions, in general, are not Galois. In particular, we derive from our Cogalois Theory results on finite Kummer extensions with few roots of unity, which are very similar to the known ones for classical finite Kummer extensions.

Chapter 8 gives a complete answer to Problem 4 within the class of G-Cogalois extensions. The main result of this chapter is the following. Let E/F be a G-Cogalois extension, let $n \in \mathbb{N}^*$, and let $(x_i)_{1 \leqslant i \leqslant n}$ be a finite family of elements of G such that $x_i F^* \neq x_j F^*$ for every $i, j \in \{1, \ldots, n\}$, $i \neq j$. Then $x_1 + \ldots + x_n$ is a primitive element of E/F if and only if $G = K^* \langle x_1, \ldots, x_n \rangle$. A series of applications to the elementary arithmetic of fields are also given.

The Kneser Criterion has nice applications not only in investigating field extensions with Cogalois correspondence, but also in proving some results in Algebraic Number Theory. Chapter 9 presents such applications. It starts with a section devoted to presenting basic notation, terminology, and facts on algebraic number fields needed in the next sections of the chapter. A series of classical theorems due to Hasse [68], Besicovitch [35], Mordell [83], and Siegel [98] related to solving Problem 2, as well as the links between them are presented in the second section of this chapter. The next section deals with a surprising application of Cogalois Theory in proving some very classical results in Algebraic Number Theory. More precisely, we apply our approach to establish very easily a classical result claimed by Hecke (but not proved) in his book [71], related to the so-called *Hecke systems of ideal numbers*.

The last chapter of this first part of the volume contains links of Cogalois Theory with graded algebras and Hopf algebras.

As is well-known, besides the classical (finite) Galois Theory there also exists its *infinite* counterpart, which involves the Krull topology on the Galois group of an infinite Galois extension. So, dually, it seems perfectly natural to ask about the possibility of extending the *Finite Cogalois Theory* to the *infinite* case. This is exactly the main purpose of the second part of this monograph: to develop an *Infinite Cogalois Theory*, that is, to investigate radical extensions E/F, not necessarily finite, with Δ-Cogalois correspondence. This theory includes as a very particular case the *classical infinite Kummer Theory*, and has nice connections with the infinite Galois cohomology and with the Pontryagin duality.

This second part of the monograph is based on the author's recent papers [8], [9], [10], [15], and [25]. In the finite case, the main tool for the investigation performed in Part I of finite field extensions with Δ-Cogalois correspondence was the *Kneser Criterion*, and the main result was a very useful characterization of these extensions in terms of a "local purity" (the n-Purity Criterion). For the infinite case we have to find the right definitions of the concepts of *infinite Kneser extension*, *infinite Cogalois extension* and *infinite strongly Kneser extension*, to prove an *Infinite Kneser*

Criterion, and to provide a similar characterization of infinite extensions with Δ-Cogalois correspondence.

Infinite G-Kneser extensions are introduced and characterized in Chapter 11. One says that a field extension E/F, which is not necessarily finite, is G-*Kneser* if G is a subgroup of E^* containing F^* such that $E = F(G)$ and the factor group G/F^* is a torsion group having a set of representatives which is linearly independent over F.

We show that the Kneser Criterion [**77**], which characterizes the finite separable G-Kneser extensions, can be generalized to infinite extensions.

Then, we introduce and investigate infinite strongly G-Kneser extensions. The key result of this chapter is a result, which roughly speaking, states that in a tower of fields $F \subseteq K \subseteq E$, whenever two subextensions are Kneser extensions, then so is the third one.

Chapter 12 is devoted to the study of infinite G-Cogalois extensions. We show that a series of results from Part I established for finite G-Cogalois extensions also hold for arbitrary infinite G-Cogalois extensions. The main tool in investigating infinite G-Cogalois extensions is the *General Purity Criterion* characterizing these extensions, which extends the *n-Purity Criterion* for finite G-Cogalois extensions.

As in the case of finite extensions, we show that if an infinite extension E/F is simultaneously G-Cogalois and H-Cogalois, then necessarily $G = H$, and so, it makes sense to define the *Kneser group* $\mathrm{Kne}(E/F)$ of a G-Cogalois extension E/F as the group G/F^*. When the Kneser group of a G-Cogalois extension E/F has finite exponent n, we say that E/F is an *n-bounded* G-Cogalois extension. For such extensions, the General Purity Criterion becomes the *Infinite n-Purity Criterion* and has the same formulation as that of the n-Purity Criterion for finite G-Cogalois extensions.

As remarked in Chapter 2, a finite extension E/F is a Cogalois extension precisely when it is $T(E/F)$-Kneser. This suggests that we define an extension E/F, which is not necessarily finite, to be *Cogalois* if it is $T(E/F)$-Kneser. A series of properties of finite Cogalois extensions are shown to hold also in the infinite case. In particular, the Greither-Harrison Criterion characterizing the finite Cogalois extensions in terms of purity remains valid for infinite extensions.

The prototype of infinite G-Cogalois extension is, by *Kummer Theory*, any *classical infinite Kummer extension*. We show in Chapter 13 that the essential part of the classical infinite Kummer Theory can be immediately deduced from our general approach developed in Chapter 12 by using the

Infinite n-Purity Criterion proved in Section 12.1. Moreover, this criterion allows us to provide large classes of infinite G-Cogalois extensions which generalize or are closely related to classical Kummer extensions: *generalized n-Kummer extensions, n-Kummer extensions with few roots of unity*, and *quasi-Kummer extensions*. The prototype of an n-Kummer extension with few roots of unity is any extension of form $\mathbb{Q}(\{\sqrt[n]{a_i} \mid i \in I\})/\mathbb{Q}$, where $\{a_i \mid i \in I\}$ is an arbitrary nonempty set of strictly positive rational numbers. We derive from our Infinite Cogalois Theory results on infinite generalized n-Kummer extensions, and, in particular, on infinite n-Kummer extensions with few roots of unity, which are very similar to the known ones for classical infinite n-Kummer extensions. Since the n-Kummer extensions with few roots of unity, in general, are not Galois if $n \geq 3$, no other approach (e.g., via Galois Theory, as in the case of classical Kummer extensions) is applicable.

Chapter 14 presents the basic facts on profinite groups, Infinite Galois Theory, character group, and Pontryagin duality which are needed in the next chapter.

Chapter 15 is devoted to the investigation of infinite Galois G-Cogalois extensions. The most relevant results are those for n-bounded extensions. It is shown that if E/F is an infinite Galois n-bounded G-Cogalois extension then the Kneser group $\mathrm{Kne}\,(E/F)$ of E/F is isomorphic to the group $Z_c^1(\mathrm{Gal}\,(E/F), \mu_n(E))$ of all *continuous* crossed homomorphisms of the compact topological group $\mathrm{Gal}\,(E/F)$ (endowed with the Krull topology) with coefficients in the discrete group $\mu_n(E)$ of all n-th roots of unity contained in E. A similar result holds for an arbitrary G-Cogalois extension, with $\mu_n(E)$ replaced by a certain subgroup $\mu_G(E)$ of the group $\mu(E)$ of all roots of unity contained in E. This allows us to generalize to infinite extensions a series of results from Chapter 6 concerning the investigation of Galois G-Cogalois extensions by means of crossed homomorphisms.

Next, we present the basic terminology, notation, and facts concerning lattice-isomorphic groups, which will be needed in the next section investigating infinite Abelian G-Cogalois extensions. A prototype of these extensions is any infinite classical n-Kummer extension. We are especially interested in finding the connection between the Kneser group $\mathrm{Kne}\,(E/F)$ and the Galois group $\mathrm{Gal}\,(E/F)$ of an arbitrary Abelian G-Cogalois extension E/F. In the case of finite extensions, by Theorem 5.2.2, these two groups are isomorphic, but not in a canonical way. We show that in the infinite case, the first one is isomorphic to the group $\mathrm{Ch}\,(\mathrm{Gal}(E/F))$ of characters of the profinite group $\mathrm{Gal}\,(E/F)$.

Each chapter ends with a section providing some historical notes, bibliographical comments, and credit for the material presented in that chapter. Also, each chapter includes a collection of exercises referring to, or extending the subject matter of the chapter. They contain many concrete examples and counterexamples. Many of the author's results are found as exercises. The exercises serve also as a way to draw attention to related results in the literature.

Part 1

FINITE COGALOIS THEORY

CHAPTER 1

PRELIMINARIES

This chapter aims to present some basic terminology, notation, and facts which will be freely used throughout this monograph, including a short review of Field Theory, the Vahlen-Capelli Criterion dealing with the irreducibility over an arbitrary field of binomials $X^n - a$, and general facts on bounded Abelian groups.

1.1. General notation and terminology

In this section we present some general notation and terminology concerning numbers, sets, groups, rings vector spaces, and modules which will be frequently used throughout this monograph.

1.1.1. Numbers and Sets

$$
\begin{aligned}
\mathbb{N} &= \{0, 1, 2, \dots\} = \text{the set of all natural numbers} \\
\mathbb{N}^* &= \mathbb{N} \setminus \{0\} \\
\mathbb{Z} &= \text{the set of all rational integers} \\
\mathbb{Q} &= \text{the set of all rational numbers} \\
\mathbb{R} &= \text{the set of all real numbers} \\
\mathbb{C} &= \text{the set of all complex numbers} \\
S^* &= S \setminus \{0\} \text{ for any } S \subseteq \mathbb{C} \\
S_+ &= \{\, x \in S \mid x \geqslant 0 \,\} \text{ for any } S \subseteq \mathbb{R} \\
S_+^* &= \{\, x \in S \mid x > 0 \,\} \text{ for any } S \subseteq \mathbb{R} \\
m \mid n \quad & m \text{ divides } n \\
\gcd(m, n) &= \text{the greatest common divisor of } m \text{ and } n \\
\mathrm{lcm}(m, n) &= \text{the least common multiple of } m \text{ and } n \\
\varphi(n) &= \text{the Euler function of } n
\end{aligned}
$$

$$\mathbb{P} \;=\; \{\, p \in \mathbb{N}^* \,|\, p \text{ prime} \,\}$$

$$\mathcal{P} \;=\; (\mathbb{P} \setminus \{2\}) \cup \{4\}$$

$$\mathbb{P}_n \;=\; \{\, p \,|\, p \in \mathbb{P},\ p \,|\, n \,\} \text{ for any } n \in \mathbb{N}$$

$$\mathbb{D}_n \;=\; \{\, m \,|\, m \in \mathbb{N},\ m \,|\, n \,\} \text{ for any } n \in \mathbb{N}$$

$$\mathcal{P}_n \;=\; \mathcal{P} \cap \mathbb{D}_n \text{ for any } n \in \mathbb{N}$$

$$|M| \;=\; \text{the cardinal number of an arbitrary set } M$$

$$1_M \;=\; \text{the identity map on the set } M$$

$$f|_A \;=\; \text{the restriction of a map } f : X \longrightarrow Y \text{ to } A \subseteq X$$

1.1.2. Groups

Unless otherwise stated, G will denote throughout this monograph a multiplicative group with identity element e.

$\mathbf{1}$	the group with only one element
$H \subseteq G$	H is a subset of G
$H \leqslant G$	H is a subgroup of G
$x \equiv y \pmod{H}$	$x^{-1}y \in H$
$H \lhd G$	H is a normal subgroup of G
$H \vee K \;=\;$	the subgroup generated by $H \cup K$
$\bigvee_{i \in I} H_i \;=\;$	the subgroup generated by $\bigcup_{i \in I} H_i$
$S^n \;=\;$	$\{\, x^n \,|\, x \in S \,\}$ for any $\varnothing \neq S \subseteq G$ and $n \in \mathbb{N}$
$\underline{\text{Subgroups}}\,(G) \;=\;$	the lattice of all subgroups of G
$\langle M \rangle \;=\;$	the subgroup of G generated by the subset $M \subseteq G$
$\langle g_1, \ldots, g_n \rangle \;=\;$	the subgroup of G generated by the subset $\{g_1, \ldots, g_n\} \subseteq G$
$(G : H) \;=\;$	the index of the subgroup H of the group G
$xH \;=\;$	the left coset $\{\, xh \,|\, h \in H \,\}$ of $x \in G$ modulo $H \leqslant G$

$$G/H \quad = \quad \text{the quotient group of the group } G \text{ modulo}$$
$$H \lhd G$$

$$\mathbb{Z}_n \quad = \quad \text{the quotient group } \mathbb{Z}/n\mathbb{Z} \text{ of integers modulo } n$$

$$\mathfrak{S}_n \quad = \quad \text{the symmetric group of degree } n$$

$$\text{ord}_G(g) \quad = \quad \text{ord}(g) \; = \; \text{the order of an element } g \in G$$

$$t(G) \quad = \quad \text{the set of all elements of } G \text{ having finite order}$$

$$t_p(G) \quad = \quad \text{the set of all elements of } G \text{ having order a power}$$
$$\text{of a prime number } p$$

$$\prod_{i \in I} G_i \quad = \quad \left\{ \, (x_i)_{i \in I} \mid x_i \in G_i, \, \forall \, i \in I \, \right\}$$

$$= \quad \text{the (external) direct product of an arbitrary family}$$
$$(G_i)_{i \in I} \text{ of groups}$$

$$\bigoplus_{i \in I} G_i \quad = \quad \left\{ \, (x_i)_{i \in I} \in \prod_{i \in I} G_i \mid x_i = e_i \text{ for all but finitely many } i \in I \, \right\}$$

$$= \quad \text{the (external) direct sum of an arbitrary family}$$
$$(G_i)_{i \in I} \text{ of groups}$$

$$\overset{\cdot}{\bigoplus_{i \in I}} H_i \quad = \quad \bigvee_{i \in I} H_i \; = \; \text{the internal direct sum of an independent}$$
$$\text{family } (H_i)_{i \in I} \text{ of normal subgroups of } G$$

A family $(H_i)_{i \in I}$ of subgroups of G is said to be *independent* if $H_i \cap \left(\bigvee_{j \in I \setminus \{i\}} H_j \right) = \{e\}$ for every $i \in I$.

For a family $(H_i)_{i \in I}$ of subgroups of G we have $G = \overset{\cdot}{\bigoplus}_{i \in I} H_i$ if and only if the following two conditions are satisfied.

(a) $h_i h_j = h_j h_i$ for all $i, j \in I$, $i \neq j$ and $h_i \in H_i$, $h_j \in H_j$.

(b) Every element $x \in G$ can be written uniquely as $x = \prod_{i \in I} x_i$, where $x_i \in H_i$ for all $i \in I$, and $x_i = e$ for all but finitely $i \in I$.

Note that the (external) direct sum $\bigoplus_{i \in I} G_i$ of a family $(G_i)_{i \in I}$ of Abelian groups is precisely the direct sum of this family in the category **Ab** of all Abelian groups, but not in the category **Gr** of all groups.

Also, note that the internal direct sum $\overset{\cdot}{\bigoplus}_{i \in I} H_i$ of an independent family $(H_i)_{i \in I}$ of normal subgroups of G is canonically isomorphic to the (external) direct sum $\bigoplus_{i \in I} H_i$ of the family $(H_i)_{i \in I}$ of groups. This allows us to identify occasionally the internal and the external direct sum of the independent family $(H_i)_{i \in I}$ of normal subgroups of a G, to denote them identically by $\bigoplus_{i \in I} H_i$, and just to call them the *direct sum* of the H_i's. The appropriate interpretation of $\bigoplus$ can always be inferred from the context.

An alternative terminology for the internal direct sum of a family of normal subgroups of a group is that of *internal direct product*, or of *restricted direct product*.

Any finite Abelian group is an internal direct sum of finitely many cyclic subgroups. Since a group is cyclic if and only if it is isomorphic to $\mathbb{Z}_n$ for some $n \in \mathbb{N}$, it follows that for any finite Abelian group G there exist $r, n_1, \ldots, n_r \in \mathbb{N}^*$ such that

$$G \cong \mathbb{Z}_{n_1} \oplus \ldots \oplus \mathbb{Z}_{n_r}.$$

Moreover, one can choose the integers $n_1, \ldots, n_r$ such that $n_1 \mid n_2 \mid \ldots \mid n_r$.

The *Chinese Remainder Theorem* states that if $m, n \in \mathbb{N}^*$ are relatively prime integers, then there exists a canonical group isomorphism (which is also a ring isomorphism)

$$\mathbb{Z}_{mn} \cong \mathbb{Z}_m \times \mathbb{Z}_n.$$

This implies that if $n = p_1^{k_1} \cdot \ldots \cdot p_s^{k_s}$ is the decomposition of an $n \in \mathbb{N}$, $n \geq 2$ into a product of distinct prime numbers $p_1, \ldots, p_s$, then

$$\mathbb{Z}_n \cong \mathbb{Z}_{p_1^{k_1}} \times \cdots \times \mathbb{Z}_{p_s^{k_s}}.$$

By a *set of representatives* of the quotient group G/H we mean any subset S of G consisting of precisely one representative of each (left) coset modulo H.

1.1.3. RINGS

All rings considered in this monograph are unital rings, all subrings of a given ring R contain the identity element of R, and all ring morphisms preserve the identity elements. By *overring* of a ring R we mean any ring which includes R as a subring.

$$
\begin{aligned}
A^* &= A \setminus \{0\} \text{ for any subset } A \text{ of a ring } R \\
U(R) &= \text{the group of all units of a ring } R \\
\mathbb{F}_q &= \text{the finite field of } q \text{ elements}
\end{aligned}
$$

$$(x_1, \ldots, x_n) \;=\; \Big\{ \sum_{1 \leqslant i \leqslant n} r_i x_i \;\Big|\; r_1, \ldots, r_n \in R \Big\} = \text{the left ideal of } R$$

generated by $x_1, \ldots, x_n \in R$

$$(x) \;=\; Rx \;=\; \text{the principal left ideal of } R$$

generated by $x \in R$

$$R[X_1, \ldots, X_n] \;=\; \text{the polynomial ring in the indeterminates}$$

$X_1, \ldots, X_n$ with coefficients in the ring R

$$\deg(f) \;=\; \text{the degree of a polynomial } f \in R[X]$$

$$Q(R) \;=\; \text{the field of quotients of the domain } R$$

$$F(X_1, \ldots, X_n) \;=\; Q(F[X_1, \ldots, X_n]) \;=\; \text{the field of rational}$$

fractions in the indeterminates $X_1, \ldots, X_n$

with coefficients in the field F

1.1.4. Vector spaces and modules

$$_F V \qquad V \text{ is a vector space over the field } F$$

$$_R M \qquad M \text{ is a left module over the ring } R$$

$$\dim_F(V) \;=\; \text{the dimension of the vector space } {}_F V$$

1.2. A short review of basic Field Theory

The aim of this section is to provide a short review of basic terminology, notation, and results in Field Theory, which will be used throughout this monograph.

Recall that a *field* is a commutative unital ring F with $0 \neq 1$, and such that any nonzero element of F is invertible, in other words, $U(F) = F^*$.

Throughout this monograph F denotes a fixed field and Ω a fixed algebraically closed field containing F as a subfield. Any algebraic extension of F is supposed to be a subfield of Ω.

For an arbitrary nonempty subset S of Ω and a natural number $n \geqslant 1$ we shall use the following notation:

$$\begin{aligned} S^* &= S \setminus \{0\}, \\ \mu_n(S) &= \{\, x \in S \mid x^n = 1 \,\}, \\ S^n &= \{\, x^n \mid x \in S \,\}. \end{aligned}$$

If $x \in \Omega^*$, then $\hat{x}$ will denote the coset xF^* of x in the quotient group Ω^*/F^*.

We are now going to review the basics of Field Theory we will currently use throughout this monograph.

1.2.1. CHARACTERISTIC OF A FIELD

Characteristic. The *characteristic* of a field F is a natural number defined as follows by means of the order $n \in \mathbb{N}^* \cup \{\infty\}$ of the identity element 1 of F in the underlying additive group $(F, +)$ of the field F:

$$\mathrm{Char}\,(F) = \begin{cases} n & \text{if } n \in \mathbb{N} \\ 0 & \text{if } n = \infty. \end{cases}$$

If $\mathrm{Char}(F) > 0$, then it is necessarily a prime number.

Characteristic exponent. The *characteristic exponent* $e(F)$ of a field F is defined by

$$e(F) = \begin{cases} 1 & \text{if } \mathrm{Char}(F) = 0 \\ p & \text{if } \mathrm{Char}(F) = p > 0. \end{cases}$$

A field F is called *perfect* if $F = F^{e(F)}$, i.e., if every element of F is an $e(F)$-th power in F.

Prime field. The *prime subfield* $P(F)$ of a field F is the intersection of all subfields of F. A *prime field* is a field having no proper subfields, or equivalently, a field which coincides with its prime subfield. The concepts of characteristic and prime subfield are related as follows:

$$\mathrm{Char}\,(F) = \begin{cases} p > 0 & \text{if } P(F) \cong \mathbb{Z}_p \\ 0 & \text{if } P(F) \cong \mathbb{Q}. \end{cases}$$

Frobenius morphism. For any field F of characteristic $p > 0$, the map

$$\varphi_F : F \longrightarrow F, \quad x \mapsto x^p$$

is a field morphism, called the *Frobenius morphism* of F. As any field morphism, it is injective. So, a field F of nonzero characteristic is perfect if and only if its Frobenius morphism φ_F is actually an isomorphism.

1.2.2. DEGREE OF A FIELD EXTENSION

Field extension. By *overfield* of a field F we mean any field which includes F as a subfield. A *field extension* is a pair (F, E) of fields, where F is a subfield of E (or E is an overfield of F), and in this case we shall write E/F. Very often, instead of "field extension" we shall use the shorter term "extension". If E is an overfield of a field F we will also say that E is an extension of F.

Of course, for a given extension E/F, F is a subgroup of the underlying Abelian group $(E, +)$ of the field E, so it makes sense to consider its quotient group modulo F. According to the standard notation, this quotient group should be also denoted by E/F, which could produce confusion. To avoid that, we will make clear in the text whenever E/F has another meaning than that of a field extension, and so, no danger of confusion concerning this notation could occur.

F-morphism, F-isomorphism, Galois group. If E/F and L/F are two extensions, then an F-*morphism* from the field E into the field L is any ring morphism $\sigma : E \longrightarrow L$ that fixes F pointwise, i.e., $\sigma(a) = a$ for all $a \in F$. Any F-morphism of fields is necessarily an injective map. An F-*isomorphism* is a surjective (hence bijective) F-morphism. An F-*automorphism* of E is any F-isomorphism from E into itself. The set of all F-automorphisms of E is a group under the binary operation of composition, called the *Galois group* of the extension E/F, and denoted by $\mathrm{Gal}\,(E/F)$.

Intermediate field, subextension, quotient extension. If E/F is an extension, then any subfield K of E, with $F \subseteq K$ is called an *intermediate field* of the extension E/F. A *subextension* (resp. *quotient extension*) of the extension E/F is any extension of the form K/F (resp. E/K), where K is an intermediate field of the extension E/F.

If E/F is any extension and K is any intermediate field of E/F, then both K/F and E/K are again extensions, and conversely, if K/F and E/K are extensions, then so is E/F.

Lattice of subextensions. By $\underline{\mathrm{Subextensions}}\,(E/F)$ we will denote the set of all subextensions of E/F. Note that $\underline{\mathrm{Subextensions}}\,(E/F)$ is a *poset*, that is, a partially ordered set, with respect to the partial order $\leqslant$ defined as follows:

$$K/F \leqslant L/F \iff K \text{ is a subfield of } L.$$

Actually, this poset is a complete lattice, where

$$\inf_{i \in I} (K_i/F) = \left(\bigcap_{i \in I} K_i \right)/F,$$

$$\sup_{i \in I} (K_i/F) = F\left(\bigcup_{i \in I} K_i \right)/F,$$

and $F\left(\bigcup_{i \in I} K_i \right)$ is the subfield of E obtained by adjoining the set $\bigcup_{i \in I} K_i$ to the field F (see 1.2.3).

Note that the lattice $\underline{\text{Subextensions}}\,(E/F)$ is essentially the same with the lattice $\underline{\text{Intermediate}}\,(E/F)$ of all intermediate fields of the extension E/F.

Degree. If E/F is an extension, then the underlying additive group $(E, +)$ of the field E may be viewed as a vector space $_F E$ via the scalar multiplication $F \times E \longrightarrow E$, $(a, x) \mapsto ax$, where ax is the product of elements $a \in F$ and $x \in E$ under the given multiplication on E. In fact, this structure of F-vector space on E together with the given ring structure on E endow E in a natural way with a structure of F-*algebra*.

We now define the *degree* of an extension E/F to be the dimension of the vector space $_F E$, and we denote it by $[E : F]$. Thus,

$$[E : F] = \dim (_F E).$$

Finite extension. One says that E/F is a *finite extension* (resp. an *infinite extension*) if $[E : F]$ is a finite (resp. an infinite) cardinal. Note that for an extension E/F one has

$$[E : F] = 1 \iff E = F.$$

An extension E/F is called *quadratic* (resp. *cubic*, *quartic*) if $[E : F] = 2$ (resp. $[E : F] = 3$, $[E : F] = 4$).

An *algebraic number field*, or for short a *number field*, is any subfield K of the field $\mathbb{C}$ of complex numbers, such that the extension $K/\mathbb{Q}$ is finite. The positive integer $[K : \mathbb{Q}]$ is called the *degree of the number field* K. The number field K is said to be *quadratic* (resp. *cubic*, *quartic*) if so is the extension $K/\mathbb{Q}$.

Tower of fields. A finite chain

$$E_0 \subseteq E_1 \subseteq E_2 \subseteq \ldots \subseteq E_n$$

of fields, where $n \geqslant 2$ and E_i is a subfield of E_{i+1} for every $i = 0, \ldots, n-1$ is called a *tower of fields*.

Tower Law. Let $F \subseteq K \subseteq E$ be a tower of fields. If B is a basis of $_F K$ and C is a basis of $_K E$, then $\{ xy \mid x \in B,\, y \in C \}$ is a basis of $_F E$, hence

$$[E : F] = [E : K] \cdot [K : F].$$

In particular, the extension E/F is finite if and only if both extensions E/K and K/F are finite.

1.2.3. ADJUNCTION

Ring adjunction. Let E be a ring, and let R be a subring of E. As in 1.2.2, E becomes in a natural way a commutative unital R-algebra. If A is any subset of E, then denote by $R[A]$ the smallest subalgebra of E containing A as a subset, that is,

$$R[A] = \bigcap_{T \in \mathcal{S}_A} T,$$

where

$$\mathcal{S}_A = \{ T \mid A \subseteq T,\, T \text{ is an } R\text{-subalgebra of } E \}.$$

Since we made the convention that any subring of a unital ring must contain the identity element of the given ring, it follows that a subset T of E is an R-subalgebra of E if and only if T is a subring of E which contains R as a subset. Thus,

$$\mathcal{S}_A = \{ T \mid A \cup R \subseteq T,\, T \text{ is a subring of } E \},$$

and so, $R[A]$ is also the smallest subring of E containing both A and R as subsets.

We call $R[A]$ the *subring of E obtained by adjoining to R the set A*, or the *R-subalgebra of E generated by A*. The procedure to obtain $R[A]$ from a ring R and a subset A of an overring E of R is called *ring adjunction*.

Finitely generated algebra. An overring E of a ring R is said to be a *finitely generated R-algebra*, or an *R-algebra of finite type*, if $E = R[A]$ for some finite subset A of E. If $A = \{a_1, \ldots, a_n\}$, then instead of $R[\{a_1, \ldots, a_n\}]$ we simply write $R[a_1, \ldots, a_n]$.

Description of $R[A]$. Clearly, $R[\varnothing] = R$. For any nonempty finite subset $\{a_1, \ldots, a_n\}$ of an overring E of a ring R, one has

$$R[a_1, \ldots, a_n] = \{ f(a_1, \ldots, a_n) \mid f \in R[X_1, \ldots, X_n] \},$$

where $f(a_1, \ldots, a_n)$ is the "value" in $(a_1, \ldots, a_n)$ of the polynomial f. Now, if A is an arbitrary subset of E, then

$$R[A] = \bigcup_{C \in \mathcal{F}_A} R[C],$$

where $\mathcal{F}_A$ denotes the set of all finite subsets of A.

Commutativity of ring adjunction. Let E be a ring, let R be a subring of E, and let A, B be subsets of E. Then

$$R[A][B] = R[B][A] = R[A \cup B].$$

Field adjunction. Let E/F be a field extension, and let A be a subset of E. We denote by $F(A)$ the smallest subfield of S containing both A and F as subsets, that is,

$$F(A) = \bigcap_{K \in \mathcal{G}_A} K,$$

where

$$\mathcal{G}_A = \{ \, K \mid A \cup F \subseteq K, \ K \text{ is a subfield of } E \, \}.$$

We call $F(A)$ the *subfield of E obtained by adjoining to F the set A*, and the extension $F(A)/F$ is called the *subextension of E/F generated by A*. The procedure of obtaining $F(A)$ from a field F and a subset A of an overfield E of F is called *field adjunction*.

Finitely generated extension. An extension E/F is said to be *finitely generated* or *of finite type* if $E = F(A)$ for some finite subset A of E. If $A = \{a_1, \ldots, a_n\}$, then instead of $F(\{a_1, \ldots, a_n\})$ we simply write $F(a_1, \ldots, a_n)$.

Simple extension. An extension E/F is said to be *simple* if there exists $a \in E$, called a *primitive element* of E/F, such that $E = F(a)$.

Description of $F(A)$. Clearly, $F(\varnothing) = F$. For any nonempty finite subset $\{a_1, \ldots, a_n\}$ of an overfield E of a field F one has

$$F(a_1, \ldots, a_n) = \{ \, f(a_1, \ldots, a_n)(g(a_1, \ldots, a_n))^{-1} \mid$$
$$f, g \in F[X_1, \ldots, X_n], \ g(a_1, \ldots, a_n) \neq 0 \, \}.$$

Now, if A is an arbitrary subset of E, then

$$F(A) = \bigcup_{C \in \mathcal{F}_A} F(C),$$

where $\mathcal{F}_A$ denotes the set of all finite subsets of A.

Connection between field adjunction and ring adjunction. For any field extension E/F and any subset A of E, $F(A)$ is the field of quotients of the domain $F[A]$.

Commutativity of field adjunction. Let E/F be an extension, and let A, B be subsets of E. Then

$$F(A)(B) = F(B)(A) = F(A \cup B).$$

Compositum. Let E/F be an extension, and let $(K_i)_{i \in I}$ be a family of intermediate fields of E/F. The *compositum* of $(K_i)_{i \in I}$ is the field $F(\bigcup_{i \in I} K_i)$, denoted by $\bigvee_{i \in I} K_i$. For the compositum of a finite family $(K_i)_{1 \leqslant i \leqslant n}$ we shall also use the notation $K_1 K_2 \cdots K_n$.

Linearly disjoint extensions. Let E/F be an extension, and let K_1/F, K_2/F be subextensions of E/F. One says that the extensions K_1/F and K_2/F are *linearly disjoint*, or that the fields K_1 and K_2 are *linearly disjoint over* F, if the canonical morphism of F-algebras

$$K_1 \otimes_F K_2 \longrightarrow E, \quad x_1 \otimes x_2 \mapsto x_1 x_2,$$

is injective. In this case we have $K_1 \cap K_2 = K$, and every linearly independent subset of K_1 (resp. K_2) over F is linearly independent over K_2 (resp. K_1). Conversely, if there exists a vector space basis of K_1 over F which is linearly independent over K_2, then the extensions K_1/F and K_2/F are linearly disjoint.

     The extensions K_1/F and K_2/F are linearly disjoint if and only if there exists a vector space basis B_i of K_i over F, $i = 1, 2$, such that $B_1 B_2 = \{ xy \mid x \in B_1, y \in B_2 \}$ is a vector space basis of $K_1 K_2$ over F.

Degree of compositum and linearly disjointness. Let E/F be an extension, and let K_1/F, K_2/F be two finite subextensions of E/F. Then, the following statements hold.

(a) $[K_1 K_2 : F] \leqslant [K_1 : F] \cdot [K_2 : F]$, with equality if and only if K_1 and K_2 are linearly disjoint over F.

(b) If $[K_1 : F]$ and $[K_2 : F]$ are relatively prime, then K_1 and K_2 are linearly disjoint over F.

1.2.4. ALGEBRAIC EXTENSION

Algebraic element, transcendental element. Let E/F be an extension, and let $u \in E$. Then u is said to be an *algebraic element over* F, or an *algebraic element of the extension* E/F if there exists a nonzero polynomial $f \in F[X]$ such that $f(u) = 0$.

The element u is said to be a *transcendental element over* F, or a *transcendental element of the extension* E/F if it is not algebraic over F, that is, if there exists no nonzero polynomial $f \in K[X]$ such that $f(u) = 0$. An *algebraic number* is any number $z \in \mathbb{C}$ which is algebraic over $\mathbb{Q}$.

Minimal polynomial. Let E/F be an extension, and let $u \in E$. Then, the *evaluation map at* u

$$\varepsilon_u : F[X] \longrightarrow E,$$

defined as

$$\varepsilon_u(f) = f(u),$$

is a ring morphism, with image

$$\mathrm{Im}(\varepsilon_u) = F[u].$$

By the Fundamental Theorem of Isomorphism for rings, ε_u induces a ring isomorphism

$$F[X]/\mathrm{Ker}(\varepsilon_u) \cong F[u].$$

Clearly, we have

$$u \text{ is transcendental over } F \Longleftrightarrow \mathrm{Ker}(\varepsilon_u) = 0 \Longleftrightarrow F[X] \cong F[u],$$

$$u \text{ is algebraic over } F \Longleftrightarrow \mathrm{Ker}(\varepsilon_u) \neq 0.$$

Consequently, since $F[X]$ is an UFD, it follows that u is algebraic over F if and only if the ideal $\mathrm{Ker}(\varepsilon_u)$ of $F[X]$ is generated by a unique monic polynomial, which is called the *minimal polynomial of* u *over* F and is denoted by $\mathrm{Min}(u, F)$.

The following statements are equivalent for an algebraic element of the extension E/F and a polynomial f.

(a) $f = \mathrm{Min}(u, F)$.
(b) f is a monic polynomial in $F[X]$ of least degree such that $f(u) = 0$.
(c) f is a monic irreducible polynomial in $F[X]$ such that $f(u) = 0$.
(d) f is a monic polynomial in $F[X]$ such that $f(u) = 0$, and $f \mid g$ for every polynomial $g \in F[X]$ with $g(u) = 0$.

Degree of an element. Let u be an algebraic element of an extension E/F, and let $n = \deg(\mathrm{Min}(u, F))$. Then

(a) $\{\, 1, u, \dots, u^{n-1} \,\}$ is a basis of the vector space $F[u]$ over F.
(b) $F[u]$ is a field, and so, $F[u] = F(u)$.
(c) $[\, F(u) : F \,] = n$, and n is called the *degree of* u *over* F.

Algebraic extension. Let E/F be an extension. We say that E/F is an *algebraic extension*, or that E is *algebraic* over F, if every element of E is algebraic over F; otherwise we say that E/F is a *transcendental extension*, or that E is *transcendental* over F.

We list below some of the main properties of algebraic extensions.

1. An extension E/F is algebraic if and only if every subring R of E with $F \subseteq R$ is a field.
2. The following assertions are equivalent for an extension E/F.
   (a) E/F is a finite extension.
   (b) E/F is algebraic and finitely generated.
   (c) There exist finitely many algebraic elements $u_1, \ldots, u_n$ of E/F such that $E = F(u_1, \ldots, u_n)$.
3. Let E/F be an extension, and let A be a subset of E consisting of algebraic elements over F. Then $F(A)/F$ is an algebraic extension, and $F[A] = F(A)$.
4. Let $F \subseteq K \subseteq E$ be a tower of fields. Then E/F is an algebraic extension if and only both E/K and K/F are algebraic extensions.
5. Any F-morphism $\sigma : E \longrightarrow E$ of an algebraic extension E/F is necessarily an F-automorphism of E.
6. For any extension E/F, denote by $\overline{F}_E$ the set of all algebraic elements of E over F. Then $\overline{F}_E$ is a subfield of E containing F, which is called the *algebraic closure of F in E*.

1.2.5. SPLITTING FIELD

Existence and uniqueness. Let F be a field, and let $f \in F[X] \setminus F$. A *splitting field of f over F* is an overfield E of F, such that the following two conditions are satisfied.

(a) f *splits over E*, i.e., we can write

$$f = c(X - x_1) \cdot \ldots \cdot (X - x_n),$$

where $x_1, \ldots, x_n \in E$ and $c \in F^*$.

(b) The field E is minimal with the property (a), i.e., there is no proper subfield E' of E containing F such that f splits over E', or equivalently, $E = F(x_1, \ldots, x_n)$.

For any field F and any polynomial $f \in F[X] \setminus F$ there exists a splitting field of f over F, which is uniquely determined up to an F-isomorphism.

Finite field. If F is a finite field, then $\text{Char}(F)$ is necessarily a prime number $p > 0$, and so, its prime subfield $P(F)$ is isomorphic to $\mathbb{Z}_p$. Since $F/P(F)$ is a finite extension, say $[F : P(F)] = n$, it follows that $|F| = p^n$.

Conversely, for every prime number $p > 0$ and every positive integer n, there exists a field F with $|F| = p^n$, namely the splitting field of the polynomial $X^{p^n} - X \in \mathbb{Z}_p[X]$, and any such field F is a splitting field of the polynomial $X^{p^n} - X$ over $P(F)$. In particular, two finite fields F and F' are isomorphic if and only if $|F| = |F'|$. Thus, for any power q of a prime number there exists a field, which we will denote by $\mathbb{F}_q$, with $|\mathbb{F}_q| = q$, and is unique up to an isomorphism. To uniform the notation, we will denote the field $\mathbb{Z}_p$ of integers modulo a prime number p by $\mathbb{F}_p$.

For any finite field $\mathbb{F}_q$ and any $n \in \mathbb{N}^*$ we always will consider that $\mathbb{F}_q$ is a subfield of $\mathbb{F}_{q^n}$.

1.2.6. ALGEBRAICALLY CLOSED EXTENSION

Algebraically closed field. A field F is said to be *algebraically closed* if it satisfies one of the following equivalent conditions.

(a) Every nonconstant polynomial in $F[X]$ splits over F.

(b) Every nonconstant polynomial in $F[X]$ has at least a root in F.

(c) Every irreducible polynomial in $F[X]$ has degree 1.

(d) The field F has no proper overfield E such that E/F is algebraic.

Steinitz's Extension Theorem. Let E/F be an algebraic extension, let Ω be an algebraically closed field, and let $\sigma : F \longrightarrow \Omega$ be a field morphism. Then σ can be extended to a field morphism $\tau : E \longrightarrow \Omega$.

Algebraic closure. An *algebraic closure* of a field F is an overfield E of E such that E is algebraically closed and E/F is an algebraic extension.

A classical result due to Steinitz asserts that any field F has an algebraic closure, which is unique up to an F-isomorphism. An algebraic closure of F will be denoted by $\overline{F}$.

From Steinitz's Extension Theorem follow immediately the following two important facts.

(1) If $\overline{F}$ is a fixed algebraic closure of F, and E/F is any algebraic extension, then there exists an F-morphism $\tau : E \longrightarrow \overline{F}$, which is necessarily injective, and extends the canonical injection $j : F \hookrightarrow \overline{F}$. Identifying E with $\tau(E)$, we can assume that any algebraic extension of F can be considered as a subfield of $\overline{F}$. This assumption will be made tacitly throughout the monograph.

(2) If E/F is an algebraic extension, then every F-automorphism of E can be extended to an F-automorphism of $\overline{F}$.

1.2.7. NORMAL EXTENSION

Conjugate elements. Let F be a field, let $\overline{F}$ be a fixed algebraic closure of F, and let $x, y \in \overline{F}$. We say that x and y are *conjugate elements over* F if one of the following equivalent conditions is satisfied.

  (a) There exists an F-automorphism σ of $\overline{F}$ such that $\sigma(x) = y$.
  (b) There exists an F-isomorphism $\tau : F(x) \to F(y)$ such that $\tau(x) = y$.
  (c) $\mathrm{Min}(x, F) = \mathrm{Min}(y, F)$.

Equivalent definitions. An extension E/F (with E contained in $\overline{F}$, as always assumed) is said to be *normal* or *quasi-Galois* if it is algebraic, and satisfies one of the following equivalent conditions.

  (a) Whenever f is an irreducible polynomial in $F[X]$, then either f splits over E, or f has no roots in E.
  (b) The minimal polynomial of each element of E splits over E.
  (c) For each $x \in E$, all the conjugates of x over F belong to E.
  (d) $\sigma(E) \subseteq E$ for every F-morphism $\tau : E \longrightarrow \overline{F}$.
  (e) $\sigma|_E \in \mathrm{Gal}(E/F)$ for every $\sigma \in \mathrm{Gal}(\overline{F}/F)$.

Properties. We list below some of the basic properties of normal extensions.

  1. Let $F \subseteq K \subseteq E$ be a tower of fields. If E/F is a normal extension, then so is also E/K.
  2. Let $(E_j/F)_{j \in I}$ be a nonempty family of normal extensions. Then $\left(\bigcap_{j \in I} E_j \right)/F$ and $F\left(\bigcup_{j \in I} E_j \right)/F$ are also normal extensions, i.e., the meet and the compositum of any family of normal extensions is again normal.
  3. For any algebraic extension E/F there exists a "least" normal extension $\widetilde{E}/F$ containing E/F as a subextension (and, of course, contained in $\overline{F}$), where $\widetilde{E}$ is the intersection of all subfields N of $\overline{F}$ containing E such that N/F is a normal extension. The extension $\widetilde{E}/F$ is called the *normal closure* of the extension E/F. It can be described as follows: let A be a subset of E which generates the extension E/F, i.e., $E = F(A)$, e.g., A could be E itself. Denote by B the set of all conjugates (in $\overline{F}$) over F of all elements in A. Then

$\widetilde{E} = F(B)$. This shows that the normal closure of a finite extension is also a finite extension.

1.2.8. SEPARABLE EXTENSION

Multiple root. Let F be a field, and let $f \in F[X] \setminus F$ have a root $a \in F$. Then

$$f = (X - a)^m \cdot g$$

for some $m \in \mathbb{N}^*$ and some $g \in F[X]$ with $g(a) \neq 0$, which are both uniquely determined. The number m is called the *multiplicity* of the root a. We say that a is a *multiple root* if $m > 1$ and a *simple root* if $m = 1$.

Derivative. For any polynomial $f = a_0 + a_1 X + \ldots + a_n X^n \in F[X]$ with coefficients in a field F we define the *derivative* Df of f by the rule

$$Df = a_1 + 2a_2 X + \ldots + na_n X^{n-1},$$

where, as usually, $ka_k = a_k + \ldots + a_k$ (k times). Thus, one obtains a map, called *differentiation*

$$D : F[X] \longrightarrow F[X], \; f \mapsto Df,$$

which has the following properties.

(1) $D(f + g) = Df + Dg$,
(2) $D(af) = aDf$,
(3) $D(fg) = (Df)g + f(Dg)$,

for every $f, g \in F[X]$ and every $a \in F$.

Differentiation provides a useful test for multiple roots: an element $a \in F$ is a multiple root of a polynomial $f \in F[X]$ if and only if $f(a) = (Df)(a) = 0$.

Let F be a field of characteristic $p \geqslant 0$, and let $f \in F[X]$. Then $Df = 0$ if and only if $f \in F[X^p]$. If $p = 0$, then $f \in F[X^p]$ means exactly that $f \in F$, and if $p > 0$, then $f \in F[X^p]$ means exactly that there exists $g(X) \in F[X]$ such that $f(X) = g(X^p)$.

Separable polynomial. Let F be a field. An irreducible polynomial $f \in F[X]$ is said to be *separable over* F if one of the following equivalent conditions is satisfied.

(a) In every splitting field of f over F, f has only simple roots.
(b) In some splitting field of f over F, f has only simple roots.
(c) $Df \neq 0$.
(d) Either $\mathrm{Char}(F) = 0$, or $\mathrm{Char}(F) = p > 0$ and $f \notin F[X^p]$.

A polynomial $f \in F[X]$ is said to be *separable* if all its irreducible factors in $F[X]$ are separable.

Separable element, separable extension. Let E/F be an extension. An algebraic element u of E/F is said to be *separable over* F if its minimal polynomial $\mathrm{Min}(u, F)$ is separable over F. An algebraic extension E/F is said to be *separable* if every element of E is separable over F.

Though the concept of *separability* is more general and applies for instance to arbitrary algebras over commutative rings, in particular to arbitrary field extensions, we shall deal in this monograph only with algebraic separable extensions. So, a separable extension E/F means for us an algebraic separable extension.

Separable degree. Let F be a field of characteristic $p > 0$, and let $f(X) \in F[X]$ be an irreducible polynomial. Then, there exists a unique number $e \in \mathbb{N}$ such that $f(X) \in F[X^{p^e}]$ but $f(X) \notin F[X^{p^{e+1}}]$, and write $f(X) = g(X^{p^e})$ for some $g(X) \in F[X]$. The following statements hold.

(a) $g(X)$ is an irreducible and separable polynomial over F.

(b) All roots of $f(X)$ in $\overline{F}$ have the same multiplicity equal to p^e, and $\deg(g(X))$ is equal to the number of distinct roots in $\overline{F}$ of f; in particular,

$$\deg(f(X)) = p^e \cdot \deg(g(X)).$$

The positive integers $\deg(g)$ and p^e are called the *separable degree* of f and the *degree of inseparability* of f, respectively.

If E/F is an extension and $u \in E$ is an algebraic element, then one defines the *separable degree* of u over F (resp. the *degree of inseparability* of u over F) as being the separable degree of $\mathrm{Min}(u, F)$. If E/F is a finite extension, then the *separable degree* of E/F is defined to be the (finite) number, denoted by $[E : F]_s$, of all F-morphisms from E into $\overline{F}$.

The connection between the numbers considered above is the following. Let F be a field of characteristic $p > 0$, let E/F be an extension, and let $u \in E$ be an algebraic element over F having the degree of inseparability p^e. Then $[F(u) : F]_s$ is equal to the separable degree of u over F, and

$$[F(u) : F] = p^e \cdot [F(u) : F]_s.$$

Properties. We list below some of the basic properties of separable extensions.

1. A field F is perfect if and only if every algebraic extension of F is separable. In particular, any algebraic extension of a field of characteristic 0 or of a finite field is a separable extension.

2. Let E/F be any extension, and let A be a subset of E consisting of algebraic separable elements over F. Then $F(A)/F$ is a separable extension.

3. Let $F \subseteq K \subseteq E$ be a tower of fields. Then E/F is a separable extension if and only if both K/F and E/K are so.

4. Let E/F be any algebraic extension and denote by E_s the set of all elements of E which are separable over F. Then E_s/F is a subextension of E/F, which is the greatest separable subextension of E/F. The subfield E_s of E is called the *separable closure of F in E*.

5. For any finite extension E/F, one has $[E : F]_s = [E_s : F]$; in particular, E/F is separable if and only if $[E : F]_s = [E : F]$. Also $[E : F]_s \mid [E : F]$, and the positive integer

$$[E : F]_i = [E : F]/[E : F]_s$$

is called the *inseparable degree* of E/F. Observe that $[E : F]_i = 1$ if $\operatorname{Char}(F) = 0$, and $[E : F]_i = p^m$ for some $m \in \mathbb{N}$ if $\operatorname{Char}(F) = p$.

6. Let $(E_j/F)_{j \in I}$ be a nonempty family of separable extensions. Then $F\left(\bigcup_{j \in I} E_j\right)/F$ is also a separable extensions, i.e., the compositum of any family of separable extensions is again separable.

7. Any finite separable extension is simple.

Purely inseparable element. Let E/F be an extension, where F is a field of characteristic $p > 0$. An element $u \in E$ is said to be *purely inseparable over F* if $u^{p^m} \in F$ for some $m \in \mathbb{N}$; if $e \in \mathbb{N}$ is the least such m, then $\operatorname{Min}(u, F) = X^{p^e} - u^{p^e}$. This immediately follows from the following classical result due to Abel: if $a \in F \setminus F^p$, then the polynomial $X^{p^n} - a$ is irreducible over F for every $n \in \mathbb{N}$.

An element $u \in E$ is simultaneously separable and purely inseparable over F if and only if $u \in F$.

Purely inseparable extension. An extension E/F is said to be *purely inseparable* if every element of E is purely inseparable over F.

For any algebraic extension E/F, the extension E_s/F is separable and the extension E/E_s is purely inseparable.

Norm and trace. Let E/F be a finite separable extension of degree n, and let $\sigma_1, \ldots, \sigma_n$ be all distinct F-morphisms of E into $\overline{F}$. For any $a \in E$

we define the following elements:

$$N_{E/F}(a) = \prod_{i=1}^{n} \sigma_i(a),$$

$$Tr_{E/F}(a) = \sum_{i=1}^{n} \sigma_i(a),$$

which are called the *norm* and the *trace* of a in the extension E/F, respectively. A priori, the elements $N_{E/F}(a)$ and $Tr_{E/F}(a)$ are in $\overline{F}$, but actually, they both belong to F.

The map $N_{E/F} : E \longrightarrow F$, called the *norm from E to F*, or the *norm of E/F*, is multiplicative, and the map $Tr_{E/F} : E \longrightarrow F$, called the *trace from E to F*, or the *trace of E/F*, is F-linear.

More generally, the concepts of "Norm" and "Trace" can be defined for any pair of commutative rings (R, S), with S an overring of R such that S is a free R-module of finite rank.

1.2.9. GALOIS EXTENSION

Equivalent definitions. An extension E/F is said to be a *Galois extension* if it is algebraic, and if one of the following equivalent conditions is satisfied.

(a) E/F is a normal and separable extension.
(b) Every element u of E such that $\sigma(u) = u$ for all $\sigma \in \text{Gal}(E/F)$, necessarily belongs to F.
(c) For every $u \in E$, $\text{Min}(u, F)$ splits over E and has only simple roots.

Properties. We list below some of the basic properties of Galois extensions.

1. Let $F \subseteq K \subseteq E$ be a tower of fields.
   - If E/F is a Galois extension, then so also is E/K.
   - $\text{Gal}(E/K) \leqslant \text{Gal}(E/F)$.
   - If E/F is a Galois extension, then K/F is also a Galois extension if and only if $\text{Gal}(E/K) \lhd \text{Gal}(E/F)$, and in this case the map $\text{Gal}(E/F) \longrightarrow \text{Gal}(K/F)$, $\sigma \mapsto \sigma|_K$ induces an isomorphism of groups

     $$\text{Gal}(E/F)/\text{Gal}(E/K) \cong \text{Gal}(K/F).$$

2. Let $(E_j/F)_{j\in I}$ be a nonempty family of Galois extensions. Then $\left(\bigcap_{j\in I} E_j\right)/F$ and $F\left(\bigcup_{j\in I} E_j\right)/F$ are also Galois extensions, i.e., the meet and the compositum of any family of Galois extensions is again Galois.

3. For any field F and any subset A of $\overline{F}$ consisting only of separable elements over F, let B be the set of all conjugates (in $\overline{F}$) over F of all elements in A. Then $F(B)/F$ is a Galois extension, which is called the *Galois extension of F generated by the subset A of $\overline{F}$*. In particular, the normal closure $\widetilde{E}/F$ of any separable extension E/F is a Galois extension.

4. A finite extension E/F is a Galois extension if and only if E is the splitting field over F of a separable polynomial in $F[X]$.

Fixed field. Let E be any field. For any nonempty subset S of the group of automorphisms of the field E we denote by $\mathrm{Fix}(S)$ or by E^S the *fixed field of S*, that is,

$$\mathrm{Fix}(S) = \{\, x \in E \mid \sigma(x) = x, \ \forall\, \sigma \in S \,\},$$

which is a subfield of E. It is also called the *field of invariants of S*, and denoted by $\mathrm{Inv}(S)$. If E/F is an extension with Galois group G and S is any nonempty subset of G, then $\mathrm{Fix}(S)$ contains F as a subfield.

With the notation above, an algebraic extension E/F is Galois if and only if $\mathrm{Fix}(\mathrm{Gal}(E/F)) = F$.

The Artin Theorem. Let E be any field and let G be any finite group of automorphisms of E. Then $E/\mathrm{Fix}(G)$ is a finite Galois extension, $G = \mathrm{Gal}(E/\mathrm{Fix}(G))$, and $[E : \mathrm{Fix}(G)] = |G|$.

Poset, lattice, isomorphism, anti-isomorphism. A *partially ordered set*, or *poset*, is a pair $(P, \leqslant)$ consisting of a nonempty set P and a binary relation $\leqslant$ on P which is reflexive, anti-symmetric, and transitive. Very often, a poset $(P, \leqslant)$ will be denoted shortly by P. The opposite poset of P will be denoted by P^{op}.

A *lattice* is a poset L in which every two elements x, y have a least upper bound $x \vee y$ (also denoted by $\sup(x, y)$) and a greatest lower bound $x \wedge y$ (also denoted by $\inf(x, y)$). A poset in which every subset A has a least upper bound $\bigvee_{x\in A} x$ (also denoted by $\sup(A)$) and a greatest lower bound $\bigwedge_{x\in A} x$ (also denoted by $\inf(A)$) is called a *complete lattice*.

A *poset morphism* (resp. a *poset anti-morphism*) from a poset P into a poset P' is a map $f : P \longrightarrow P'$ which is *increasing*, or *order-preserving* (resp. *decreasing*, or *order-reversing*), i.e., satisfies the following condition:

$$\forall\, x \leqslant y \ \text{in} \ P \Longrightarrow f(x) \leqslant f(y)$$

$$(\text{resp.} \ \forall\, x \leqslant y \ \text{in} \ P \Longrightarrow f(y) \leqslant f(x)).$$

A *poset isomorphism* (resp. a *poset anti-isomorphism*) from a poset P into a poset P' is a bijective poset morphism (resp. poset anti-morphism) $f : P \longrightarrow P'$ such that its inverse $f^{-1} : P' \longrightarrow P$ is a poset morphism (resp. poset anti-morphism).

If L and L' are two lattices, then a *lattice morphism* (resp. a *lattice anti-morphism*) from into L into L' is a map $f : L \longrightarrow L'$ satisfying the following conditions:

$$f(x \vee y) = f(x) \vee f(y) \quad \text{and} \quad f(x \wedge y) = f(x) \wedge f(y), \ \forall x, y \in L$$

$$(\text{resp.} \ f(x \vee y) = f(x) \wedge f(y) \quad \text{and} \quad f(x \wedge y) = f(x) \vee f(y), \ \forall x, y \in L).$$

Any lattice isomorphism between complete lattices commutes with arbitrary meets and joins.

A bijection $f : L \longrightarrow L'$ between two lattices L and L' is a lattice isomorphism (resp. lattice anti-isomorphism) if and only if f and its inverse f^{-1} are both order-preserving (resp. order-reversing) maps, i.e., if and only if f is a poset isomorphism (resp. poset anti-isomorphism).

The Fundamental Theorem of Finite Galois Theory. The following statements hold for a finite Galois extension E/F with Galois group G.

(a) The maps

$$\alpha : \underline{\text{Intermediate}} \, (E/F) \longrightarrow \underline{\text{Subgroups}} \, (G), \ \ \alpha(K) = \text{Gal}(E/K),$$

and

$$\beta : \underline{\text{Subgroups}} \, (G) \longrightarrow \underline{\text{Intermediate}} \, (E/F), \ \ \beta(H) = \text{Fix}(H),$$

establish anti-isomorphisms of lattices, inverse to one another, between the lattice $\underline{\text{Intermediate}} \, (E/F)$ of all intermediate fields of the extension E/F and the lattice $\underline{\text{Subgroups}} \, (G)$ of all subgroups of G.

(b) For any $H_1 \leqslant H_2 \leqslant G$ one has

$$(H_2 : H_1) = [\, \text{Fix}(H_1) : \text{Fix}(H_2)\,].$$

In particular, $|G| = [\, E : F\,]$, and

$$|H| = [\, E : \text{Fix}(H)\,] \quad \text{and} \quad (G : H) = [\, \text{Fix}(H) : F\,]$$

for any $H \leqslant G$.

(c) If $K \in$ Intermediate(E/F), then K/F is a Galois extension if and
   only if Gal$(E/K) \lhd$ Gal(E/F), and in this case

$$\mathrm{Gal}(E/F)/\mathrm{Gal}(E/K) \cong \mathrm{Gal}(K/F).$$

Lattice-theoretical aspects of Galois Theory. Let E/F be a finite
Galois extension with Galois group G, let $(H_i)_{i \in I}$ be a nonempty family
of subgroups of G, and let $(K_j)_{j \in J}$ be a nonempty family of intermediate
fields of the extension E/F. Then

(a) Fix$\left(\bigcap_{i \in I} H_i\right) = \bigvee_{i \in I}$ Fix(H_i).
(b) Fix$\left(\bigvee_{i \in I} H_i\right) = \bigcap_{i \in I}$ Fix(H_i).
(c) Gal$(E/(\bigvee_{j \in J} K_j)) = \bigcap_{j \in J}$ Gal(E/K_j).
(d) Gal$(E/(\bigcap_{j \in J} K_j)) = \bigvee_{j \in J}$ Gal(E/K_j).

Base field change. Let E/F and K/F be subextensions of an extension
L/F. If $E/(E \cap K)$ is a finite Galois extension, then EK/K is also a finite
Galois extension, and the map

$$\mathrm{Gal}(EK/K) \longrightarrow \mathrm{Gal}(E/(E \cap K)), \ \sigma \mapsto \sigma|_E,$$

is an isomorphism of groups.

Compositum of Galois extensions. Let E_1/F and E_2/F be finite Galois
extensions, and assume that E_1, E_2 are subfields of some other field. Then
$E_1 E_2/F$ is a finite Galois extension, and the map

$$\mathrm{Gal}(E_1 E_2/F) \longrightarrow \mathrm{Gal}(E_1/F) \times \mathrm{Gal}(E_2/F), \ \sigma \mapsto (\sigma|_{E_1}, \sigma|_{E_2}),$$

is an monomorphism of groups, which is an isomorphism when $E_1 \cap E_2 = F$.

More generally, let $(E_i/F)_{1 \leqslant i \leqslant n}$ be a finite family of finite Galois extensions. Assume that all $E_1, \dots, E_n$ are subfields of some other field,
and

$$E_{i+1} \cap (E_1 \cdots E_i) = F$$

for every i, $1 \leqslant i \leqslant n - 1$. Then

$$\mathrm{Gal}(E_1 \cdots E_n/F) \cong \mathrm{Gal}(E_1/F) \times \dots \times \mathrm{Gal}(E_n/F).$$

Conversely, whenever the Galois group G of a finite Galois extension
E/F can be decomposed into a finite internal finite direct sum of subgroups

$$G = G_1 \oplus \dots \oplus G_n,$$

then the field E can be expressed as the compositum

$$E = E_1 \cdots E_n$$

of intermediate fields E_i of E/F, $i = 1, \ldots, n$, where E_i is the fixed field of the subgroup

$$G_1 \oplus \ldots \oplus \{e\} \oplus \ldots \oplus G_n$$

of G, where the identity element e of G occurs in the i-th position. Moreover, every extension E_i/F, $1 \leqslant i \leqslant n$ is Galois, and

$$E_{i+1} \cap (E_1 \cdots E_i) = F$$

for every i, $1 \leqslant i \leqslant n - 1$.

1.2.10. CYCLOTOMIC EXTENSION

Root of unity. Let F be a field, and let $n \in \mathbb{N}^*$. An element $\zeta \in F$ is said to be an *n-th root of unity* (or of 1) if $\zeta^n = 1$. A *root of unity* is an n-th root of unity for some $n \in \mathbb{N}^*$.

We denote by $\mu_n(F)$ the set of all n-th roots of unity in F, and by $\mu(F)$ the set of all roots of unity in F. Then

$$\mu_n(F) \leqslant \mu(F) \leqslant F^*$$

for all $n \in \mathbb{N}^*$. We also have

$$\mu(F) = \bigcup_{k \geqslant 1} \mu_k(F) \text{ and } \mu_m(F) \subseteq \mu_n(F) \text{ if } m \mid n.$$

Let $n \in \mathbb{N}^*$. The group $\mu_n(F)$ is a cyclic group having order a divisor of n. In case the polynomial $X^n - 1$ splits over F (in particular, this happens whenever F is algebraically closed), then $\mu_n(F)$ is a cyclic group of order n if and only if $\gcd(n, e(F)) = 1$. Recall that $e(F)$ denotes the characteristic exponent of F (see 1.2.1). By a *primitive n-th root of unity over F* we mean throughout this monograph any generator of the cyclic group $\mu_n(\overline{F})$, and ζ_n will always denote such an element. If $\gcd(n, e(F)) = 1$, then there exist precisely $\varphi(n)$ primitive n-th roots of unity over F, where $\varphi(n)$ is the Euler function of n. If there is no danger of ambiguity about the ground field F, we will simply say *primitive n-th root of unity*.

Note that the standard definition of the notion of primitive root of unity is somewhat different: usually, by a primitive n-th root of unity over F one understands any element in $\overline{F}$ having order n in the group $\overline{F}^*$. Clearly, when $\gcd(n, e(F)) = 1$, then the two definitions coincide, but if $\mathrm{Char}(F) = p > 0$ and $p \mid n$, then no primitive n-root of unity in the standard definition exists, while, in our definition, always exists a primitive n-th root of unity.

Cyclotomic extension, cyclotomic polynomial. Let F be a field, let $n \in \mathbb{N}^*$ be such that $\gcd(n, e(F)) = 1$, and let ζ_n be a primitive n-th root of unity over F. The field $F(\zeta_n)$ is called the *n-th cyclotomic field over F*, and the extension $F(\zeta_n)/F$ is called the *n-th cyclotomic extension* over F.

Since $F(\zeta_n)$ is the splitting field over F of the separable polynomial $X^n - 1$, it follows that any cyclotomic extension is a Galois extension.

Let $\varepsilon_1, \ldots, \varepsilon_r$, $r = \varphi(n)$, be all primitive n-th roots of unity over F. Then

$$\Phi_n = \prod_{i=1}^{r} (X - \varepsilon_i)$$

is a monic polynomial in $F[X]$, and is called the *n-th cyclotomic polynomial over F*.

Galois group of a cyclotomic extension. Let F be a field, and let $n \in \mathbb{N}^*$ be such that $\gcd(n, e(F)) = 1$. Then $\mathrm{Gal}(F(\zeta_n)/F)$ is isomorphic to a subgroup of the group of units $U(\mathbb{Z}_n)$ of the ring $\mathbb{Z}_n$. Moreover, the following conditions are equivalent.

(a) $\mathrm{Gal}(F(\zeta_n)/F) \cong U(\mathbb{Z}_n)$.
(b) Φ_n is irreducible over F.
(c) $[F(\zeta_n) : F] = \varphi(n)$.

Cyclotomic extension of $\mathbb{Q}$. Let $n \in \mathbb{N}^*$, let ζ_n be a primitive n-root of unity over $\mathbb{Q}$, e.g., $\zeta_n = \cos(2\pi/n) + i\sin(2\pi/n)$, and let Φ_n be the n-th cyclotomic polynomial over $\mathbb{Q}$. Then

(a) $\mathrm{Gal}(\mathbb{Q}(\zeta_n)/\mathbb{Q}) \cong U(\mathbb{Z}_n)$.
(b) Φ_n is irreducible over $\mathbb{Q}$.
(c) $[\mathbb{Q}(\zeta_n) : \mathbb{Q}] = \varphi(n)$.
(d) $X^n - 1 = \prod_{d|n} \Phi_d$.
(e) If n is prime, then $\Phi_n = X^{n-1} + X^{n-2} + \ldots + 1$.

1.2.11. ABELIAN EXTENSION, CYCLIC EXTENSION

Abelian extension. An extension E/F, which is not necessarily finite, is said to be *Abelian* if E/F is a Galois extension and $\mathrm{Gal}(E/F)$ is an Abelian group. Any cyclotomic extension is Abelian. Also, any subextension and any quotient extension of a cyclotomic extension is Abelian.

Cyclic extension. An extension E/F, which is not necessarily finite is said to be *cyclic* if E/F is a Galois extension and $\mathrm{Gal}(E/F)$ is a cyclic group.

Let $n \in \mathbb{N}^*$, and let E/F be a finite extension such that $\gcd(n, e(F)) = 1$ and $\zeta_n \in F$. Then E/F is a cyclic extension with $[E : F] \mid n$ if and only if there exists $x \in E^*$ such that $x^n \in F$ and $E = F(x)$.

1.3. The Vahlen-Capelli Criterion

The aim of this section is to prove a criterion for the irreducibility of binomials $X^n - a$ over an arbitrary field, known as the Vahlen-Capelli Criterion. A special case of this criterion, which is mainly applicable to subfields of $\mathbb{R}$, is also presented.

Throughout this section F will denote an arbitrary field and Ω a fixed algebraically closed field containing F as a subfield.

We start with some preparatory results which are needed in the proof of the Vahlen-Capelli Criterion.

LEMMA 1.3.1. *Let $a \in F$, and let $m, n \in \mathbb{N}^*$ be relatively prime integers. Then, the polynomial $X^{mn} - a$ is irreducible in $F[X]$ if and only if both polynomials $X^m - a$ and $X^n - a$ are irreducible in $F[X]$.*

PROOF. Since $X^{mn} - a = (X^n)^m - a = (X^m)^n - a$, we deduce that if $X^{mn} - a$ is irreducible in $F[X]$, then so are $X^m - a$ and $X^n - a$.

Conversely, assume that both $X^m - a$ and $X^n - a$ are irreducible in $F[X]$. Let $u \in \Omega$ be a root of $X^{mn} - a$. Then u^m is a root of the irreducible polynomial $X^n - a$, and u^n is a root of the irreducible polynomial $X^m - a$, hence

$$[F(u^m) : F] = n \text{ and } [F(u^n) : F] = m.$$

Using the Tower Law for the following two towers of fields

$$F \subseteq F(u^m) \subseteq F(u) \text{ and } F \subseteq F(u^n) \subseteq F(u),$$

we deduce that $n \mid [F(u) : F]$ and $m \mid [F(u) : F]$, hence $mn \mid [F(u) : F]$ since $\gcd(m, n) = 1$. But u is a root of the polynomial $X^{mn} - a$, so $[F(u) : F] \leqslant mn$. It follows that $[F(u) : F] = mn$, which shows that $\mathrm{Min}(u, F) = X^{mn} - a$. This proves that $X^{mn} - a$ is irreducible in $F[X]$. $\square$

LEMMA 1.3.2. *Let $p \in \mathbb{P}$ and $a \in F$. Then $X^p - a$ is irreducible in $F[X]$ if and only if $a \notin F^p$.*

PROOF. If $a \in F^p$, then $a = b^p$ for some $b \in F$, and then $X^p - a = X^p - b^p$ is clearly reducible in $F[X]$.

Conversely, assume that $a \notin F^p$, and prove that $X^p - a$ is irreducible in $F[X]$. Suppose the contrary, and let f be an irreducible factor of degree n, $1 \leqslant n < p$, of $X^p - a$. If u is a fixed root of f in Ω, then the roots of f in Ω all have the form ζu, where $\zeta \in \Omega$ is a p-th root of unity. Let c be the constant term of f. Since $\pm c$ is the product of the roots of f, we have $\pm c = \xi u^n$, where $\xi \in \Omega$ is a p-th root of unity. Since n and p are relatively prime integers, there exist $r, s \in \mathbb{Z}$ such that $rn + sp = 1$, and so,

$$u = u^{rn}u^{sp} = (\pm c\xi^{-1})^r a^s.$$

Thus $u\xi^r \in F$, hence $a = u^p = (u\xi^r)^p \in F^p$, which contradicts our assumption. $\square$

LEMMA 1.3.3. *Let $p \in \mathbb{P}$ and $a \in F$ be such that $X^p - a$ is irreducible in $F[X]$, and let $u \in \Omega$ be a root of $X^p - a$.*

(1) *If $p > 2$, or if $p = 2$ and $\mathrm{Char}(F) = 2$, then $u \notin F(u)^p$.*
(2) *If $p = 2$ and $\mathrm{Char}(F) \neq 2$, then $u \in F(u)^2$ if and only if $-4a \in F^4$.*

PROOF. (1) Assume that $u = v^p$ for some $v \in F(u)$. Then we have $v = \sum_{k=0}^{p-1} c_k u^k$ for some $c_0, \ldots, c_{k-1} \in F$.

If $\mathrm{Char}(F) = p$, then we have

$$u = v^p = \sum_{k=0}^{p-1} c_k^p (u^p)^k = \sum_{k=0}^{p-1} c_k^p a^k \in F,$$

hence $a = u^p \in F^p$, which contradicts Lemma 1.3.2. In particular, if $\mathrm{Char}(F) = p = 2$, then we have $u \notin F(u)^p$.

If $\mathrm{Char}(F) \neq p$, then consider the field $E = F(u, \zeta_p)$. Since E is clearly the splitting field of the polynomial $X^p - a \in F[X]$, E/F is a normal extension. Any $\sigma \in \mathrm{Gal}(E/F)$ sends u into some $\zeta_p^i u$, $0 \leqslant i \leqslant p - 1$. Conversely, for every $i \in \{0, \ldots, p - 1\}$, $\zeta_p^i u$ is a root of the irreducible polynomial $X^p - a$, hence it is a conjugate of u over F. Therefore, there exists $\sigma_i \in \mathrm{Gal}(E/F)$ such that $\sigma_i(u) = \zeta_p^i u$. Set $v_i := \sigma_i(v)$, $1 \leqslant i \leqslant p-1$. Then $\sigma_i(v^p) = v_i^p = \sigma_i(u) = \zeta_p^i u$.

By Lemma 1.3.2, $v \in F(u) \setminus F$, hence $F(v) = F(u)$. If we set $f := \mathrm{Min}(v, F)$, then

$$\deg(f) = [\, F(v) : F \,] = [\, F(u) : F \,] = \deg(\mathrm{Min}(u, F)) = \deg(X^p - a) = p.$$

Now observe that $\sigma_i(u) \neq \sigma_j(u)$ for every $i \neq j$ in $\{0, \ldots, p-1\}$ since ζ_p has order p in Ω^*. This implies that $\sigma_i(v) \neq \sigma_j(v)$ for every $i \neq j$ in $\{0, \ldots, p-1\}$, so the polynomial f of degree p has at least p distinct roots $v_0, \ldots, v_{p-1}$ in E. Consequently, these elements are exactly all the roots in Ω of f, and then $w = \prod_{k=0}^{p-1} v_k \in F$. We now multiply together the equalities $\zeta_p^k u = v_k^p$, $k = 0, \ldots, p-1$, and obtain $\eta u^p = \eta a = w^p$, where

$$\eta = \prod_{k=0}^{p-1} \zeta_p^k = \zeta_p^{p(p-1)/2}.$$

If p is odd, then $\eta = 1$, and so, $a = w^p \in F^p$, which contradicts Lemma 1.3.2. This proves (1).

(2) Assume that $u \in F(u)^2$. Then $u = v^2$ with $v = \alpha + \beta u$, $\alpha, \beta \in F$. Thus

$$u = (\alpha + \beta u)^2 = \alpha^2 + \beta^2 a + 2\alpha\beta u.$$

Since $\{1, u\}$ is a basis of the vector space $F(u)$ over F, we deduce that

$$\alpha^2 + \beta^2 a = 0 \quad \text{and} \quad 2\alpha\beta = 1.$$

Eliminating β, we obtain $a = -4\alpha^4$, and so

$$-4a = 16\alpha^4 = (2\alpha)^4 \in F^4.$$

Conversely, if $-4a \in F^4$, then we can write $-4a = \gamma^4 = (2\alpha)^4$ for some $\gamma, \alpha \in F^*$. Now, if we take $\beta = (2\alpha)^{-1}$, then it is easily verified that $u = (\alpha + \beta u)^2 \in F(u)^2$. $\qquad\square$

LEMMA 1.3.4. *Let $p \in \mathbb{P}$, $n \in \mathbb{N}$, $n \geqslant 2$, and $a \in F$.*

(1) *If $p > 2$, or if $p = 2$ and $\mathrm{Char}(F) = 2$, then $X^{p^n} - a$ is irreducible in $F[X]$ if and only if $a \notin F^p$.*

(2) *If $p = 2$ and $\mathrm{Char}(F) \neq 2$, then $X^{2^n} - a$ is irreducible in $F[X]$ if and only if $a \notin F^2$ and $-4a \notin F^4$.*

PROOF. (1) If $a \in F^p$, then $a = b^p$ for some $b \in F$, hence

$$X^{p^n} - a = (X^{p^{n-1}})^p - b^p$$

is divisible by $X^{p^{n-1}} - b$.

Conversely, assume that $a \notin F^p$. Let $v \in \Omega$ be a root of $X^{p^n} - a$, and set $u := v^{p^{n-1}}$. Then $u^p = a$, hence $[F(u) : F] = p$ by Lemma 1.3.2. Clearly $X^{p^n} - a$ is irreducible if and only if v has degree p^n over F, which will be proved by induction on n. This is true if $n = 1$ by Lemma 1.3.2. By Lemma 1.3.3 (1), $u \notin F(u)^p$, hence the polynomial $X^{p^{n-1}} - u \in F(u)[X]$ is irreducible over $F(u)$ by the inductive hypothesis. Consequently, the root v of $X^{p^{n-1}} - u$ has degree p^{n-1} over $F(u)$. Then, by the Tower Law, it follows that v has degree p^n over F, and we are done.

(2) If $a \in F^2$, then clearly $X^{2^n} - a$ is reducible over F. Now assume that $-4a \in F^4$. Then $-4a = c^4$ for some $c \in F$. Since $\mathrm{Char}(F) \neq 2$, we have $c = 2b$ for some $b \in F$, hence

$$a = -4^{-1}c^4 = -4^{-1}2^4 b^4 = -4b^4.$$

If we set $Y = X^{2^{n-2}}$, then

$$X^{2^n} - a = Y^4 + 4b^4 = Y^4 + 4b^2 Y^2 + 4b^4 - 4b^2 Y^2$$

$$= (Y^2 + 2b^2)^2 - (2bY)^2 = (Y^2 + 2bY + 2b^2)(Y^2 - 2bY + 2b^2),$$

hence $X^{2^n} - a$ is reducible in $F[X]$.

Conversely, assume that $\mathrm{Char}(F) \neq 2$, $a \notin F^2$ and $-4a \notin F^4$. We prove that $X^{2^n} - a$ is irreducible over F by induction on n. Let $v \in \Omega$ be a root of $X^{2^n} - a$, and set $u := v^{2^{n-1}}$. Then $u^2 = a$, hence $[F(u) : F] = 2$ by Lemma 1.3.2. Therefore, as in the proof of (1), the fact that $X^{2^n} - a$ is irreducible over F means precisely that $X^{2^{n-1}} - u$ is irreducible over $F(u)$.

If $n = 2$, then $v^2 = u \notin F(u)^2$ by Lemma 1.3.3 (2), hence $X^2 - u$ is irreducible over $F(u)$.

If $n > 2$, then by the inductive hypothesis, $X^{2^{n-1}} - u \in F(u)[X]$ is irreducible over $F(u)$ if and only if $u \notin F(u)^2$ and $-4u \notin F(u)^4$. Now observe that

$$-4u \in F(u)^4 \implies -u \in F(u)^2$$

since $0 \neq 4 = 2^2 \in F(u)^2$, and

$$-u \in F(u)^2 \iff u \in F(u)^2$$

since the map defined by $u \mapsto -u$ is an element of $\mathrm{Gal}(F(u)/F)$.
Consequently, we have

$$u \notin F(u)^2 \text{ and } -4u \notin F(u)^4 \iff u \notin F(u)^2.$$

But $-4a \notin F^4$, hence $u \notin F(u)^2$ by Lemma 1.3.3 (2). This proves that $X^{2^{n-1}} - u$ is irreducible over $F(u)$, and we are done. $\square$

We are now in a position to state and prove the main result of this section. We shall use the notation $-4F^4 := \{ -4b^4 \mid b \in F \}$.

THEOREM 1.3.5 (THE VAHLEN-CAPELLI CRITERION). *Let F be an arbitrary field, let $n \in \mathbb{N}^*$, and let $a \in F$. Then, the following assertions are equivalent.*

(1) *The polynomial $X^n - a$ is irreducible in $F[X]$.*

(2) *$a \notin F^p$ for all $p \in \mathbb{P}_n$, and $a \notin -4F^4$ whenever $4 \mid n$.*

PROOF. Of course, the result is vacuously true for $n = 1$, so we can assume that $n \geqslant 2$. Let p be a prime divisor of n. If $a \in F^p$, then $a = b^p$ for some $b \in F$, and $X^n - a$ is divisible in $F[X]$ by $X^m - b$, where $m = n/p$, hence $X^n - a$ is reducible in $F[X]$.

If $4 \mid n$ and $a = -4b^4$ for some $b \in F$, then $X^n - a$ is a reducible polynomial in $F[X]$ since

$$X^n - a = X^n + 4b^4 = (X^{n/2} - 2bX^{n/4} + 2b^2)(X^{n/2} + 2bX^{n/4} + 2b^2)$$

This proves the implication (1) $\implies$ (2).

Conversely, assume that condition (2) is satisfied. Let

$$n = p_1^{k_1} \cdot \ldots \cdot p_r^{k_r}$$

be the decomposition of n as a product of mutually distinct prime numbers $p_1, \ldots, p_r$, with $r, k_1, \ldots, k_r \in \mathbb{N}^*$. By Lemma 1.3.1, we only need to check that $X^{p_i^{r_i}} - a$ is irreducible in $F[X]$ for every i, $1 \leqslant i \leqslant r$. This follows from Lemma 1.3.2, Lemma 1.3.4, and the following simple fact: if $\text{Char}(F) \neq 2$ and $a \notin -4F^4$, then $-4a \notin F^4$. □

The next result is an alternative version of the Vahlen-Capelli Criterion.

THEOREM 1.3.6 (A VARIANT OF THE VAHLEN-CAPELLI CRITERION). *Let F be an arbitrary field, let $n \in \mathbb{N}$, $n \geqslant 2$, and let $a \in F^*$. Then, the polynomial $X^n - a$ is reducible in $F[X]$ if and only if one of the following two conditions is satisfied.*

(1) *$a \in F^s$ for some $s \in \mathbb{N}$, $s \geqslant 2$, $s \mid n$, or*

(2) *$4 \mid n$ and $-4a \in F^{*4}$.*

PROOF. First, observe that for any $a \in F^*$, one has

$$-4a \in F^{*4} \iff a \in -4F^{*4}.$$

Indeed, if $-4a \in F^{*4}$, then $-4a = b^4$ for some $b \in F^*$ and $4 \neq 0$, hence $\text{Char}(F) \neq 2$. Thus $b = 2c$ for some $c \in F^*$, and so, $-4a = 16c^4$. Multiplying the last equality by $(-4)^{-1}$ we find that $a = -4c^4 \in -4F^{*4}$. The implications above can be clearly reversed.

By negation, Theorem 1.3.5 is equivalent to the following statement: $X^n - a$ is reducible in $F[X]$ if and only if either $a \in F^p$ for some prime divisor p of n, or $-4a \in F^{*4}$ whenever $4 \mid n$. To conclude, observe that the divisor s in (1) can be chosen to be prime. $\square$

REMARK 1.3.7. Occasionally, the Vahlen-Capelli Criterion is wrongly stated in the literature, as for instance in Bastida [**33**, p. 154]:

Let F be an arbitrary field, let $n \in \mathbb{N}^*$, and let $a \in F$. Then, $X^n - a$ is a reducible polynomial in $F[X]$ if and only if either $a \in F^p$ for some $p \in \mathbb{P}_n$, or $4 \mid n$ and $-4a \in F^4$.

To see that the statement above is wrong, take as F any non perfect field of characteristic 2, e.g., $\mathbb{F}_2(Y)$, and as a any element in $F \setminus F^2$. Then, the polynomial $X^4 - a$ is irreducible in $F[X]$ by Theorem 1.3.5 since $a \notin -4F^4 = \{0\}$, but $-4a = 0 \in F^4$. $\square$

Next we will show that, imposing additional conditions on F or on the binomial $X^n - a$, the somewhat uncomfortable condition

$$\text{``} a \notin -4F^4 \text{ whenever } 4 \mid n \text{''}$$

in the Vahlen-Capelli Criterion can be removed.

DEFINITIONS 1.3.8. *The field F is said to satisfy the condition $C_0(n; a)$ (resp. $C_1(n; a)$), where $n \in \mathbb{N}^*$ and $a \in F^*$, if the binomial $X^n - a$ has a root in Ω, say $\sqrt[n]{a}$, such that $\mu_n(F(\sqrt[n]{a})) \subseteq F$ (resp. $\mu_n(F(\sqrt[n]{a})) \subseteq \{-1, 1\}$).* $\square$

Whenever the field F satisfies the condition $C_0(n; a)$ (resp. $C_1(n; a)$), then we shall always denote in the sequel by $\sqrt[n]{a}$ a specified root in Ω of the polynomial $X^n - a$ such that $\mu_n(F(\sqrt[n]{a})) \subseteq F$ (resp. $\mu_n(F(\sqrt[n]{a})) \subseteq \{-1, 1\}$).

EXAMPLES 1.3.9. (1) Any field satisfying the condition $C_1(n; a)$ clearly satisfies the condition $C_0(n; a)$. The condition $C_1(n; a)$ is however stronger than the condition $C_0(n; a)$: indeed, the field $\mathbb{C}$ satisfies the condition $C_0(n; a)$ for any $n \in \mathbb{N}$, $n \geqslant 2$ and any $a \in \mathbb{C}^*$, but it does not satisfy the condition $C_1(2, -1)$.

(2) If $\zeta_n \in F$ for some $n \in \mathbb{N}^*$, then obviously F satisfies the condition $C_0(n; a)$ for any $a \in F^*$.

(3) Any subfield F of $\mathbb{R}$ satisfies the condition $C_1(n; a)$ for any odd number $n \in \mathbb{N}^*$ and any $a \in F^*$, as well as for any $n \in \mathbb{N}^*$ and any $a \in \mathbb{R}_+^*$.

(4) The field $\mathbb{Q}$ does not satisfy the condition $C_0(4; -4)$. Indeed the complex roots of the polynomial $X^4 + 4$ are $1 + i$, $1 - i$, $-1 + i$, $-1 - i$, hence, if $\sqrt[4]{-4}$ denotes any of these roots, then $\mathbb{Q}(\sqrt[4]{-4}) = \mathbb{Q}(i)$, and so, $i \in \mu_4(\mathbb{Q}(\sqrt[4]{-4})) \setminus \mathbb{Q}$. $\qquad \square$

PROPOSITION 1.3.10. *The following assertions hold for a field F satisfying the condition $C_0(n; a)$.*

(1) *The polynomial $X^n - a$ is reducible in $F[X]$ if and only if $a \in F^s$ for some divisor $s > 1$ of n.*

(2) $\mathrm{Min}(\sqrt[n]{a}, F) = X^m - b$, *where* $m = \mathrm{ord}(\widehat{\sqrt[n]{a}})$, $m \mid n$, *and* $b = \sqrt[n]{a}^m$.

PROOF. (1) Let $f = \mathrm{Min}(\sqrt[n]{a}, F)$ and $m = \deg(f)$. Then f is a product in $\Omega[X]$ of m linear polynomials of type $X - \zeta_n^j \sqrt[n]{a}$, hence the constant term b_0 of f has the form

$$b_0 = \pm \zeta_n^r \sqrt[n]{a}^m.$$

It follows that

$$\zeta_n^r = \pm b_0 \sqrt[n]{a}^{-m} \in \mu_n(\Omega) \cap F(\sqrt[n]{a}) = \mu_n(F(\sqrt[n]{a})) \subseteq F$$

since F satisfies the condition $C_0(n; a)$. So, $\sqrt[n]{a}$ is a root of the polynomial $X^m - b \in F[X]$, where $b = \pm b_0 \zeta_n^{-r}$. This proves that $f = \mathrm{Min}(\sqrt[n]{a}, F) = X^m - b$.

Clearly, if $a \in F^s$ for some divisor $s > 1$ of n, then $X^n - a$ is reducible in $F[X]$. Conversely, suppose that $X^n - a$ is reducible in $F[X]$. Then $1 \leqslant m < n$. If $d = \gcd(m, n)$, then we can write $n = ds$ and $m = dt$ for some $m, n \in \mathbb{N}^*$. Since $\gcd(s, t) = 1$, there exist $u, v \in \mathbb{Z}$ such that $ut + vs = 1$.

But $a = \sqrt[n]{a}^{ds}$ and $b = \sqrt[n]{a}^{dt}$, hence $a^t = b^s$, and so

$$a = a^{ut+vs} = a^{vs} b^{us} = (a^v b^u)^s \in F^s.$$

Notice that $s > 1$ since $m < n$, and $s \mid n$.

(2) We have already proved that $\mathrm{Min}(\sqrt[n]{a}, F) = X^m - b$, hence it remains only to show that $m = \mathrm{ord}(\widehat{\sqrt[n]{a}})$ and $m \mid n$. If we set $k = \mathrm{ord}(\widehat{\sqrt[n]{a}})$, then $c := \sqrt[n]{a}^k \in F$, and k is the least $r \in \mathbb{N}^*$ such that $\sqrt[n]{a}^r \in F$. Since $\sqrt[n]{a}^n = a \in F$, we deduce that $k \mid n$ by the definition of k.

We claim that $\mathrm{Min}(\sqrt[n]{a}, F) = X^k - c$, in other words, $X^k - c$ is irreducible in $F[X]$. Indeed, suppose the contrary, i.e., suppose that $X^k - c$ is reducible in $F[X]$. Since $k \mid n$, we have $\mu_k(F(\sqrt[n]{a})) \subseteq \mu_n(F(\sqrt[n]{a})) \subseteq F$, hence F satisfies the condition $C_0(k; c)$. By (1), applied to the polynomial

$X^k - c$, we deduce that $c = e^i$ for some $e \in F^*$ and $i \in \mathbb{N}$, $i > 1$, $i \mid k$. Then $k = ij > j$ for some $j \in \mathbb{N}^*$, and

$$c = \sqrt[n]{a}^k = \sqrt[n]{a}^{ij} = \left(\sqrt[n]{a}^j \right)^i = e^i,$$

so $\sqrt[n]{a}^j \cdot e^{-1} \in \mu_i(\Omega) \cap F(\sqrt[n]{a}) = \mu_i(F(\sqrt[n]{a})) \subseteq \mu_n(F(\sqrt[n]{a})) \subseteq F$. Thus $\sqrt[n]{a}^j \in F^*$ with $1 \leqslant j < k$, which contradicts the definition of k.

Consequently, $\mathrm{Min}(\sqrt[n]{a}, F) = X^k - c = X^m - b$, hence $m = k$, $m \mid n$, and $m = \mathrm{ord}(\widehat{\sqrt[n]{a}})$. The proof is now complete.   □

REMARK 1.3.11. Assertion (1) in Proposition 1.3.10 can be restated as follows:

(1') *The polynomial* $X^n - a$ *is irreducible in* $F[X]$ *if and only if* $a \notin F^p$ *for every* $p \in \mathbb{P}_n$.   □

PROPOSITION 1.3.12. *Any field* F *satisfies the condition* $C_0(n; a)$ *for any* $n \in \mathbb{P}$ *and any* $a \in F^*$.

PROOF. If $n = e(F)$, then $\mu_n(\Omega) = \{1\}$ and we have nothing to prove. We can therefore assume that $n \neq e(F)$. Since $\mu_n(F) = \mu_n(\Omega) \cap F^*$ is a subgroup of the group $\mu_n(\Omega)$ of prime order n, we have either $\mu_n(F) = \mu_n(\Omega)$ or $\mu_n(F) = \{1\}$. If $\mu_n(F) = \mu_n(\Omega)$ then clearly $\mu_n(\Omega) \subseteq F$.

Now assume that $\mu_n(F) = \{1\}$. Let $\sqrt[n]{a} \in \Omega$ be an arbitrary root in Ω of $X^n - a$ if $a \notin F^n$, and let $\sqrt[n]{a} = b$ if $a = b^n$ for some $b \in F$. We claim that $\mu_n(F(\sqrt[n]{a})) \neq \mu_n(\Omega)$.

This is clear when $a \in F^n$ because $F(\sqrt[n]{a}) = F(b) = F$, hence $\mu_n(F(\sqrt[n]{a})) = \mu_n(F) = \{1\} \neq \mu_n(\Omega)$.

Now consider the case when $a \notin F^n$, and suppose that $\mu_n(F(\sqrt[n]{a})) = \mu_n(\Omega)$. Then $\zeta_n \in F(\sqrt[n]{a})$, hence $F(\zeta_n) \subseteq F(\sqrt[n]{a})$, and so

$$[F(\zeta_n) : F] \mid [F(\sqrt[n]{a}) : F].$$

On the other hand, $[F(\sqrt[n]{a}) : F] = n$ by Lemma 1.3.2, and $[F(\zeta_n) : F] \leqslant \varphi(n) = n - 1$, where φ is the Euler function. Consequently, $[F(\zeta_n) : F] = 1$, i.e., $\zeta_n \in F$. It follows that $\zeta_n \in \mu_n(F) = \{1\}$, which is a contradiction.

This proves our claim: $\mu_n(F(\sqrt[n]{a})) \neq \mu_n(\Omega)$, and then, necessarily $\mu_n(F(\sqrt[n]{a})) = \{1\}$. Thus F satisfies the condition $C_0(n; a)$.   □

1.4. Bounded Abelian groups

In this section we present some basic properties of Abelian groups of bounded order which will be frequently used in the sequel.

Throughout this section G will denote an arbitrary multiplicative group with identity element e. If $n \in \mathbb{N}^*$, we set $G^n = \{\, x^n \mid x \in G \,\}$. For any torsion group G we will use the following notation:

$$\mathcal{O}_G = \{\, \mathrm{ord}(x) \mid x \in G \,\}.$$

Recall that a group G is said to be a *torsion group* if every element of G has finite order.

If $a, b \in \mathbb{N}$, then $\gcd(a, b)$ (resp. $\mathrm{lcm}(a, b)$) will denote the greatest common divisor (resp. the least common multiple) of a and b. For a nonempty finite set A of natural numbers, $\mathrm{lcm}(A)$ will denote the least common multiple of all numbers of A, and $\max(A)$ will denote the greatest number of A.

DEFINITION 1.4.1. *A group G is said to be a* group of bounded order *if G is a torsion group and the subset $\mathcal{O}_G$ of $\mathbb{N}$ is a bounded set, or equivalently, a finite set.* $\qquad\square$

Clearly, any finite group is a group of bounded order, and any direct sum or direct product of infinitely many copies of a finite Abelian group of order $n > 1$ is an infinite group of bounded order.

The structure of Abelian groups of bounded order is given by the following classical result.

THEOREM 1.4.2. *Any Abelian group of bounded order is isomorphic to a direct sum of finite cyclic groups.*

PROOF. See Kaplansky [**73**, Theorem 6]. $\qquad\square$

As is well-known, any torsion Abelian group G is the internal direct sum of its p-primary components $t_p(G)$, $p \in \mathbb{P}$. Using this fact, Theorem 1.4.2 implies that an Abelian group G is a group of bounded order if and only if there exists a finite subset $\{p_1, \dots, p_r\}$ of the set $\mathbb{P}$ of all positive prime numbers and a finite family $(G_i)_{1 \leqslant i \leqslant r}$ of Abelian groups, such that

$$G \cong \bigoplus_{i=1}^{r} G_i \quad \text{and} \quad G_i \cong \bigoplus_{j \in A_i} \mathbb{Z}_{p_i^{n_{ij}}},$$

where $A_1, \dots, A_r$ are arbitrary sets, $n_{ij} \in \mathbb{N}^*$, and for each $i = 1, \dots, r$, the set $\{\, n_{ij} \mid j \in A_i \,\}$ is a bounded subset of $\mathbb{N}^*$.

Using this description, the results of this section about orders and exponents become almost obvious. In fact, they are the same as if the considered Abelian groups of bounded order were finite. However, we prefer a more elementary approach which does not refer to Theorem 1.4.2.

LEMMA 1.4.3. *The following statements hold for an arbitrary Abelian group G.*

(1) *For any $a \in \mathcal{O}_G$ and any $d \in \mathbb{N}$ with $d \,|\, a$ one has $d \in \mathcal{O}_G$.*

(2) *For any $a, b \in \mathcal{O}_G$ one has $\mathrm{lcm}(m, n) \in \mathcal{O}_G$, and so, $\mathrm{lcm}(A) \in \mathcal{O}_G$ for any finite nonempty subset A of $\mathcal{O}_G$.*

(3) *If G is a group of bounded order and $m = \max(\mathcal{O}_G)$, then every element of $\mathcal{O}_G$ divides m.*

PROOF. (1) If $a \in \mathcal{O}_G$ and $d \,|\, a$, $d \in \mathbb{N}$, then $a = \mathrm{ord}(x)$ for some $x \in G$, hence $d = \mathrm{ord}(x^{a/d})$, and so $d \in \mathcal{O}_G$.

(2) First, we observe that if $a, b \in \mathcal{O}_G$, with $\gcd(a, b) = 1$, then $ab \in \mathcal{O}_G$. Indeed, if $a = \mathrm{ord}(x)$, $b = \mathrm{ord}(y)$, with $x, y \in G$, it is known that $ab = \mathrm{ord}(xy)$, so $ab \in \mathcal{O}_G$. By induction one proves that if $a_1, \ldots, a_r \in \mathcal{O}_G$, with $\gcd(a_i, a_j) = 1$ for all $i \neq j$ in $\{1, \ldots, r\}$, then $a_1 \cdot \ldots \cdot a_r \in \mathcal{O}_G$.

We are now going to prove that if $a, b \in \mathcal{O}_G$, then $\mathrm{lcm}(a, b) \in \mathcal{O}_G$. Clearly, we can suppose that $a > 1$ and $b > 1$. If

$$a = p_1^{\alpha_1} \cdot \ldots \cdot p_s^{\alpha_s} \text{ and } b = p_1^{\beta_1} \cdot \ldots \cdot p_s^{\beta_s},$$

with $p_1, \ldots, p_s$ mutually distinct primes and $\alpha_1, \ldots, \alpha_s, \beta_1, \ldots, \beta_s \in \mathbb{N}$, then $p_i^{\max(\alpha_i, \beta_i)}$ divides a or b, so $p_i^{\max(\alpha_i, \beta_i)} \in \mathcal{O}_G$, $i = 1, \ldots, s$, by (1). Consequently, $\mathrm{lcm}(a, b) = p_1^{\max(\alpha_1, \beta_1)} \cdot \ldots \cdot p_s^{\max(\alpha_s, \beta_s)} \in \mathcal{O}_G$.

If A is an arbitrary nonempty finite subset of $\mathcal{O}_G$, then the fact that $\mathrm{lcm}(A) \in \mathcal{O}_G$ follows by induction on $|A|$.

(3) Let $a \in \mathcal{O}_G$. By (2), we have $\mathrm{lcm}(a, m) \in \mathcal{O}_G$, so

$$\mathrm{lcm}(a, m) \leqslant \max(\mathcal{O}_G) = m.$$

But $m \leqslant \mathrm{lcm}(a, m)$, hence $m = \mathrm{lcm}(a, m)$. Thus $a \,|\, m$, as desired.   □

COROLLARY 1.4.4. If G is an Abelian group of bounded order, and $m = \max(\mathcal{O}_G)$, then $G^m = \{e\}$ and $\mathrm{lcm}(\mathcal{O}_G) \,|\, m$.

PROOF. Apply Lemma 1.4.3.   □

DEFINITION 1.4.5. *Let G be a group of bounded order. The least number $n \in \mathbb{N}^*$ with the property that $G^n = \{e\}$ is called the* exponent *of G*

and is denoted by $\exp(G)$. *The group* G *is said to be* n-bounded *if* G *is a group of bounded order and* $\exp(G) = n$. □

LEMMA 1.4.6. *Let* G *be an* n-*bounded Abelian group. If* $G^k = \{e\}$ *for some* $k \geqslant 1$, *then* $n \mid k$.

PROOF. By the Division Theorem we can write $k = nq + r$, with $q, r \in \mathbb{N}$, $0 \leqslant r < n$. Since $G^k = \{e\}$ it follows that $G^{nq+r} = \{e\}$, hence $G^r = \{e\}$. But $n \in \mathbb{N}^*$ is minimal with $G^n = \{e\}$, so necessarily $r = 0$. Thus $k = nq$, i.e., $n \mid k$. □

PROPOSITION 1.4.7. *Let* G *be an* n-*bounded Abelian group, and let* $m = \max(\mathcal{O}_G)$, $c = \operatorname{lcm}(\mathcal{O}_G)$. *Then* $n = m = c$.

PROOF. We have $c \mid m$ by Corollary 1.4.4. But $m \in \mathcal{O}_G$, so $m \mid c$ by the definition of $c = \operatorname{lcm}(\mathcal{O}_G)$. We deduce that $m = c$.

By Corollary 1.4.4 we have $G^m = \{e\}$, hence $n \mid m$ by Lemma 1.4.6. Since $m \in \mathcal{O}_G$ there exists $x_0 \in G$ with $m = \operatorname{ord}(x_0)$. But $x_0^n = e$ and consequently $\operatorname{ord}(x_0) \mid n$, i.e., $m \mid n$. From $n \mid m$ and $m \mid n$ it follows that $m = n$. □

Recall that for a given $n \in \mathbb{N}$ we denote by $\mathbb{D}_n$ the set of all divisors $d \in \mathbb{N}$ of n, and by $\mathcal{P}_n$ the set of all divisors $p > 0$ of n, with p odd prime or 4.

PROPOSITION 1.4.8. *Let* G *be an* n-*bounded Abelian group. Then, for any* $d \in \mathbb{N}$, $d \mid n$, *there exists* $x_d \in G$ *such that* $d = \operatorname{ord}(x_d)$. *In particular, for every* $p \in \mathcal{P}_n$ *there exists* $x_p \in G$ *such that* $p = \operatorname{ord}(x_p)$.

PROOF. By Proposition 1.4.7 we have $n = m = \max(\mathcal{O}_G)$, and by Lemma 1.4.3 (1) we have $d \in \mathcal{O}_G$ for every $a \in \mathcal{O}_G$ and $d \mid a$. Since $m \in \mathcal{O}_G$, we deduce that $d \in \mathcal{O}_G$ for every divisor d of $m = n$. Thus $d = \operatorname{ord}(x_d)$ for some $x_d \in G$. □

REMARK 1.4.9. Proposition 1.4.8 can be reformulated as follows: *If* G *is any* n-*bounded Abelian group, then* $\mathcal{O}_G = \mathbb{D}_n$. □

The connection between the exponent and the order of a finite Abelian group is given by the next result.

PROPOSITION 1.4.10. *For any finite Abelian group* G, $\exp(G)$ *divides* $|G|$, *and* $|G|$ *divides a power of* $\exp(G)$. *In particular,* $|G|$ *and* $\exp(G)$ *have the same prime divisors.*

PROOF. Set $n = |G|$ and $m = \exp(G)$. Since clearly $G^n = \{e\}$, one has $m \mid n$ by Lemma 1.4.6.

Now we shall prove that n divides a power of m by induction on $n = |G|$. The statement is clear for $n = 1$ and $n = 2$. Let G be an Abelian group of order $n \geqslant 3$, and assume that the statement is true for any Abelian group of order $< n$. Pick an element $g \in G$, $g \neq e$ and set $H = \langle g \rangle$. Then $|H| = \mathrm{ord}(g)$ divides $m = \exp(G)$ since $g^m = e$. Since G is an Abelian group, we can consider the quotient Abelian group G/H which has order $< n$. By the induction hypothesis, $|G/H|$ divides a power of $\exp(G/H)$.

On the other hand, $\exp(G/H)$ divides $\exp(G)$ by Lemma 1.4.6, $|H|$ divides $\exp(G)$, and $n = |G| = |G/H| \cdot |H|$, hence n divides a power of m, as desired.   □

COROLLARY 1.4.11 (CAUCHY'S LEMMA). *For any finite Abelian group G and any prime divisor p of $|G|$ there exists at least an element $x_p \in G$ with $\mathrm{ord}(x_p) = p$.*

PROOF. Apply Proposition 1.4.10 and Proposition 1.4.8.   □

1.5. Exercises to Chapter 1

1. Show that the field $\mathbb{Q}$ does not satisfy the condition $C_1(4; -9)$, but satisfies the condition $C_1(4; -2)$.

2. Let F be a field satisfying the condition $C_0(n; a)$. If $\widehat{\widehat{a}}$ denotes the coset of a in the quotient group F^*/F^{*n} and d is the greatest divisor of n such that $a \in F^{*d}$, then prove that

$$\mathrm{ord}(\widehat{\sqrt[n]{a}}) = \mathrm{ord}(\widehat{\widehat{a}}) = n/d.$$

3. Find $[\mathbb{Q}(\sqrt[60]{67500}) : \mathbb{Q}]$.

4. Let F be a field satisfying the condition $C_0(n; a)$, and let $m = [F(\sqrt[n]{a}) : F]$. Show that the map
   $$\alpha : \mathbb{D}_m \longrightarrow \underline{\mathrm{Intermediate}}\,(F(\sqrt[n]{a})/F)), \quad \alpha(d) = F(\sqrt[n]{a}^{\,d}),$$
   establishes an anti-isomorphism of lattices.

5. Find all subfields of the field $\mathbb{Q}(\sqrt[n]{a})$, where $n \in \mathbb{N}^*$ and $a \in \mathbb{Q}_+^*$.

6. Let $p \in \mathbb{P}$ and let F be a field such that $\mu_p(F) = \{1\}$. Prove that F satisfies the condition $C_1(p; a)$ for any $a \in F^*$.

7. Prove that any finite field $\mathbb{F}_q$ satisfies the condition $C_1(p;a)$ for any $a \in \mathbb{F}_q^*$ and any $p \in \mathbb{P}$ with $\gcd(p, q-1) = 1$.

8. Let F be a field satisfying the condition $C_1(n;a)$, with $n \geqslant 3$. Prove that for any Abelian extension E/F which is not necessarily finite, the following assertions hold.
   (a) If n is odd, then $\sqrt[n]{a} \in E \iff \sqrt[n]{a} \in F$.
   (b) If $n = 4$, then $\sqrt[4]{a} \in E \implies \sqrt[4]{a}^2 \in F$.

9. Show that the implication " $\impliedby$ " in Exercise 8 (b) does not hold in general.

10. With notation and hypotheses of Exercise 8, show that if $n = 4$ and $\sqrt[4]{a} \in E$ then we may have $\sqrt[4]{a} \notin F$.

11. Show that both statements (a) and (b) in Exercise 8 may fail for fields which do not satisfy the condition $C_1(n;a)$.

12. (*Bourbaki* [40]). Let F be a field of characteristic p, let $a \in F^*$, let $n \in \mathbb{N}^*$ with $p \nmid n$, and let N be a splitting field of the polynomial $X^n - a$.
   (a) Prove that the group $\mathrm{Gal}(N/F)$ is isomorphic to a subgroup of the group Γ of all matrices $\left(\begin{smallmatrix} x & y \\ 0 & 1 \end{smallmatrix}\right)$ with $x \in U(\mathbb{Z}_n)$ and $y \in \mathbb{Z}_n$.
   (b) Show that if n is a prime number, then the group $\mathrm{Gal}(N/F)$ is Abelian if and only if $\zeta_n \in F$ or $a \in F^n$.
   (c) Show that if $F = \mathbb{Q}$, $n \in \mathbb{P}$, and $a \notin \mathbb{Q}^n$, then $\mathrm{Gal}(N/K)$ is isomorphic to the whole group Γ.

13. With notation and hypotheses of Exercise 12, prove that if F satisfies the condition $C_1(n;a)$ and $n \neq 4$, then $\mathrm{Gal}(N/F)$ is Abelian if and only if $a \in F^n$. (*Hint*: Use Exercise 8.)

14. Use Exercise 13 and Proposition 1.3.12 to prove Exercise 12 (b).

15. Find $\mathcal{O}_{\mathfrak{S}_n}$ for $n \leqslant 4$.

16. Show that if G is a finite cyclic group of order n, then $\mathcal{O}_G = \mathbb{D}_n$.

17. Show that Lemma 1.4.3 may fail for a non Abelian finite group G.

18. Let G be a finite group of order n, and consider the canonical map
$$\omega_G : \underline{\mathrm{Subgroups}}(G) \longrightarrow \mathbb{D}_n, \quad H \mapsto |H|.$$
   Prove that the following assertions are equivalent.
   (a) G is a cyclic group.

    (b) The map ω_G is injective.

    (c) The map ω_G is a lattice isomorphism.

19. Using Lemma 1.4.3 or Exercise 18, prove that any finite subgroup of the multiplicative group F^* of any field F is cyclic. In particular, for any finite field $\mathbb{F}_q$, the group $\mathbb{F}_q^*$ is cyclic.

20. (*Darbi* [**53**], [**60**]). Let F be a field of characteristic 0, and let $f = X^n - a \in F[X]$ be an irreducible binomial with root $u \in \Omega$. Let $k = \max\{\, m \in \mathbb{D}_n \mid u^{n/m} \in F(\zeta_n)\,\}$. Prove that the degree of the splitting field of f over F is $[\, F(\zeta_n) : F\,] \cdot n/k$.

21. Prove that the following statements hold for a finite extension E/F.

    (a) $|\mathrm{Gal}(E/F)| \leqslant [\,E : F\,]$.

    (b) $|\mathrm{Gal}(E/F)| = [\,E : F\,]$ if and only if the extension E/F is Galois.

1.6. Bibliographical comments to Chapter 1

Section 1.1. The basic terminology and notation we use in this monograph are the standard ones, as those from Bourbaki [**40**], Kaplansky[**74**], Karpilovsky[**76**], and Lang [**80**].

Section 1.2. The review presented in this section is mainly based on Bourbaki [**40**], Kaplansky [**74**], Karpilovsky [**76**], and Lang [**80**].

Section 1.3. Lemma 1.3.2 is due to Abel. The Vahlen-Capelli Criterion was first proved in 1895 by Vahlen [**107**] for $\mathbb{Q}$, and it was extended in 1897, 1898 by Capelli [**43**], [**44**] and in 1900 by Wendt [**112**] to arbitrary fields of characteristic zero. A summary of Capelli's results is given in his short Note [**45**] which appeared in 1901. The general result for fields of arbitrary characteristic was proved only in 1959 by Rédei [**88**, §172]. The proof of the Vahlen-Capelli Criterion we gave in this section basically follows the outline in Kaplansky [**74**], although there are several additions, simplifications, and improvements. The conditions $C_0(n; a)$ and $C_1(n; a)$ applied for fields were introduced by Albu [**3**] under different names.

Section 1.4. Most of the results in this section are essentially folklore.

CHAPTER 2

KNESER EXTENSIONS

In this chapter we introduce and investigate two basic concepts of Cogalois Theory, that of *G-radical extension* and that of *G-Kneser extension*. Then we prove a criterion due to Kneser [**77**], which characterizes Kneser extensions and plays a fundamental role in this monograph.

Roughly speaking, a radical extension is an extension E/F such that E is obtained by adjoining to the base field F an arbitrary set of "radicals" over F, i.e., of elements $x \in E$ such that $x^n = a \in F$ for some $n \in \mathbb{N}^*$. Such an x will be denoted by $\sqrt[n]{a}$ and will be called an *n*-th *radical* of x. Notice that this terminology is somewhat different from that used in Galois Theory (see e.g., Kaplansky [**74**], Karpilovsky [**76**], Lang [**80**]), but coincides for simple extensions. So, E/F is a radical extension when $E = F(R)$, where R is a set of radicals over F. Clearly, one can replace R by the subgroup $G = F^* \langle R \rangle$ of the multiplicative group E^* of E generated by F^* and R. Thus, any radical extension E/F has the form $E = F(G)$, where G is a subgroup of E^* containing F^*, with G/F^* a torsion group. Such an extension is called *G-radical*. A finite extension E/F is said to be *G-Kneser* when it is G-radical and $|G/F^*| = [E : F]$.

2.1. *G*-Radical and *G*-Kneser extensions

In this section we introduce and investigate two basic notions of this monograph, that of G-radical extension and that of G-Kneser extension.

For any extension E/F we shall use throughout this monograph the following notation.

$$T(E/F) := \{\, x \in E^* \mid x^n \in F^* \text{ for some } n \in \mathbb{N}^* \,\}.$$

Clearly $F^* \leqslant T(E/F)$, so it makes sense to consider the quotient group $T(E/F)/F^*$. Remember that for any $x \in E$ we denote by $\widehat{x}$ the coset xF^* of x modulo F^* in the group E^*/F^*.

Observe that the quotient group $T(E/F)/F^*$ is precisely the torsion group $t(E^*/F^*)$ of the quotient group E^*/F^*. This group, playing a major role in this monograph, is somewhat dual to the Galois group of E/F, which explains the terminology below.

DEFINITIONS 2.1.1. *The* Cogalois group *of an arbitrary field extension* E/F, *denoted by* $\text{Cog}\,(E/F)$, *is the quotient group* $T(E/F)/F^*$.

A finite extension E/F *is said to be a* Cogalois extension *if* $E = F(T(E/F))$ *and* $|\text{Cog}\,(E/F)| = [\,E : F\,]$. □

In the next chapter we will characterize and investigate Cogalois extensions and will compute the Cogalois group of some extensions.

Observe that for every element in $x \in T(E/F)$ there exists an $n \in \mathbb{N}^*$ such that $x^n = a \in F$. Such an x is usually denoted by $\sqrt[n]{a}$ and is called an n-th *radical* of a. Thus, $T(E/F)$ is precisely the set of all "radicals" belonging to E of elements of F. This observation suggests to define a *radical extension* as being an extension E/F such that E is obtained by adjoining to the base field F an arbitrary set of "radicals" over F. More precisely, we have the following definition.

DEFINITION 2.1.2. *An extension* E/F *is said to be* radical *if there exists* $A \subseteq T(E/F)$ *such that* $E = F(A)$. *We say that* E/F *is a* simple radical extension *if there exists an* $a \in T(E/F)$ *such that* $E = F(a)$. □

Clearly, one can replace the set A in Definition 2.1.2 by the subgroup $G = F^*\langle A \rangle$ of the multiplicative group E^* of E generated by F^* and A. So, any radical extension E/F has the form $E = F(G)$, where G is a subgroup of E^* containing F^*, with G/F^* a torsion group. For such extensions we will use a special name.

DEFINITION 2.1.3. *Let* E/F *be an extension and let* G *be a group. Then,* E/F *is said to be a* G-radical extension *if* $F^* \leqslant G \leqslant T(E/F)$ *and* $E = F(G)$. □

REMARKS 2.1.4. (1) Clearly, a G-radical extension E/F is also G'-radical for any G' with $G \leqslant G' \leqslant T(E/F)$.

(2) Let $F \subseteq K \subseteq E$ be a tower of fields. If E/F is a radical extension, then clearly so is also the extension E/K. We will see in Section 3.2 that, in general, the extension K/F is not radical, and if K/F and E/K are both radical extensions, then E/F is not necessarily so.

(3) Any radical extension is clearly algebraic. This implies that if E/F is a G-radical extension, then $E = F(G) = F[G]$. □

EXAMPLES 2.1.5. The examples below show that the Cogalois group of a finite radical extension could be infinite.

(1) Consider the extension $\mathbb{F}_2(X)/\mathbb{F}_2(X^2)$ of degree 2. We claim that $\mathrm{Cog}\,(\mathbb{F}_2(X)/\mathbb{F}_2(X^2)) = \mathbb{F}_2(X)^*/\mathbb{F}_2(X^2)^*$ is a countably infinite group isomorphic to a countably infinite direct sum of copies of $\mathbb{Z}_2$. Since each element of this group has order $\leqslant 2$, our claim will follow once we prove that this group is infinite. To do that, observe that there exist infinitely many irreducible polynomials in the ring $\mathbb{F}_2[X]$, and for any distinct irreducible polynomials p_1, p_2 in $\mathbb{F}_2[X]$, their cosets $\widehat{p_1}, \widehat{p_2}$ in $\mathrm{Cog}\,(\mathbb{F}_2(X)/\mathbb{F}_2(X^2))$ are distinct. Indeed, assume that $\widehat{p_1} = \widehat{p_2}$. Then, there exist $f, g \in \mathbb{F}_2[X]$ such that $\gcd(f,g) = 1$ and $p_1/p_2 = f^2/g^2$, that is, $p_1 g^2 = p_2 f^2$. It follows that both polynomials f and g are nonconstant, so, decomposing them into irreducible polynomials, and using the fact that $\mathbb{F}_2[X]$ is an UFD, we deduce that necessarily $p_1 = p_2$, which proves our claim.

(2) Similar arguments show that $\mathbb{F}_p(X)/\mathbb{F}_p(X^p)$, where $p > 0$ is any prime number, is an extension of degree p with Cogalois group isomorphic to a countably infinite direct sum of copies of the cyclic group $\mathbb{Z}_p$. $\qquad \square$

LEMMA 2.1.6. *Let E/F be a G-radical extension, which is not necessarily finite. Then any set of representatives of the factor group G/F^* is a set of generators of the F-vector space E. In particular, one has*

$$|G/F^*| \geqslant [E : F].$$

PROOF. Let $T = \{\, t_i \,|\, i \in I \,\}$ be a set of representatives of the factor group G/F^*. Then $G/F^* = \{\, \widehat{t_i} \,|\, i \in I \,\}$. For each $x \in E = F(G) = F[G]$ there exist $\lambda_1, \dots, \lambda_s \in F$ and $g_1, \dots, g_s \in G$ such that $x = \sum_{k=1}^{s} \lambda_k g_k$. But $\widehat{g_k} \in G/F^*$, so $\widehat{g_1} = \widehat{t_{i_1}}, \dots, \widehat{g_s} = \widehat{t_{i_s}}$, where $i_1, \dots, i_s \in I$. It follows that $g_k = \mu_k t_{i_k}$ for some $\mu_k \in F^*$. Consequently

$$x = \sum_{k=1}^{s} (\lambda_k \mu_k) t_{i_k} = \sum_{k=1}^{s} \nu_k t_{i_k},$$

with $\nu_k = \lambda_k \mu_k \in F$. Thus T is a set of generators of the F-vector space E. $\qquad \square$

COROLLARY 2.1.7. *Let E/F be a finite G-radical extension. Then there exists a subgroup H of G such that H/F^* is a finite group and E/F is H-radical.*

PROOF. Let S be a fixed set of representatives of the factor group G/F^*. By Lemma 2.1.6, S is a set of generators of the F-vector space E. Now extract from S a finite basis $B = \{b_1, \ldots, b_n\}$ of this vector space, and set $H = F^*\langle B \rangle$. Since the Abelian group H/F^* is generated by the finite set $\{\widehat{b}_1, \ldots, \widehat{b}_r\}$ consisting of elements of finite order, we deduce that

$$H/F^* = \{ \widehat{b}_1^{\,j_1} \cdot \ldots \cdot \widehat{b}_r^{\,j_r} \mid 0 \leqslant j_1 < n_1, \ldots, 0 \leqslant j_r < n_r \}$$

is a finite group of order $\leqslant n_1 \cdot \ldots \cdot n_r$, where $n_i = \operatorname{ord}(\widehat{b}_i)$ $1 \leqslant i \leqslant r$. Since $E = F[B] = F[H] = F(H)$, we conclude that the extension E/F is H-radical. $\square$

The next result works for G-radical extensions which are not necessarily finite, and will be also used in Part 2 for infinite extensions.

PROPOSITION 2.1.8. *The following assertions are equivalent for an arbitrary G-radical extension E/F.*

(1) *There exists a set of representatives of the factor group G/F^* which is linearly independent over F.*

(2) *Every set of representatives of G/F^* is linearly independent over F.*

(3) *Every set of representatives of G/F^* is a vector space basis of E over F.*

(4) *There exists a set of representatives of G/F^* which is a vector space basis of E over F.*

(5) *Every subset of G consisted of elements having distinct cosets in the group G/F^* is linearly independent over F.*

(6) *Every finite subset $\{x_1, \ldots, x_n\} \subseteq G$ such that $\widehat{x}_i \neq \widehat{x}_j$ for each $i, j \in \{1, \ldots, n\}$, $i \neq j$, is linearly independent over F.*

(7) *For every subgroup H of G such that $F^* \leqslant H$ and H/F^* is a finite group, $|H/F^*| \leqslant [F(H) : F]$.*

PROOF. (1) $\Longrightarrow$ (2): Suppose that there exists a set of representatives T of G/F^* which is linearly independent over F. Let S be an arbitrary set of representatives of G/F^*. Clearly, for each $s \in S$ there exists $t \in T$ with $\widehat{s} = \widehat{t}$, i.e., $s = \lambda t$ for some $\lambda \in F^*$. This implies that S is linearly independent over F because T is so.

(2) $\Longrightarrow$ (3) follows from Lemma 2.1.6.

(3) $\Longrightarrow$ (4), (4) $\Longrightarrow$ (1), and (5) $\Longleftrightarrow$ (6) are obvious.

(3) $\Longrightarrow$ (5): Let $T = \{ t_i \mid i \in I \}$ be a set of representatives of G/F^*. Then T is a basis for E over F. Let $M = \{ x_j \mid j \in J \} \subseteq G$ be such

that $\widehat{x_{j_1}} \neq \widehat{x_{j_2}}$ for each $j_1 \neq j_2$ in J. If $\{x_1, \ldots, x_n\}$ is a finite subset of M, then we have $\widehat{x_1} = \widehat{t_{i_1}}, \ldots, \widehat{x_n} = \widehat{t_{i_n}}$, with $i_1, \ldots, i_n \in I$. It follows that $x_1 = \lambda_1 t_{i_1}, \ldots, x_n = \lambda_n t_{i_n}$ for some $\lambda_1, \ldots, \lambda_n \in F^*$. The set $\{t_{i_1}, \ldots, t_{i_n}\}$ is clearly linearly independent over F, since it is a subset of the basis T. It follows that $\{x_1, \ldots, x_n\}$ is linearly independent over F, so $M = \{\, x_j \mid j \in J\}$ is linearly independent over F.

(5) $\Longrightarrow$ (7): Let H be a subgroup of G such that $F^* \leqslant H$ and H/F^* is a finite group, and let S be a fixed set of representatives of the factor group H/F^*. The set S is a subset of G consisting of elements with distinct classes, and by assumption, it follows that S is linearly independent over F. This shows that $|H/F^*| \leqslant [F(H) : F]$.

(7) $\Longrightarrow$ (2): Let $T = \{\, t_i \mid i \in I\}$ be a set of representatives of the factor group G/F^* and let $\{t_1, \ldots, t_n\}$ be a finite subset of T. Consider the subgroup $\langle \widehat{t_1}, \ldots, \widehat{t_n} \rangle$ of G/F^* generated by $\{\widehat{t_1}, \ldots, \widehat{t_n}\}$. This subgroup is finite since each $\widehat{t_1}, \ldots, \widehat{t_n}$ has finite order in G/F^*. Clearly $\langle \widehat{t_1}, \ldots, \widehat{t_n} \rangle = H/F^*$ for some $H \leqslant G$ with $F^* \leqslant H$. By assumption, we have $|H/F^*| \leqslant [F(H) : F]$, and by Lemma 2.1.6, applied to the H-radical extension $F(H)/F$, the opposite inequality also holds. Thus $|H/F^*| = [F(H) : F]$, hence any set of representatives of the finite group H/F^* is a vector space basis of $F(H)$ over F. Since $\{t_1, \ldots, t_n\}$ is a subset of such a set of representatives, it follows that it is linearly independent over F. This proves that T is linearly independent over F. $\qquad\square$

As we showed in Lemma 2.1.6, for any G-radical extension E/F we have the inequality $|G/F^*| \geqslant [E : F]$. The finite extensions for which we have the opposite inequality deserve a special name.

DEFINITIONS 2.1.9. *A finite extension E/F is said to be G-Kneser if it is a G-radical extension such that $|G/F^*| \leqslant [E : F]$. The extension E/F is called* Kneser *if it is G-Kneser for some group G.* $\qquad\square$

COROLLARY 2.1.10. *The following assertions are equivalent for a finite G-radical extension E/F.*

(1) *E/F is G-Kneser.*
(2) *$|G/F^*| = [E : F]$.*
(3) *Every set of representatives of G/F^* is a vector space basis of E over F.*
(4) *There exists a set of representatives of G/F^* which is linearly independent over F.*

(5) *Every finite subset* $\{x_1, \ldots, x_n\} \subseteq G$ *such that* $\widehat{x_i} \neq \widehat{x_j}$ *for each* $i, j \in \{1, \ldots, n\}$, $i \neq j$, *is linearly independent over* F.

(6) *The extension* $F(H)/F$ *is* H-*Kneser for every* H, $F^* \leqslant H \leqslant G$.

PROOF. Apply Lemma 2.1.6 and Proposition 2.1.8. □

PROPOSITION 2.1.11. *Let* E/F *be a finite* G-*Kneser extension. Then, the extension* $F(H)/F$ *is* H-*Kneser and* $F(H) \cap G = H$ *for every* H *with* $F^* \leqslant H \leqslant G$.

PROOF. The extension $F(H)/F$ is H-Kneser by Corollary 2.1.10. We are going to prove the equality $F(H) \cap G = H$.

The inclusion $H \subseteq F(H) \cap G$ is obvious. Now let $x \in F(H) \cap G$. Then $x = \lambda_1 h_1 + \ldots + \lambda_s h_s$ for some $\lambda_i \in F$, $h_i \in H$. We can assume that the classes $\widehat{h_1}, \ldots, \widehat{h_s}$ are mutually distinct in the group H/F^* and so, also in the group G/F^*. Since $x, h_1, \ldots, h_s$ are linearly dependent over F, it follows by Corollary 2.1.10 that there exist at least two equal cosets among $\widehat{x}, \widehat{h_1}, \ldots, \widehat{h_s}$. This is possible only if $\widehat{x} = \widehat{h_i}$ for some $i \in \{1, \ldots, s\}$, i.e., $x = \lambda h_i$ with $\lambda \in F^*$, and so $x \in H$. This proves the inclusion $F(H) \cap G \subseteq H$. □

COROLLARY 2.1.12. *Let* E/F *be a finite extension, which is simultaneously* G-*Kneser and* H-*Kneser. If* $H \leqslant G$, *then* $H = G$.

PROOF. We have $G = E \cap G - F(H) \cap G = H$ according to Proposition 2.1.11. □

EXAMPLE 2.1.13. The next example shows the non-uniqueness of G for a given G-Kneser extension. Consider the extension $\mathbb{Q}(\sqrt[4]{-9})/\mathbb{Q}$, where $\sqrt[4]{-9}$ is one of the complex roots of the irreducible polynomial $X^4 + 9 \in \mathbb{Q}[X]$, say $\sqrt{6}(1 + i)/2$. Since $(\sqrt[4]{-9})^2 = 3i$, we have $i = \frac{1}{3}(\sqrt[4]{-9})^2 \in \mathbb{Q}(\sqrt[4]{-9})$, and so, $\sqrt{6} = 2\sqrt[4]{-9}/(1 + i) \in \mathbb{Q}(\sqrt[4]{-9})$.

Now consider the groups $G = \mathbb{Q}^* \langle \sqrt[4]{-9} \rangle$ and $H = \mathbb{Q}^* \langle i, \sqrt{6} \rangle$. Then

$$G/\mathbb{Q}^* = \left\{ \widehat{1}, \widehat{\sqrt[4]{-9}}, \widehat{\sqrt[4]{-9}}^2, \widehat{\sqrt[4]{-9}}^3 \right\}$$

and

$$H/\mathbb{Q}^* = \left\{ \widehat{1}, \widehat{i}, \widehat{\sqrt{6}}, \widehat{i\sqrt{6}} \right\}.$$

But $\mathbb{Q}(\sqrt[4]{-9}) = \mathbb{Q}(i, \sqrt{6})$, hence

$$|G/\mathbb{Q}^*| = |H/\mathbb{Q}^*| = 4 = [\mathbb{Q}(\sqrt[4]{-9}) : \mathbb{Q}],$$

which shows that the extension $\mathbb{Q}(\sqrt[4]{-9})/\mathbb{Q}$ is simultaneously *G*-Kneser and *H*-Kneser. However, $G \neq H$ since $\sqrt{6} \in H \setminus G$. $\qquad \square$

The uniqueness of the group *G* for the so called *G-Cogalois extensions*, which are separable *G*-Kneser extensions possessing a certain inheritance property, will be discussed in Section 4.4.

REMARKS 2.1.14. (1) A subextension of a Kneser extension is not necessarily Kneser. Indeed, we will see in Section 5.3 that the extension $\mathbb{Q}(\sqrt{2+\sqrt{2}})/\mathbb{Q}$ is not radical, and so, it is not Kneser. On the other hand, it is easily seen that $\mathbb{Q}(\sqrt{2+\sqrt{2}}) \subseteq \mathbb{Q}(\zeta_{16})$, and by Exercise 16, $\mathbb{Q}(\zeta_{16})/\mathbb{Q}$ is a $\mathbb{Q}^* \langle \zeta_{16} \rangle$-Kneser extension.

(2) Also, we will see in Section 3.2 that the property of an extension being Kneser is, in general, not transitive. $\qquad \square$

EXAMPLES 2.1.15. (1) The extension $\mathbb{Q}(\sqrt{-3})/\mathbb{Q}$ is $\mathbb{Q}^* \langle \sqrt{-3} \rangle$-Kneser but it is not $\mathbb{Q}^* \langle \zeta_3 \rangle$-Kneser. Note that $\mathbb{Q}(\sqrt{-3}) = \mathbb{Q}(\zeta_3)$.

Indeed, since $\sqrt{-3}^2 = -3 \in \mathbb{Q}^*$, it follows that $(\widehat{\sqrt{-3}})^2 = \widehat{1}$, that is, $\mathbb{Q}^* \langle \sqrt{-3} \rangle / \mathbb{Q}^* = \{\widehat{1}, \widehat{\sqrt{-3}}\}$, hence

$$| \mathbb{Q}^* \langle \sqrt{-3} \rangle / \mathbb{Q}^* | = 2 = [\mathbb{Q}(\sqrt{-3}) : \mathbb{Q}].$$

On the other hand, since $\zeta_3, \zeta_3^2, \zeta_3^2/\zeta_3 \notin \mathbb{Q}^*$ and $\zeta_3^3 = 1 \in \mathbb{Q}^*$, we have

$$\mathbb{Q}^* \langle \zeta_3 \rangle / \mathbb{Q}^* = \{\widehat{1}, \widehat{\zeta_3}, \widehat{\zeta_3^2}\} \quad \text{and} \quad |\mathbb{Q}^* \langle \zeta_3 \rangle / \mathbb{Q}^*| = 3 > [\mathbb{Q}(\zeta_3) : \mathbb{Q}] = 2.$$

(2) A *Cogalois* extension E/F is nothing else than a $T(E/F)$-Kneser extension. These extensions will be discussed in the next chapter.

(3) Any Cogalois extension is a Kneser extension as mentioned in the previous example. The converse of this statement does not hold: the extension $\mathbb{Q}(\sqrt{-3})/\mathbb{Q}$ is $\mathbb{Q}^* \langle \sqrt{-3} \rangle$-Kneser by (1), but it is not Cogalois.

Indeed, we have shown at (1) that the group $\mathbb{Q}^* \langle \zeta_3 \rangle / \mathbb{Q}^*$ has order 3. Since this group is a subgroup of the group $\mathrm{Cog}\,(\mathbb{Q}(\sqrt{-3})/\mathbb{Q})$, it follows that

$$|\mathrm{Cog}\,(\mathbb{Q}(\sqrt{-3})/\mathbb{Q})| \geqslant 3 > [\mathbb{Q}(\sqrt{-3}) : \mathbb{Q}] = 2.$$

This means that the extension $\mathbb{Q}(\sqrt{-3})/\mathbb{Q}^*$ is not Cogalois.

We will show in Section 3.3 that

$$\mathrm{Cog}\,(\mathbb{Q}(\sqrt{-3})/\mathbb{Q}) = \langle \widehat{\sqrt{-3}}, \widehat{\zeta_3} \rangle \cong \mathbb{Z}_6,$$

that is, the Cogalois group of this extension is a cyclic group of order 6. $\square$

2.2. The Kneser Criterion

In this section we prove a nice result, due to Kneser [**77**], which characterizes finite separable G-Kneser extensions E/F according to whether or not certain roots of unity belonging to G are in F.

THEOREM 2.2.1 (THE KNESER CRITERION). *The following assertions are equivalent for a separable G-radical extension E/F with finite G/F^*.*

(1) *E/F is a G-Kneser extension.*

(2) *For every odd prime p, $\zeta_p \in G \implies \zeta_p \in F$, and $1 \pm \zeta_4 \in G \implies \zeta_4 \in F$.*

(3) *$\mu_p(G) = \mu_p(F)$ for every odd prime p, and $1 \pm \zeta_4 \in G \implies \zeta_4 \in F$.*

PROOF. First of all, note that E/F is necessarily a finite extension, since $E = F(G)$ and G/F^* is a finite group.

$(2) \iff (3)$: Assume that $\mu_p(G) = \mu_p(F)$ for every odd prime p. Then clearly $\zeta_p \in G \implies \zeta_p \in F$. Conversely, assume that $\zeta_p \in G \implies \zeta_p \in F$ for every odd prime p, and let $g \in \mu_p(G)$. Then $g^p = 1$, hence the order $\operatorname{ord}(g)$ of g in G is either 1 or p. If $\operatorname{ord}(g) = 1$ then $g = 1 \in F$, as desired. If $\operatorname{ord}(g) = p$ then g generates the subgroup $\langle g \rangle = \langle \zeta_p \rangle$ of order p of G. Thus $\zeta_p \in \langle g \rangle \leqslant G$, hence $\zeta_p \in F$ by our assumption. It follows that $g \subset \langle \zeta_p \rangle \leqslant F^*$, as desired.

$(1) \implies (2)$: Let us specify that the condition $1 \pm \zeta_4 \in G$ means that either $1 + \zeta_4 \in G$ or $1 - \zeta_4 \in G$.

Observe that $1 + \zeta_4 \in G \iff 1 - \zeta_4 \in G$. Indeed, this is clear whenever $\operatorname{Char}(F) = 2$. If $\operatorname{Char}(F) \neq 2$ and $1 + \zeta_4 \in G$, then $(1 + \zeta_4)^2 = 2\zeta_4 \in G$, hence $\zeta_4 \in G$ since $2 \in F^* \subseteq G$. Thus $\zeta_4(1 + \zeta_4) = -1 + \zeta_4 \in G$, and then $1 - \zeta_4 = (-1)(-1 + \zeta_4) \in G$.

Suppose that the extension E/F is G-Kneser and $\zeta_p \in G$, where p is an odd prime. Then $\widehat{\zeta_p}^{\,p} = \widehat{1}$, hence $\operatorname{ord}(\widehat{\zeta_p})$ in the group G/F^* divides p, that is, $\operatorname{ord}(\widehat{\zeta_p}) \in \{1, p\}$.

If we would have $\operatorname{ord}(\widehat{\zeta_p}) = p$, then the cosets $\widehat{1}, \widehat{\zeta_p}, \ldots, \widehat{\zeta_p}^{\,p-1}$ would be distinct in G/F^*, hence the subset $\{1, \zeta_p, \ldots, \zeta_p^{p-1}\}$ of G would be linearly independent over F by Corollary 2.1.10, which would contradict the fact that $1 + \zeta_p + \cdots + \zeta_p^{p-1} = 0$. Consequently $\operatorname{ord}(\widehat{\zeta_p}) = 1$, and this implies that $\zeta_p \in F^*$, as desired.

If $1 + \zeta_4 \in G$, we investigate two cases:

i) $\mathrm{Char}(F) = 2$. Then $\zeta_4^4 = 1$, hence $\zeta_4 = 1 \in F$.

ii) $\mathrm{Char}(F) \neq 2$. Since $(1 + \zeta_4)^4 = -4 \in F^*$ it follows that the element $\widehat{1 + \zeta_4} \in G/F^*$ has order 1 or 2 or 4.

If $\mathrm{ord}(\widehat{1 + \zeta_4}) = 1$, then $1 + \zeta_4 \in F^*$, so $\zeta_4 \in F$.

If $\mathrm{ord}(\widehat{1 + \zeta_4}) = 2$, then $(1 + \zeta_4)^2 \in F^*$, so $2\zeta_4 \in F^*$; since $\mathrm{Char}(F) \neq 2$ it follows that $\zeta_4 \in F$.

If $\mathrm{ord}(\widehat{1 + \zeta_4}) = 4$, then the elements

$$\widehat{1}, \ \widehat{1 + \zeta_4}, \ \widehat{(1 + \zeta_4)}^2, \ \widehat{(1 + \zeta_4)}^3$$

are distinct in G/F^*. By Corollary 2.1.10, the elements $1, 1 + \zeta_4, (1 + \zeta_4)^2$, $(1 + \zeta_4)^3$ are linearly independent over F, which is a contradiction because $(1 + \zeta_4)^2 = 2\zeta_4$, $2 \in F^*$, and

$$2 \cdot 1 + (-2) \cdot (1 + \zeta_4) + (1 + \zeta_4)^2 + 0 \cdot (1 + \zeta_4)^3 = 0.$$

This shows that we cannot have $\mathrm{ord}(\widehat{1 + \zeta_4}) = 4$, so necessarily we must have $\mathrm{ord}(\widehat{1 + \zeta_4}) \in \{1, 2\}$, and then $\zeta_4 \in F$, as shown above.

$(2) \Longrightarrow (1)$: Suppose that the conditions in (2) are satisfied, and show that

$$|G/F^*| \leqslant [E : F].$$

Of course, we can assume that $|G/F^*| \geqslant 2$.

To do that, it is sufficient to prove the implication only when the group G/F^* is a p-group, that is, G/F^* has order a power of a prime number p.

Indeed, if $|G/F^*| = p_1^{t_1} \cdot \ldots \cdot p_r^{t_r}$, with p_i mutually distinct prime numbers and $t_i \in \mathbb{N}^*$, $1 \leqslant i \leqslant r$, let H_i/F^* be a p_i-Sylow subgroup of G/F^*, i.e., $|H_i/F^*| = p_i^{t_i}$, $i = 1, \ldots, r$. Assuming that the result holds for p-groups, we deduce that $p_i^{t_i} = [F(H_i) : F]$. But $[F(H_i) : F]$ divides $[E : F]$, hence $p_i^{t_i}$ divides $[E : F]$, and so, $\prod_{i=1}^{r} p_i^{t_i}$ divides $[E : F]$. It follows that $|G/F^*|$ divides $[E : F]$. Thus

$$|G/F^*| \leqslant [E : F].$$

Consequently, without loss of generality we can assume that $|G/F^*| = p^t$, with $p \in \mathbb{P}$ and $t \in \mathbb{N}^*$. There exists then an ascending chain

$$F^* = H_0 \leqslant H_1 \leqslant \ldots \leqslant H_t = G$$

of subgroups of G such that $|H_s/H_{s-1}| = p$, $s = 1, \ldots, t$. We are going to prove by induction on s, $0 \leqslant s \leqslant t$, the following two statements.

$(\mathcal{A}_s)$ \quad $[F(H_s) : F(H_{s-1})] = p$.

$(\mathcal{B}_s)$ If $p > 2$, and $c \in F(H_s)$ with $c^p \in H_s$, then $c \in H_s$.
             If $p = 2$, and $c \in F(H_s)$ with $c^2 \in H_s$, and either
             $c \in G$ or $\zeta_4 \notin F(H_s)$, then $c \in H_s$.

Observe that $(\mathcal{B}_0)$ is obvious and $(\mathcal{A}_0)$ has no meaning. Let $s \geqslant 1$. First, we show that $(\mathcal{B}_{s-1})$ implies $(\mathcal{A}_s)$. Let $a \in H_s$ be such that the coset $\overline{a} = aH_{s-1}$ is a generator of the cyclic group H_s/H_{s-1} of order p. Then

$$a \in H_s \setminus H_{s-1}, \ H_s = H_{s-1}\langle a \rangle, \ \text{and} \ a^p \in H_{s-1}.$$

Clearly, a is a root of the polynomial $f = X^p - a^p \in F(H_{s-1})[X]$. We claim that this polynomial is irreducible over $F(H_{s-1})$, and then it will be the minimal polynomial $\mathrm{Min}(a, F(H_{s-1}))$ of a over $F(H_{s-1})$, which will imply that $[F(H_s) : F(H_{s-1})] = p$, as desired. If f would be reducible, then, by Lemma 1.3.2, it would have a root $b \in F(H_{s-1})$. Then $b^p = a^p \in H_{s-1}$.

If $p > 2$, then we have $b \in H_{s-1}$ by $(\mathcal{B}_{s-1})$. Since $b = a\zeta_p$, $a \in H_s$, and $b \in H_{s-1}$, we have $\zeta_p = ba^{-1} \in H_s \leqslant G$. In view of (2), this would imply that $\zeta_p \in F$ and so, $a = b\zeta_p^{-1} \in H_{s-1}$, contradicting the choice of a.

If $p = 2$, then $a^2 = b^2$, hence $a = \pm b$, and so, $b \in H_s \leqslant G$. As before, by $(\mathcal{B}_{s-1})$, we have $b \in H_{s-1}$, hence $a \in H_{s-1}$, which is a contradiction.

Thus, for any prime p, the polynomial $X^p - a^p$ is irreducible over $F(H_{s-1})$, i.e., $(\mathcal{A}_s)$ holds.

Now, we are going to prove that $(\mathcal{B}_{s-1})$ and $(\mathcal{A}_s)$ together imply $(\mathcal{B}_s)$. We will examine separately the cases when p is odd or even.

Case 1: If $p > 2$, let $c \in F(H_s)$ be such that $c^p \in H_s$. Then $\overline{c}^p = \overline{a}^q$, with $0 \leqslant q \leqslant p - 1$, hence $c^p = a^q d$ for some $d \in H_{s-1}$. We want to show that $c \in H_s$. Assume first that $q > 0$, and denote by N the norm of $F(H_s)$ over $F(H_{s-1})$. Since $\mathrm{Min}(a, F(H_{s-1})) = X^p - a^p$, we have $N(a) = (-1)^{p-1}a^p$. Taking the norm N in both sides of the equality

$$a^q = c^p d^{-1},$$

we obtain

(†) $\left((-1)^{p-1}a^p\right)^q = N(c)^p d^{-p}.$

But p has been supposed to be odd, hence (†) becomes

$$(a^p)^q = (N(c)d^{-1})^p.$$

Since $1 \leqslant q < p$, we deduce that p and q are relatively prime, hence there exist $u, v \in \mathbb{Z}$ such that $up + vq = 1$. Consequently,

$$a^p = (a^p)^{up+vq} = ((a^p)^u)^p \left((N(c)d^{-1})^v\right)^p.$$

Since a^p, $N(c)d^{-1} \in F(H_{s-1})$, we deduce that a^p is a p-th power of an element from $F(H_{s-1})$, which contradicts the irreducibility of the polynomial $f = X^p - a^p \in F(H_{s-1})[X]$.

Thus, we must have $q = 0$, and then $c^p = d \in H_{s-1}$. For simplicity, set $K = F(H_{s-1})$, and observe that $F(H_s) = K(a)$ since $H_s = H_{s-1}\langle a \rangle$. Let L be the normal closure (contained in Ω) of $K(a)/K$. Since the extension E/F is separable, we deduce that L/K is a finite Galois extension, hence $a \in L \setminus K$ implies that there exists $\varphi \in \mathrm{Gal}\,(L/K)$ such that $\varphi(a) \neq a$.

Since $a^p \in H_{s-1}$, we have $\varphi(a^p) = a^p = \varphi(a)^p$, hence $\varphi(a) = a\zeta_p$. Similarly, $c^p \in H_{s-1}$ implies that $\varphi(c^p) = c^p = \varphi(c)^p$, hence $\varphi(c) = c\zeta_p^m$ for some $0 \leqslant m \leqslant p-1$. Then

$$\varphi(a^{-m}c) = \varphi(a)^{-m}\varphi(c) = a^{-m}\zeta_p^{-m}c\zeta_p^m = a^{-m}c.$$

We claim that $a^{-m}c \in F(H_{s-1}) = K$. Indeed, set $w = a^{-m}c$. Since $\mathrm{Min}(a, K) = X^p - a^p$, there exist $\lambda_i \in K$, $0 \leqslant i \leqslant p-1$, such that

$$w = \sum_{0 \leqslant i \leqslant p-1} \lambda_i a^i.$$

Since $\varphi(a) = \zeta_p a$, we deduce that

$$\sum_{0 \leqslant i \leqslant p-1} \lambda_i a^i = w = \varphi(w) = \sum_{0 \leqslant i \leqslant p-1} (\lambda_i \zeta_p^i) a^i.$$

But $[K(\zeta_p) : K] \leqslant p-1$ and $[K(a) : K] = p$, hence these degrees are relatively prime. Then

$$[K(a, \zeta_p) : K] = [K(\zeta_p) : K] \cdot [K(a) : K],$$

and so,

$$[K(\zeta_p)(a) : K(\zeta_p)] = [K(a) : K] = p.$$

In particular, it follows that the set $\{1, a, \dots, a^{p-1}\}$ is linearly independent over $K(\zeta_p)$. Therefore, $\lambda_i = \lambda_i \zeta_p^i$ for all i, $0 \leqslant i \leqslant p-1$.

If we assume that $w \notin K$, then there would exist $1 \leqslant j \leqslant p-1$ such that $\lambda_j \neq 0$, which would imply that $\zeta_p^j = 1$. But this contradicts the fact that $\zeta_p \neq 1$.

Thus, we have proved that $w = a^{-m}c \in K = F(H_{s-1})$. Since $(a^{-m}c)^p = (a^p)^{-m}c^p \in H_{s-1}$, the statement $(\mathcal{B}_{s-1})$ implies that $a^{-m}c \in H_{s-1}$, hence $c \in H_s$.

Case 2: If $p = 2$, let $c \in F(H_s)$ be such that $c^2 \in H_s$, and either $c \in G$ or $\zeta_4 \notin F(H_s)$. We have to prove that $c \in H_s$. As in Case 1, we have, $c^2 = a^q d$ for some $0 \leqslant q < 2$ and $d \in H_{s-1}$.

If $q = 1$ then $c^2 = ad$, hence (†), which also holds for $p = 2$, implies that $-a^2 = z^2$ for some $z \in F(H_{s-1})$, and then $a = \pm\zeta_4 z$. This implies that $\zeta_4 \notin F(H_{s-1})$, $F(H_s) = F(H_{s-1})(\zeta_4)$, and $c^2 = ad = \pm\zeta_4 zd$. It follows that $c = x + \zeta_4 y$ for some $x, y \in F(H_{s-1})$. Thus

$$c^2 = (x^2 - y^2) + 2xy\zeta_4 = \pm\zeta_4 zd,$$

so $x^2 = y^2$, i.e., $x = \pm y$, hence $c = (1 \pm \zeta_4)x$. Observe that necessarily $\mathrm{Char}(F) \neq 2$, for otherwise, it would follow that $\zeta_4 = 1$, and then $a = z \in F(H_{s-1})$, which is a contradiction.

On the other hand, $c^2 \in H_s$ and $|H_s/H_{s-1}| = 2$, hence $c^4 = (c^2)^2 \in H_{s-1}$. Then $x^4 = (-c^4) \cdot 4^{-1} \in H_{s-1}$. By $(\mathcal{B}_{s-1})$, it follows that $x^2 \in H_{s-1}$, and applying $(\mathcal{B}_{s-1})$ again, we deduce that $x \in H_{s-1}$. Then $1\pm\zeta_4 = cx^{-1} \in G$, and by our assumption in (2) we deduce that $\zeta_4 \in F \subseteq F(H_{s-1})$, which is a contradiction. Hence q is necessarily 0, and then $c^2 = d \in H_{s-1}$.

Now we will proceed as for $p > 2$ by taking an automorphism $\varphi \in \mathrm{Gal}\,(F(H_s)/F(H_{s-1}))$ such that $\varphi(a) = -a$. We deduce that $\varphi(a^j c) = a^j c$ for a suitable integer j, and $a^j c \in F(H_{s-1})$. Since $a^{2j}c^2 \in H_{s-1}$, the statement $(\mathcal{B}_{s-1})$ implies that $a^j c \in H_{s-1}$, and so, $c \in H_s$.

This completes the inductive proof of $(\mathcal{A}_s)$ and $(\mathcal{B}_s)$. From $(\mathcal{A}_s)$ we now deduce that

$$[E : F] = [F(H_t) : F(H_{t-1})] \cdot \ldots \cdot [F(H_1) : F(H_0)] = p^t.$$

which proves the implication $(2) \Longrightarrow (1)$, and we are done. $\qquad\square$

REMARKS 2.2.2. (1) The condition "G/F^* is a finite group" in the statement of the Kneser Criterion can be replaced by the condition "E/F is a finite extension". Indeed, since E/F is a finite G-radical extension, $E = F(H)$ for a suitable group H, with $F^* \leqslant H \leqslant G$ and H/F^* finite, by Corollary 2.1.7. Then E/F is H-Kneser by the Kneser Criterion applied to the H radical extension E/F with finite H/F^*. Let $g \in G$, and set $H' = H\langle g \rangle$. Then E/F is also H'-Kneser, again by the Kneser Criterion, and H'/F^* is a finite group. By Corollary 2.1.12, we have $H = H'$, hence

$g \in H$. Since g was an arbitrary element of G, we deduce that $H = G$. This proves that $G/F^* = H/F^*$ is a finite group.

(2) The next example shows that the *separability* condition cannot be dropped from the Kneser Criterion. Take the extension E/F considered in Example 2.1.5 (1), that is $F = \mathbb{F}_2(X^2)$ and $E = \mathbb{F}_2(X)$. We have seen that $\mathrm{Cog}(E/F) = T(E/F)/F^* = E^*/F^*$ is an infinite group. Observe that E/F is $T(E/F)$-radical since $E = F(X)$ and $X^2 \in F$. Further, the condition (2) from the Kneser Criterion is satisfied for $G = T(E/F)$ since $\mu_n(E) = \{1\}$ for every $n \in \mathbb{N}^*$. Indeed, let $n \in \mathbb{N}^*$ and $f \in E$ be such that $f^n = 1$. This means that the element $f \in E$ is an integral element over $\mathbb{F}_2[X]$. Since $\mathbb{F}_2[X]$ is an integrally closed domain, we deduce that necessarily $f \in \mathbb{F}_2[X]$, and moreover, f is a unit of the domain $\mathbb{F}_2[X]$. Then f is a nonzero constant polynomial in $\mathbb{F}_2[X]$, that is, $f = 1$, as desired.

However, the finite extension E/F is not G-Kneser, since

$$2 = [\, E : F \,] < |G/F^*| = \aleph_0.$$

(3) We do not know whether or not the separability condition in the Kneser Criterion can be weakened. $\qquad\qquad\qquad\qquad\qquad\qquad\qquad\square$

Applications of the Kneser Criterion to algebraic number fields will be given in Chapter 9.

2.3. Exercises to Chapter 2

1. Prove that $\mu_n(F(X_1, \dots, X_m)) = \mu_n(F)$ for any field F and any $m, n \in \mathbb{N}^*$.

2. Let p be any positive prime number, and let n be any positive integer. Prove that the extension $\mathbb{F}_p(X_1, \dots, X_n)/\mathbb{F}_p(X_1^p, \dots, X_n^p)$ is an extension of degree p^n, with Cogalois group isomorphic to a countably infinite direct sum of copies of the cyclic group $\mathbb{Z}_p$.

3. Show that $\mathrm{Cog}(\mathbb{F}_4/\mathbb{F}_2) \cong \mathbb{F}_4^*$.

4. Prove that for any extension $\mathbb{F}_{q^n}/\mathbb{F}_q$ of finite fields, one has

$$\mathrm{Cog}(\mathbb{F}_{q^n}/\mathbb{F}_q) \cong \mathbb{Z}_m,$$

where $m = 1 + q + \dots + q^{n-1}$.

5. (*Greither and Harrison* [**63**]). Let $E = \mathbb{Q}(\{\,\zeta_{2^n} \mid n \in \mathbb{N}^*\,\})$ and $F = E \cap \mathbb{R}$. Prove that $E = F(i)$, $[\,E : F\,] = 2$, and $\mathrm{Cog}\,(E/F)$ is a countably infinite group.

6. (*Dummit* [**55**]). Let p be an odd prime, let $E = \mathbb{Q}(\zeta_p)$, and let $F = E \cap \mathbb{R}$. Show that $[\,E : F\,] = 2$ and $|\mathrm{Cog}\,(E/F)| = 2p$.

7. (*May* [**81**]). Let F be a field, and let p be a prime different from the characteristic of F. Let $u \in \Omega$ with $u^p \in F \setminus F^p$, and let $E = F(u)$. If $p = 2$, further assume that $E \neq F(\zeta_4)$. Prove that $\mathrm{Cog}(E/F) = (\langle u \rangle t(E^*)F^*)/F^*$.

8. (*May* [**81**]). Let $F = \mathbb{Q}$, $p = 2$, $u = \zeta_4 = i$, and $E = F(u)$. Show that $1 + i \in \mathrm{Cog}(E/F) \setminus (\langle u \rangle t(E^*)F^*)/F^*$.

9. (*Brandis* [**41**]). Let F be an infinite field, and let E be an overfield of F with $E \neq F$. Prove that the group E^*/F^* is not finitely generated.

10. Let $E = \mathbb{F}_2(X)$ and $F = \mathbb{F}_2(X^2)$. Show that the extension E/F is simultaneously $F^*\langle X \rangle$-Kneser and $F^*\langle X + 1 \rangle$-Kneser, but the groups $F^*\langle X \rangle$ and $F^*\langle X + 1 \rangle$ are distinct.

11. Prove that any quadratic extension E/F, with $\mathrm{Char}(F) \neq 2$ is Kneser.

12. Prove that the extension $\mathbb{F}_4/\mathbb{F}_2$ is not Kneser.

13. Investigate whether or not the quotient extension of a Kneser extension is also Kneser.

14. Is the compositum of two Kneser (resp. radical) extensions again a Kneser (resp. radical) extension?

15. Give an example of a radical extension which is not a Kneser extension.

16. Let $n \in \mathbb{N}^*$. Prove that $\mathbb{Q}(\zeta_n)/\mathbb{Q}$ is a $\mathbb{Q}^*\langle \zeta_n \rangle$-Kneser extension if and only if $n = 2^r$ for some $r \in \mathbb{N}$.

17. For what $n \in \mathbb{N}^*$ is $\mathbb{Q}(\zeta_n)/\mathbb{Q}$ a Kneser extension?

18. Let F be an arbitrary field, let $u \in \Omega$ be any root of an irreducible binomial $X^n - a \in F[X]$, and let $m = \mathrm{ord}(\widehat{u})$ in the group Ω^*/F^*, where Ω in an algebraically closed overfield of F. Prove that the extension $F(u)/F$ is $F^*\langle u \rangle$-Kneser.

19. (*Gay and Vélez* [**61**]). Let F be any field, let $u \in \Omega$ be a root of a binomial $X^n - a \in F[X]$, and let $m = \mathrm{ord}(\hat{u})$ in the group Ω^*/F^*, where Ω in an algebraically closed overfield of F. If $\gcd(n, e(F)) = 1$ and $\zeta_{2p} \notin F(u) \setminus F$ for every prime divisor p of n, then prove that $[F(u) : F] = m$.

20. (*Risman*). Let E/F be a separable extension, and let $u \in T(E/F)$. If $m = \mathrm{ord}(\hat{u})$ in the group E^*/F^*, then prove that $m = nt$, where $\gcd(n, t) = 1$, $n \mid [F(u) : F]$, and $\zeta_{2p} \notin F(u) \setminus F$ for every prime divisor p of t. (*Hint:* See [**29**] and [**61**].)

21. (*Schinzel* [**92**]). Let F be any field. Prove that the Galois group of the splitting field of a binomial $X^n - a \in F[X]$ is Abelian if and only if $a^{w_n} \in F^n$, where $w_n = |\mu_n(F)|$.

22. (*Schinzel* [**92**]). Let F be any field, and let $f = X^n - a \in F[X]$ be an irreducible binomial such that the Galois group G of the splitting field of f is Abelian. Prove that if $4 \mid n$ and $\zeta_4 \notin F$, then $G \cong \mathbb{Z}_2 \times \mathbb{Z}_{n/2}$, otherwise G is cyclic.

23. (*Halter-Koch* [**65**]). Let E/F be a finite separable extension, let $n \in \mathbb{N}^*$ be such that $\gcd(n, e(F)) = 1$, and let G be a group such that $F^* \leqslant G \leqslant E^*$ and $G^n \subseteq F$. Prove the following facts.
    (a) $(G^n : F^{*n}) \mid [E : F]$
    (b) G/F^* is a finite group.

2.4. Bibliographical comments to Chapter 2

Section 2.1. The concept of *radical extension* is rather basic and well-known in Galois Theory. However, our terminology is somewhat different from that used in Galois Theory (see e.g., Kaplansky [**74**], Karpilovsky [**76**], Lang [**80**]), but coincides for simple extensions. Note that radical extensions are called *coseparable* by Greither and Harrison [**63**]. The concept of *G*-radical extension is due to Albu and Nicolae [**19**]. Radical extensions were studied among others by Acosta de Orozco and Vélez [**1**], Barrera-Mora and Vélez [**32**], Gay and Vélez [**60**], Halter-Koch [**65**], [**66**], Norris and Vélez [**86**] [**109**], Vélez [**108**], [**110**].

The torsion group $t(E^*/F^*)$ of an extension E/F was intensively investigated by Acosta de Orozco and Vélez [**2**], Gay and Vélez [**61**]. To the best of our knowledge, the name of *Cogalois group* of E/F for the group $t(E^*/F^*)$ appeared for the first time in the literature in the fundamental paper [63] of Greither and Harrison. The term of "coGalois group" was also used by Enochs, Rozas, and Oyonarte [56], [57], but with a completely different meaning, involving the concept of $\mathcal{F}$-cover of a module.

The concept of *Kneser extension* was introduced by Albu and Nicolae [19].

This section basically follows the outline in Albu and Nicolae [19] and Albu and Țena [25], although there are several extensions, simplifications, and improvements.

Section 2.2. The Kneser Criterion which, of course, is due to Kneser [77] was published in 1975. Actually, Kneser proved only the difficult implication (2) $\implies$ (1) in Theorem 2.2.1. The other implication (1) $\implies$ (2) in Theorem 2.2.1 is mentioned by Schinzel [91], Gay and Vélez [61], and it was explicitly proved for the first time in 1982 by Schinzel [93].

The very thorough proof of the Kneser Criterion presented in this section details the original ideas from the very concise Kneser's paper [77], and follows the more accurate proof given in Schinzel [93] (see also Schinzel [94]), which corrects a little gap in the original proof of Kneser [77].

The Kneser Criterion is not only a basic tool in the whole Cogalois Theory, but it has nice applications to classical Algebraic Number Theory (see Chapter 9) and also to Gröbner bases (see Becker, Grobe, and Niermann [34]).

CHAPTER 3

COGALOIS EXTENSIONS

The aim of this chapter is to investigate the Cogalois extensions, introduced into the literature in 1986 by Greither and Harrison [**63**]. Using the concept of Kneser extension, a finite extension E/F is a *Cogalois extension* precisely when it is $T(E/F)$-Kneser. We defined in Section 2.1 the *Cogalois group* of an extension E/F as being the group $T(E/F)/F^*$ and we denoted it by $\mathrm{Cog}(E/F)$. Thus, a finite radical extension E/F is Cogalois if and only if $|\mathrm{Cog}\,(E/F)| = [E : F]$.

Using the Kneser Criterion we provide a short proof of the *Greither-Harrison Criterion* characterizing Cogalois extensions. We present then the *Gay-Vélez Criterion*, which is an equivalent form of the Greither-Harrison Criterion. Some simple properties of Cogalois extensions are also given. Finally, we calculate explicitly the Cogalois group of any quadratic extension of $\mathbb{Q}$.

3.1. The Greither-Harrison Criterion

The aim of this section is to state and prove a criterion characterizing Cogalois extensions in terms of purity, due to Greither and Harrison [**63**]. Our simple proof is based on the Kneser Criterion. We also relate the Kneser Criterion to the Gay-Vélez Criterion.

Recall that throughout this monograph we shall use the following notation:

$$\mathbb{P} = \{\, p \in \mathbb{N}^* \,|\, p \text{ prime }\},$$
$$\mathcal{P} = (\mathbb{P} \setminus \{2\}) \cup \{4\}.$$

Recall also that Ω is a fixed algebraically closed field containing the fixed base field F as a subfield; any considered overfield of F is supposed to be a subfield of Ω. For any $n \in \mathbb{N}^*$, ζ_n will denote a primitive n-th root of unity over F, i.e., a generator of the cyclic group $\mu_n(\Omega)$.

A fundamental concept in the theory of radical extensions is that of *purity*, which is somewhat related to that used in Group Theory. Recall that a subgroup H of an Abelian multiplicative group G is called *pure* if $G^n \cap H = H^n$ for every $n \in \mathbb{N}^*$.

DEFINITION 3.1.1. *An extension E/F is said to be* pure *if $\mu_p(E) \subseteq F$ for every $p \in \mathcal{P}$.*   □

LEMMA 3.1.2. *The following assertions are equivalent for an extension E/F.*

(1) *E/F is pure.*
(2) *$\mu_p(E) = \mu_p(F)$ for every $p \in \mathcal{P}$.*
(3) *$\zeta_p \in E \implies \zeta_p \in F$ for every $p \in \mathcal{P}$.*
(4) *$\zeta_{2p} \notin E \setminus F$ for every $p \in \mathbb{P}$.*

PROOF. (1) $\iff$ (2) and (2) $\implies$ (3) are obvious.

(3) $\implies$ (1): Assume that (3) holds, and let $\zeta \in \mu_p(E)$. Then $\zeta^p = 1$, hence the order $\mathrm{ord}(\zeta)$ of ζ in E^* is a divisor of p.

If p is an odd prime, $\mathrm{ord}(\zeta)$ is either 1 or p. If $\mathrm{ord}(\zeta) = 1$, then $\zeta = 1 \in F$, as desired. If $\mathrm{ord}(\zeta) = p$, then ζ generates the subgroup $\langle \zeta \rangle = \langle \zeta_p \rangle$ of order p of Ω^*. Thus $\zeta_p \in \langle \zeta \rangle \leqslant E^*$, hence $\zeta_p \in F$ by our assumption. It follows that $\zeta \in \langle \zeta_p \rangle \leqslant F^*$, as desired.

Now assume that $p = 4$. Then $\zeta^4 = 1$, hence $\mathrm{ord}(\zeta) \in \{1, 2, 4\}$. If $\mathrm{ord}(\zeta) = 1$ then $\zeta = 1 \in F$, and if $\mathrm{ord}(\zeta) = 2$ then $\zeta = -1 \in F$. If $\mathrm{ord}(\zeta) = 4$, then as in the case of an odd prime, we have $\zeta \in \langle \zeta_4 \rangle \leqslant F^*$. Thus, E/F is a pure extension.

(3) $\iff$ (4): First of all, observe that if $\mathrm{Char}(F) = 2$, then clearly $\zeta_4 = 1$ and ζ_{2p} is a primitive p-th root of unity for every odd prime p. Indeed, since $\mathrm{Char}(F) = 2$, the equality $\zeta_{2p}^{2p} = 1$ implies that $\zeta_{2p}^p = 1$, hence the order of ζ_{2p} in Ω^* is a divisor of p. Since clearly $\zeta_{2p} \neq 1$, we deduce that $\mathrm{ord}(\zeta_{2p}) = p$, that is, ζ_{2p} is a primitive p-th root of unity.

If $\mathrm{Char}(F) \neq 2$, we claim that $-\zeta_{2p}$ is a primitive p-th root of unity for every odd prime p. Indeed, we have $(\zeta_{2p}^p)^2 = 1$, hence $\zeta_{2p}^p = -1$, and then $(-\zeta_{2p})^p = (-1)^p \zeta_{2p}^p = (-1)(-1) = 1$. If $\mathrm{Char}(F) = p$, then $-\zeta_{2p} = \zeta_p = 1$. If $\mathrm{Char}(F) \neq p$, then $-\zeta_{2p} \neq 1$ since $\mathrm{ord}(\zeta_{2p}) = 2p > 2$, hence $\mathrm{ord}(-\zeta_{2p}) = p$. This proves our claim.

Assume that (3) holds and $\zeta_{2p} \in E \setminus F$ for some $p \in \mathbb{P}$. Then necessarily p is odd since, by (3), $\zeta_4 \in E \implies \zeta_4 \in F$. If $\mathrm{Char}(F) = 2$ then we have $\zeta_p \in E \setminus F$, which contradicts (3). If $\mathrm{Char}(F) \neq 2$, then we have

$-\zeta_{2p} \in E \setminus F$, that is $\zeta_p \in E \setminus F$, which again contradicts (3). This proves the implication (3) $\implies$ (4).

Now assume that (4) holds. Since $\zeta_4 \notin E \setminus F$ by hypothesis, it is obvious that $\zeta_4 \in E \implies \zeta_4 \in F$. Let p be an odd prime, and assume that $\zeta_p \in E$. Then $\pm\zeta_{2p} \in E$ in view of the considerations above. If we would have $\zeta_p \notin F$, this would imply that $\pm\zeta_{2p} \notin F$, i.e., $\zeta_{2p} \in E \setminus F$, which contradicts (4). This proves the implication (4) $\implies$ (3). $\square$

REMARK 3.1.3. Consider the extension E/F with $E = \mathbb{Q}(\zeta_p)$ and $F = \mathbb{Q}$, where p is an odd prime. Then $\zeta_{p^2} \notin E$, for otherwise, it would follow that $\mathbb{Q}(\zeta_{p^2}) = \mathbb{Q}(\zeta_p)$, hence

$$p^2 - p = \varphi(p^2) = [\mathbb{Q}(\zeta_{p^2}) : \mathbb{Q}] = [\mathbb{Q}(\zeta_p) : \mathbb{Q}] = p - 1,$$

which is a contradiction. Hence $\zeta_{p^2} \in E \implies \zeta_{p^2} \in F$ holds vacuously. On the other hand, $\langle \zeta_p \rangle = \mu_{p^2}(E) \neq \mu_{p^2}(F) = \{1\}$. This shows that implication (3) $\implies$ (2) in Lemma 3.1.2 may fail when p is not in $\mathcal{P}$. $\square$

EXAMPLES 3.1.4. (1) A field F is said to be a *field with few n-th roots of unity*, where $n \in \mathbb{N}^*$, if $\mu_n(F) \subseteq \{-1, 1\}$; when this holds for every $n \in \mathbb{N}^*$, i.e., if $\mu(F) \subseteq \{-1, 1\}$, then F is said to be a *field with few roots of unity*. Clearly, any extension E/F with E any field with few roots of unity is a pure extension, and any subfield of $\mathbb{R}$ is a field with few roots of unity. In particular, any extension E/F, where E is any subfield of $\mathbb{R}$ is pure.

More generally, any extension E/F, with $\mu_n(E) \subseteq F$ for every $n \in \mathbb{N}^*$ is obviously pure. Thus, for any field F and any $m \in \mathbb{N}^*$, the extension $F(X_1, \ldots, X_m)/F$ is pure, by Exercise 1, Chapter 2. In particular, the extension $\mathbb{F}_2(X)/\mathbb{F}_2(X^2)$ is pure.

(2) A quadratic extension $\mathbb{Q}(\sqrt{d})/\mathbb{Q}$ where d is a square-free integer is pure if and only if $d \neq -1, -3$ (see Corollary 3.3.3). $\square$

PROPOSITION 3.1.5. *Let $F \subseteq K \subseteq E$ be a tower of fields. Then E/F is pure if and only if both K/F and E/K are pure.*

PROOF. Clearly $\mu_p(F) \subseteq \mu_p(K) \subseteq \mu_p(E)$ for every $p \in \mathcal{P}$, hence

$$\mu_p(F) = \mu_p(E) \iff (\mu_p(F) = \mu_p(K) \text{ and } \mu_p(K) = \mu_p(E)),$$

which proves the proposition. $\square$

LEMMA 3.1.6. *Let E/F be a finite separable G-radical extension. If E/F is pure, then E/F is G-Kneser, and $G = T(E/F)$.*

PROOF. First of all, note that the extension E/F is also $T(E/F)$-radical. Let p be an odd prime with $\zeta_p \in T(E/F)$. Then $\zeta_p \in E$, hence $\zeta_p \in F$ by purity. If $1 + \zeta_4 \in T(E/F)$, then $1 + \zeta_4 \in E$, hence $\zeta_4 \in E$, and so, $\zeta_4 \in F$ again by purity. Now apply Theorem 2.2.1 to deduce that E/F is $T(E/F)$-Kneser. The same argument, or Corollary 2.1.10, shows that E/F is also G-Kneser. Then $G = T(E/F)$ by Corollary 2.1.12. □

Recall that a finite extension E/F is said to be *Cogalois* if E/F is a radical extension such that $|\mathrm{Cog}\,(E/F)| = [E : F]$.

THEOREM 3.1.7 (THE GREITHER-HARRISON CRITERION). *The following assertions are equivalent for a finite extension E/F.*

(1) E/F is Cogalois.
(2) E/F is radical, separable, and pure.

PROOF. (1) $\Longrightarrow$ (2): Assume that E/F is a Cogalois extension. Then, by definition, it is radical.

We are going to show that E/F is separable. Assume that E/F is not separable. Then necessarily $\mathrm{Char}(F) = p > 0$ and $[E : F]$ is divisible by p, because the degree of any finite nonseparable extension E/F is the product between its separable degree and a power of the characteristic of F (see 1.2.8). Since E/F is Cogalois, we have $|\mathrm{Cog}(E/F)| = [E : F]$, hence $|\mathrm{Cog}(E/F)|$ is divisible by p. Then, by Lemma 1.4.11, the finite group $\mathrm{Cog}(E/F)$ contains an element $\hat{\lambda} \in \mathrm{Cog}(E/F)$ of order p, i.e., $\lambda \notin F^*$ and $\lambda^p \in F^*$. The field F is necessarily infinite, for otherwise, the extension E/F would be separable, which contradicts our assumption.

Since $\lambda \notin F$, we have $\mu + \lambda \notin F$ for all $\mu \in F$, but $(\mu + \lambda)^p = \mu^p + \lambda^p \in F$. Observe that $\widehat{\mu_1 + \lambda} \neq \widehat{\mu_2 + \lambda}$ for every $\mu_1, \mu_2 \in F$ with $\mu_1 \neq \mu_2$. Indeed, $\widehat{\mu_1 + \lambda} = \widehat{\mu_2 + \lambda}$ implies that $\mu_1 + \lambda = a(\mu_2 + \lambda)$ for some $a \in F^*$. If $a \neq 1$, then $\lambda = (\mu_1 - a\mu_2)(a - 1)^{-1} \in F$, which is impossible. Hence $a = 1$, and so, $\mu_1 + \lambda = \mu_2 + \lambda$, i.e., $\mu_1 = \mu_2$.

Thus, we showed that the finite group $\mathrm{Cog}(E/F)$ contains infinitely many elements of the form $\widehat{\mu + \lambda}$ with $\mu \in F$, which is a contradiction. This proves that the extension E/F is separable.

Now we are going to show that the extension E/F is pure. Let $\zeta \in \mu_q(E)$ with $q \in \mathcal{P}$. We have to show that $\zeta \in F$. Assume first that q is an odd prime. If $q = p = \mathrm{Char}(F)$, then $\zeta^p = 1$ implies that $\zeta = 1 \in F$. If $q \neq p = \mathrm{Char}(F)$, we can assume that $\zeta \neq 1$ (since $1 \in F$), and then $1 + \zeta + \cdots + \zeta^{q-1} = 0$, i.e., $1, \zeta, \ldots, \zeta^{q-1}$ are linearly dependent over F.

Since $\widehat{\zeta}^q = \widehat{1}$, we deduce that $\mathrm{ord}(\widehat{\zeta})$ is 1 or q. If we would have $\mathrm{ord}(\widehat{\zeta}) = q$, then the elements $\widehat{1}, \widehat{\zeta}, \ldots, \widehat{\zeta}^{q-1}$ of the group $\mathrm{Cog}(E/F)$ would be distinct. By Corollary 2.1.10, this would imply that the elements $1, \zeta, \ldots, \zeta^{q-1}$ of E would be linearly independent over F, which is a contradiction. Therefore, we must have $\mathrm{ord}(\widehat{\zeta}) = 1$, i.e., $\zeta \in F^*$.

If $q = 4$, we may again assume that $\mathrm{Char}(F) \neq 2$. Since $1+\zeta-(1+\zeta) = 0$, the elements $1, \zeta, 1 + \zeta$ of E are linearly dependent over F, hence, by Corollary 2.1.10, the elements $\widehat{1}, \widehat{\zeta}, \widehat{1+\zeta}$ of $\mathrm{Cog}(E/F)$ cannot be distinct. But, any equality between two such cosets implies that $\zeta \in F$, as desired.

Conversely, suppose that E/F is radical, separable and pure. Then Lemma 3.1.6 implies that E/F is $T(E/F)$-Kneser, i.e., Cogalois.   □

EXAMPLES 3.1.8. (1) The quadratic extension $\mathbb{F}_2(X)/\mathbb{F}_2(X^2)$ is radical, pure by Examples 3.1.4 (1), but not separable, so it is not Cogalois. We have seen in Examples 2.1.5 (1) that the Cogalois group of this extension is a countably infinite group.

(2) The quartic extension $\mathbb{Q}(\sqrt{1 + \sqrt{2}})/\mathbb{Q}$ is pure and separable, but it is neither radical nor Cogalois (see Proposition 3.2.6 (e)).

(3) The quadratic extension $\mathbb{Q}(i)/\mathbb{Q}$ is clearly a separable radical extension which is not pure, hence it is not Cogalois. We will show in Section 3.3 that $\mathrm{Cog}\,(\mathbb{Q}(i)/\mathbb{Q}) = \{\,\widehat{1}, \widehat{i}, \widehat{1+i}, \widehat{1-i}\,\} = \langle 1 + i \rangle \cong \mathbb{Z}_4$.   □

THEOREM 3.1.9 (THE GAY-VÉLEZ CRITERION). *The following assertions are equivalent for a finite extension E/F.*

(1) *E/F is Cogalois.*
(2) *E/F is radical, separable, and satisfies the following conditions: for every odd prime p, $\zeta_p \in F$ whenever $\zeta_p \in T(E/F)$, and $\zeta_4 \in F$ whenever $1 \pm \zeta_4 \in T(E/F)$.*
(3) *E/F is radical, separable, and $\zeta_{2p} \notin E \setminus F$ for every $p \in \mathbb{P}$.*

PROOF. (1) $\Longleftrightarrow$ (3) follows immediately from the Greither-Harrison Criterion and Lemma 3.1.2.

(1) $\Longrightarrow$ (2): If E/F is a Cogalois extension, then it is $T(E/F)$-Kneser. Now apply the Kneser Criterion.

(2) $\Longrightarrow$ (1): By Remarks 2.2.2 (1) and the Kneser Criterion, we deduce that the extension E/F is $T(E/F)$-Kneser, i.e., it is Cogalois.   □

REMARK 3.1.10. By the Greither-Harrison Criterion, any Cogalois extension is separable. However, a Kneser extensions is not necessarily separable, as the following example shows: $F = \mathbb{F}_2(X^2)$, $E = \mathbb{F}_2(X)$, $G = F^*\langle X \rangle$. The extension E/F is G-Kneser since $|G/F^*| = \operatorname{ord}(\widehat{X}) = 2 = [E:F]$, but it is not separable.   □

3.2. Examples and properties of Cogalois extensions

The aim of this section is to present a series of examples, and to establish the basic properties of Cogalois extensions.

EXAMPLES 3.2.1. (1) Any finite G-radical extension E/F with E a subfield of $\mathbb{R}$ is pure by Examples 3.1.4 (1), hence it is Cogalois by the Greither-Harrison Criterion. Notice that for such an extension E/F we have $\operatorname{Cog}(E/F) = G/F^*$ in view of Lemma 3.1.6. For example consider the extension

$$\mathbb{Q}(\sqrt[n_1]{a_1}, \dots, \sqrt[n_r]{a_r})/\mathbb{Q},$$

where $r \in \mathbb{N}^*$, $n_1, \dots, n_r, a_1, \dots, a_r \in \mathbb{N}^*$, and $\sqrt[n_i]{a_i}$ is the positive real n_i-th root of a_i for every $i, 1 \leqslant i \leqslant r$. This is a G-radical Cogalois extension, where $G = \mathbb{Q}^*\langle \sqrt[n_1]{a_1}, \dots, \sqrt[n_r]{a_r}\rangle$, hence its Cogalois group is precisely $\mathbb{Q}^*\langle \sqrt[n_1]{a_1}, \dots, \sqrt[n_r]{a_r}\rangle/\mathbb{Q}^*$.

(2) A quadratic extension $\mathbb{Q}(\sqrt{d})/\mathbb{Q}$ where $d \neq 1$ is a square-free integer is Cogalois if and only if $d \neq -1, -3$ (see Corollary 3.3.3).

(3) By Exercise 1, for any odd prime $p > 0$ and any $n \in \mathbb{N}^*$, the radical extension $\mathbb{Q}(\zeta_{p^n})/\mathbb{Q}(\zeta_p)$ is pure, hence it is Cogalois by the Greither-Harrison Criterion.

(4) By Exercise 4, $\mathbb{Q}(\zeta_9, \sqrt[9]{5})/\mathbb{Q}(\zeta_3)$ is a Galois and Cogalois extension which is not Abelian.   □

Next, we will investigate the property of an extension being Cogalois in a tower of fields.

PROPOSITION 3.2.2. *The following assertions hold for a tower of fields* $F \subseteq K \subseteq E$.

(1) *There exists a canonical exact sequence of Abelian groups*

$$1 \longrightarrow \operatorname{Cog}(K/F) \longrightarrow \operatorname{Cog}(E/F) \longrightarrow \operatorname{Cog}(E/K).$$

(2) *If E/F is a Cogalois extension, then E/K and K/F are both Cogalois extensions.*

(3) *If E/F is a radical extension, and E/K, K/F are both Cogalois extensions, then E/F is a Cogalois extension.*

(4) *If E/F is a Cogalois extension, then the groups $\mathrm{Cog}(E/K)$ and $\mathrm{Cog}(E/F)/\mathrm{Cog}(K/F)$ are canonically isomorphic.*

PROOF. (1) Clearly, the canonical map

$$\mathrm{Cog}\,(E/F) \longrightarrow \mathrm{Cog}\,(E/K),\ xF^* \mapsto xK^*$$

is a group morphism with kernel $\mathrm{Cog}\,(K/F)$. This shows that the sequence

$$1 \longrightarrow \mathrm{Cog}\,(K/F) \longrightarrow \mathrm{Cog}\,(E/F) \longrightarrow \mathrm{Cog}\,(E/K)$$

of Abelian groups is exact.

(2) Suppose that E/F is Cogalois. Then E/F is radical, separable and pure by Theorem 3.1.7, hence E/K is also radical, separable and pure. Again by Theorem 3.1.7, we conclude that E/K is a Cogalois extension.

Now, we are going to show that K/F is also Cogalois. Since E/F is a Cogalois extension, the group $\mathrm{Cog}(E/F)$ is finite, hence its subgroup $\mathrm{Cog}(K/F)$ is also finite. Let $r = |\mathrm{Cog}(K/F)|$. If $\widehat{x_1},\dots,\widehat{x_r}$ are all the elements of $\mathrm{Cog}(K/F)$, then using again the fact that E/F is a Cogalois extension, we deduce by Corollary 2.1.10 that the subset $\{x_1,\dots,x_r\}$ of K is linearly independent over F, hence

$$|\mathrm{Cog}(K/F)| \leqslant [\,K:F\,].$$

By (1), there exists a canonical monomorphism of groups

$$\mathrm{Cog}(E/F)/\mathrm{Cog}(K/F) \rightarrowtail \mathrm{Cog}(E/K),$$

hence

$$|\mathrm{Cog}(E/F)|/|\mathrm{Cog}(K/F)| \leqslant |\mathrm{Cog}(E/K)|.$$

Using the facts that E/F and E/K are Cogalois extensions, we obtain

$$[\,E:F\,]/|\mathrm{Cog}(K/F)| \leqslant [\,E:K\,] = [\,E:F\,]/[\,K:F\,].$$

This implies that

$$1/|\mathrm{Cog}(K/F)| \leqslant 1/[\,K:F\,],$$

i.e.,

$$|\mathrm{Cog}(K/F)| \geqslant [\,K:F\,].$$

Since we have shown that the opposite inequality also holds, we deduce that

$$|\mathrm{Cog}(K/F)| = [\,K:F\,].$$

This shows that $\{\widehat{x_1}, \ldots, \widehat{x_r}\}$ is a vector space basis of K over F. In particular $K = F(x_1, \ldots, x_r)$, and since $\{x_1, \ldots, x_r\} \subseteq T(K/F)$, it follows that K/F is a radical extension. Consequently, K/F is a Cogalois extension by the Greither-Harrison Criterion.

We present below an alternative proof of the fact that K/F is a Cogalois extension. A piece of this proof will be used in the proof of Theorem 3.2.3 (1). For simplicity, set $G = T(E/F)$. Since E/K is Cogalois, it is $T(E/K)$-Kneser. But $GK^* = T(E/F)K^* \leqslant T(E/K)$, and $E = F(G) = K(GK^*)$, hence the extension E/K is GK^*-radical, separable and pure. Consequently, $GK^* = T(E/K)$ by Lemma 3.1.6, i.e., E/K is GK^*-Kneser, hence

$$[E : K] = |(GK^*)/K^*| = |G/(K^* \cap G)|.$$

Thus

$$[K : F] = [E : F]/[E : K] = |G/F^*|/|G/(K^* \cap G)| = |(K^* \cap G)/F^*|.$$

On the other hand, $F(K^* \cap G)/F$ is $K^* \cap G$-Kneser by Corollary 2.1.10, so $[F(K^* \cap G) : F] = |(K^* \cap G)/F^*|$. Since $F(K^* \cap G) \subseteq K$ we deduce that $K = F(K^* \cap G)$. Consequently, K/F is $K^* \cap G$-Kneser. But $K^* \cap G = K^* \cap T(E/F) = T(K/F)$. Thus K/F is $T(K/F)$-Kneser, i.e., K/F is a Cogalois extension.

(3) Suppose that E/K and K/F are both Cogalois extensions. Since E/K and K/F are separable extensions, so is E/F. Since E/K and K/F are pure, so is also E/F by Proposition 3.1.5. But E/F is a radical extension by hypothesis, hence E/F is Cogalois by Theorem 3.1.7.

(4) By (1), there exists a canonical monomorphism

$$\varphi : \mathrm{Cog}\,(E/F)/\mathrm{Cog}\,(K/F) \rightarrowtail \mathrm{Cog}\,(E/K)$$

of Abelian groups. Observe that since all extensions E/F, K/F, E/K are Cogalois, one has

$$|\mathrm{Cog}\,(E/F)/\mathrm{Cog}\,(K/F)| = |\mathrm{Cog}\,(E/F)|/|\mathrm{Cog}\,(K/F)|$$

$$= [E : F]/[K : F] = [E : K] = |\mathrm{Cog}\,(E/K)|,$$

hence φ is also surjective. Thus φ is an isomorphism of groups. $\square$

An easy consequence of Proposition 3.2.2 is the following fundamental result, which essentially says that any Cogalois extension is an extension with Cogalois correspondence. A much shorter proof, involving the n-Purity Criterion will be given in Section 7.5.

THEOREM 3.2.3. *The following statements hold for a finite Cogalois extension E/F.*

(1) *The maps* $- \cap T(E/F) : \mathcal{E} \longrightarrow \mathcal{C}$ *and* $F(-) : \mathcal{C} \longrightarrow \mathcal{E}$ *are isomorphisms of lattices, inverse to one another, where* $\mathcal{E} = \{\, K \mid F \subseteq K,\ K\ subfield\ of\ E \,\}$ *and* $\mathcal{C} = \{\, H \mid F^* \leqslant H \leqslant T(E/F) \,\}.$

(2) *For every intermediate field $K \in \mathcal{E}$ one has $K = F(T(K/F))$.*

(3) *For every subgroup $H \in \mathcal{C}$ one has $\mathrm{Cog}(F(H)/F) = H/F^*$.*

PROOF. (1) For every $H \in \mathcal{C}$ and $K \in \mathcal{E}$ we have

$$F(H) \cap T(E/F) = H$$

by Proposition 2.1.11, and

$$F(K \cap T(E/F)) = F(K^* \cap T(E/F)) = K$$

by the alternative proof of point (2) in Proposition 3.2.2. These two relations mean precisely that the order-preserving maps $- \cap T(E/F) : \mathcal{E} \longrightarrow \mathcal{C}$ and $F(-) : \mathcal{C} \longrightarrow \mathcal{E}$ are bijections, hence isomorphisms of lattices, inverse to one another.

(2) and (3) are consequences of (1) and of the following simple facts: for every $H \in \mathcal{C}$ and every $K \in \mathcal{E}$ one has

$$K \cap T(E/F) = T(K/F) \quad \text{and} \quad H = F(H) \cap T(E/F) = T(F(H)/F).$$

$\square$

PROPOSITION 3.2.4. *Let E_1/F and E_2/F be Cogalois extensions which are subextensions of an extension E/F. Then, the extension E_1E_2/F is Cogalois if and only if it is pure. In this case, the following statements are equivalent.*

(1) $T(E_1/F) \cap T(E_2/F) = F^*$.

(2) *The canonical map*

$$\alpha : \mathrm{Cog}(E_1/F) \times \mathrm{Cog}(E_2/F) \longrightarrow \mathrm{Cog}(E_1E_2/F)$$

defined by $\alpha(x_1 F^*, x_2 F^*) = x_1 x_2 F^*$, $x_1 \in E_1$, $x_2 \in E_2/F$, *is a monomorphism of groups.*

(3) *The canonical map α in (2) is an isomorphism of groups.*

(4) *The fields E_1 and E_2 are linearly disjoint over F.*

(5) $E_1 \cap E_2 = F$.

PROOF. We have $E_1 = F(T(E_1/F))$ and $E_2 = F(T(E_2/F))$, hence $E_1 E_2 = F(T(E_1/F) \cup T(E_2/F))$. Thus E_1E_2/F is a radical extension. Since E_1/F and E_2/F are separable, so is also E_1E_2/F. Therefore, by the

Greither-Harrison Criterion, the extension $E_1 E_2/F$ is Cogalois if and only if it is pure.

(1) $\implies$ (2): We are going to show that the group morphism α is injective. As always throughout this monograph, for any $x \in E$ we denote by $\widehat{x}$ the coset xF^* in the quotient group E^*/F^*. If $(\widehat{x}, \widehat{y}) \in \mathrm{Ker}(\alpha)$, then $\widehat{xy} = \widehat{1}$, hence $xy \in F^*$. Since $x \in T(E_1/F)$ it follows that $y \in T(E_1/F)$. But $y \in T(E_2/F)$, so $y \in T(E_1/F) \cap T(E_2/F) = F^*$. This implies that $x \in F^*$. Consequently, $(\widehat{x}, \widehat{y}) = (\widehat{1}, \widehat{1})$, hence $\mathrm{Ker}(\alpha) = \{(\widehat{1}, \widehat{1})\}$. This shows that α is injective.

(2) $\implies$ (3): We have to prove that α is surjective. To do that, observe that since $E_1 E_2/F$ is Cogalois, one has

$$|\mathrm{Cog}\,(E_1 E_2/F)| = [E_1 E_2 : F] \leqslant [E_1 : F] \cdot [E_2 : F]$$
$$= |\mathrm{Cog}\,(E_1/F)| \cdot |\mathrm{Cog}\,(E_2/F)|.$$

But α is a monomorphism, hence

$$|\mathrm{Cog}\,(E_1/F) \times \mathrm{Cog}\,(E_2/F)| = |\mathrm{Cog}\,(E_1/F)| \cdot |\mathrm{Cog}\,(E_2/F)|$$
$$\leqslant |\mathrm{Cog}\,(E_1 E_2/F)|.$$

Therefore α is surjective, hence an isomorphism.

(3) $\implies$ (4): By the definition of linearly disjoint extensions (see 1.2.3), we have to show that there exists a vector space basis B_i of E_i over F, $i = 1, 2$, such that $B_1 B_2 = \{\, x_1 x_2 \,|\, x_1 \in B_1,\, x_2 \subset B_2 \,\}$ is a vector space basis of $E_1 E_2$ over F.

Let B_i be a set of representatives in $T(E_i/F)$ of the factor group $\mathrm{Cog}(E_i/F) = T(E_i/F)/F^*$, $i = 1, 2$. Then B_i is a vector space basis of E_i over F, $i = 1, 2$, by Corollary 2.1.10.

Since $\mathrm{Cog}(E_i/F) = \{\, \widehat{x}_i \,|\, x_i \in B_i \,\}$, $i = 1, 2$, and α is an isomorphism we deduce that

$$\mathrm{Cog}(E_1 E_2/F) = \mathrm{Im}(\alpha) = \{\, \widehat{x_1 x_2} \,|\, x_1 \in B_1,\, x_2 \in B_2 \,\},$$

and

$$\widehat{x_1 x_2} = \widehat{y_1 y_2} \iff (\widehat{x_1}, \widehat{x_2}) = (\widehat{y_1}, \widehat{y_2}) \iff (x_1, x_2) = (y_1, y_2)$$

for every $x_1, y_1 \in B_1$, $x_2, y_2 \in B_2$. In other words, $B_1 B_2$ is a set of representatives of $\mathrm{Cog}(E_1 E_2/F)$. By Proposition 2.1.8, it follows that $B_1 B_2$ is a basis of the vector space $E_1 E_2$ over F, as desired.

(4) $\implies$ (5) and (5) $\implies$ (1) are obvious. $\square$

REMARKS 3.2.5. (1) The compositum of two Cogalois subextensions of an extension is not necessarily Cogalois, by Exercise 5.

(2) By Exercise 6, the conditions (1)-(5) in Proposition 3.2.4 are also equivalent to the following one:

The groups $\mathrm{Cog}(E_1E_2/F)$ *and* $\mathrm{Cog}(E_1/F) \times \mathrm{Cog}(E_2/F)$ *are isomorphic, but not necessarily via the canonical isomorphism* α. $\square$

We end this section by showing that the property of an extension being radical, Kneser, or Cogalois is, in general, not transitive. This will follow immediately from the next proposition investigating the quartic extension $\mathbb{Q}(\sqrt{1+\sqrt{2}})/\mathbb{Q}$.

PROPOSITION 3.2.6. *The following assertions hold.*

(a) $\mathbb{Q}(\sqrt{2})$ *is a subfield of the field* $\mathbb{Q}(\sqrt{1+\sqrt{2}})$.

(b) $[\mathbb{Q}(\sqrt{1+\sqrt{2}}) : \mathbb{Q}(\sqrt{2})] = 2$ *and* $[\mathbb{Q}(\sqrt{1+\sqrt{2}}) : \mathbb{Q}] = 4$.

(c) $\mathbb{Q}(\sqrt{1+\sqrt{2}})/\mathbb{Q}$ *is not a Galois extension.*

(d) $\mathbb{Q}(\sqrt{1+\sqrt{2}})/\mathbb{Q}(\sqrt{2})$ *and* $\mathbb{Q}(\sqrt{2})/\mathbb{Q}$ *are both Cogalois extensions.*

(e) $\mathbb{Q}(\sqrt{1+\sqrt{2}})/\mathbb{Q}$ *is neither a radical extension, nor a Kneser extension, nor a Cogalois extension.*

(f) $\mathrm{Cog}\left(\mathbb{Q}(\sqrt{1+\sqrt{2}})/\mathbb{Q}\right) = \{\,\widehat{1},\ \widehat{\sqrt{2}}\,\}$.

(g) *The element* $\widehat{\sqrt{1+\sqrt{2}}}$ *of the group* $\mathbb{Q}(\sqrt{1+\sqrt{2}})^*/\mathbb{Q}^*$ *has infinite order.*

PROOF. For simplicity, set

$$F = \mathbb{Q},\ K = \mathbb{Q}(\sqrt{2}),\ E = \mathbb{Q}\left(\sqrt{1+\sqrt{2}}\right),\ \text{and}\ \theta = \sqrt{1+\sqrt{2}}.$$

(a) We have $\sqrt{2} = \theta^2 - 1 \in \mathbb{Q}(\theta) = E$, hence $K = \mathbb{Q}(\sqrt{2}) \subseteq \mathbb{Q}(\theta) = E$, and $E = K(\theta)$.

(b) Since θ is a root of the irreducible polynomial $X^4 - 2X^2 - 1 \in \mathbb{Q}[X]$, we deduce that $[E : F] = 4$, and this clearly implies that $[E : K] = 2$.

(c) The conjugates of θ over $\mathbb{Q}$ are the roots of its minimal polynomial $X^4 - 2X^2 - 1$ over $\mathbb{Q}$. Two of them are the nonreal complex roots of the equation $x^2 = 1 - \sqrt{2}$, so they do not belong to $E \subseteq \mathbb{R}$. This shows that the extension E/F is not normal, so it is not Galois.

(d) Since E is a subfield of the field $\mathbb{R}$ of real numbers, it follows that the extensions K/F and E/K are both pure by Examples 3.1.4 (1). On the other hand, the extensions K/F and E/K are both separable and radical. So, in view of the Greither-Harison Criterion, K/F and E/K are both Cogalois extensions.

(e) The extension E/F is pure, again by Example 3.1.4 (1). Since E/F is clearly a separable extension, then E/F is a Cogalois extension if and only if it is radical by the Greither-Harison Criterion. Since any Kneser extension is necessarily radical, it is sufficient to prove only that E/F is not a Cogalois extension.

Assume that E/F is a Cogalois extension, and look for a contradiction. Since $[E : F] = 4$, then $\mathrm{Cog}(E/F)$ is a group of order 4 by the definition of the concept of Cogalois extension, hence this group is isomorphic either to $\mathbb{Z}_2 \times \mathbb{Z}_2$ or to $\mathbb{Z}_4$.

Case 1: If $\mathrm{Cog}(E/F) \cong \mathbb{Z}_2 \times \mathbb{Z}_2$, then there exist $\beta, \gamma \in \mathbb{Q}_+^*$ such that

$$\mathrm{Cog}(E/F) = \mathbb{Q}^* \langle \sqrt{\beta}, \sqrt{\gamma} \rangle / \mathbb{Q}^* = \{ \widehat{1}, \widehat{\sqrt{\beta}}, \widehat{\sqrt{\gamma}}, \widehat{\sqrt{\beta\gamma}} \}.$$

Then, by Corollary 2.1.10, we deduce that $\{ 1, \sqrt{\beta}, \sqrt{\gamma}, \sqrt{\beta\gamma} \}$ is a vector space basis of E over $\mathbb{Q}$. In particular, it follows that

$$E = \mathbb{Q}\left(\sqrt{1 + \sqrt{2}} \right) = \mathbb{Q}(\sqrt{\beta}, \sqrt{\gamma}).$$

But $\mathbb{Q}(\sqrt{\beta}, \sqrt{\gamma})/\mathbb{Q}$ is a Galois extension, hence so is $\mathbb{Q}(\sqrt{1 + \sqrt{2}})/\mathbb{Q}$, which contradicts (c). Thus, this case cannot occur.

Case 2: If $\mathrm{Cog}(E/F) \cong \mathbb{Z}_4$, then there exists $\alpha \in \mathbb{Q}_+^*$ such that

$$\mathrm{Cog}(E/F) = \mathbb{Q}^* \langle \sqrt[4]{\alpha} \rangle / \mathbb{Q}^* = \{ \widehat{1}, \widehat{\sqrt[4]{\alpha}}, \widehat{\sqrt[4]{\alpha}^2}, \widehat{\sqrt[4]{\alpha}^3} \}.$$

Again by Corollary 2.1.10, $\{ 1, \sqrt[4]{\alpha}, \sqrt[4]{\alpha}^2, \sqrt[4]{\alpha}^3 \}$ is a vector space basis of E over $\mathbb{Q}$. In particular, it follows that

$$E = \mathbb{Q}\left(\sqrt{1 + \sqrt{2}} \right) = \mathbb{Q}(\sqrt[4]{\alpha}).$$

Since $[E : F] = 4$, we have $\sqrt{\alpha} \notin \mathbb{Q}$.

Clearly, $\mathbb{Q}(\sqrt{\alpha})$ is a proper subfield of $\mathbb{Q}(\sqrt[4]{\alpha}) = E$. Since the group $\mathrm{Cog}(E/F)$ has a unique proper subgroup, we deduce by Theorem 3.2.3 (1) that E/F has a unique proper intermediate field. Since $\mathbb{Q}(\sqrt{2})$ is also such an intermediate field, this implies that necessarily $\mathbb{Q}(\sqrt{\alpha}) = \mathbb{Q}(\sqrt{2})$, hence $\sqrt{2} = k\sqrt{\alpha}$ for some $k \in \mathbb{Q}_+^*$, and then $2 = \alpha k^2$.

Since $E = \mathbb{Q}(\theta) = \mathbb{Q}(\sqrt[4]{\alpha})$, there exist $a, b, c, d \in \mathbb{Q}$ such that

$$(1) \qquad \theta = \sqrt{1 + \sqrt{2}} = a + b\sqrt[4]{\alpha} + c\sqrt[4]{\alpha}^2 + d\sqrt[4]{\alpha}^3.$$

We can also write (1) as

$$(2) \qquad \theta - (a + c\sqrt{\alpha}) = \sqrt[4]{\alpha}\,(b + d\sqrt{\alpha}).$$

Squaring (2), we obtain

$$\theta^2 - 2\theta(a + c\sqrt{\alpha}) + (a + c\sqrt{\alpha})^2 = \sqrt{\alpha}\,(b + d\sqrt{\alpha})^2.$$

Since $\theta^2 \in \mathbb{Q}(\sqrt{2}) = \mathbb{Q}(\sqrt{\alpha})$, we deduce that we must have $a + c\sqrt{\alpha} = 0$, for otherwise, it would follow that $\theta \in \mathbb{Q}(\sqrt{\alpha}) = \mathbb{Q}(\sqrt{2})$, which is a contradiction. Thus, (2) becomes

$$(3) \qquad \theta = \sqrt[4]{\alpha}\,(b + d\sqrt{\alpha}),$$

which can be also written as

$$\frac{b + d\sqrt{\alpha}}{\theta} = \frac{1}{\sqrt[4]{\alpha}},$$

or

$$(4) \qquad \frac{bk + d\sqrt{2}}{\theta} = \frac{k}{\sqrt[4]{\alpha}},$$

since $\sqrt{\alpha} = \sqrt{2}/k$.

From (4) we deduce that

$$\left(\frac{bk + d\sqrt{2}}{\theta}\right)^4 \in \mathbb{Q},$$

i.e.,

$$u := \frac{\left(bk + d\sqrt{2}\right)^4}{(1 + \sqrt{2})^2} \in \mathbb{Q}.$$

Then, u coincides with its conjugate in the quadratic extension $\mathbb{Q}(\sqrt{2})/\mathbb{Q}$, i.e.,

$$\frac{\left(bk + d\sqrt{2}\right)^4}{(1 + \sqrt{2})^2} = \frac{\left(bk - d\sqrt{2}\right)^4}{(1 - \sqrt{2})^2}.$$

This implies that

$$\frac{\left(bk + d\sqrt{2}\right)^2}{1 + \sqrt{2}} = \frac{\left(bk - d\sqrt{2}\right)^2}{\sqrt{2} - 1},$$

hence

$$\left(bk + d\sqrt{2}\right)^2(\sqrt{2} - 1) = \left(bk - d\sqrt{2}\right)^2(\sqrt{2} + 1).$$

After easy calculations we obtain

$$b^2k^2 - 4bdk + 2d^2 = 0,$$

which can be also written as

(5) $$(bk - 2d)^2 = 2d^2.$$

We claim that $d = 0$, for otherwise, (5) would imply that $\sqrt{2} \in \mathbb{Q}$, which is a contradiction. Then, (5) yields $bk = 0$, hence $b = 0$. Thus, (3) becomes $\theta = 0$, which is a contradiction. This completes the proof of the fact that E/F is not a Cogalois extension.

(f) The inclusion

$$\left\{ \widehat{1}, \ \widehat{\sqrt{2}} \right\} \subseteq \mathrm{Cog}\left(\mathbb{Q}\!\left(\sqrt{1 + \sqrt{2}} \right) \middle/ \mathbb{Q} \right)$$

is clear. To prove the equality it is sufficient to show that $|\mathrm{Cog}\,(E/F)| = 2$. To do that, observe that, by Proposition 3.2.2 (1), there exists a canonical monomorphism of groups

$$\mathrm{Cog}\,(E/F)/\mathrm{Cog}\,(K/F) \rightarrowtail \mathrm{Cog}\,(E/K).$$

By (d), the quadratic extensions E/K and K/F are both Cogalois. Then

$$|\mathrm{Cog}\,(E/K)| = |\mathrm{Cog}\,(K/F)| = 2,$$

hence necessarily $|\mathrm{Cog}\,(E/F)| \in \{2, 4\}$. Since E/F is not a Cogalois extension by (e), we are going to show that we cannot have $|\mathrm{Cog}\,(E/F)| = 4$. This will imply that $|\mathrm{Cog}\,(E/F)| = 2$, and we will be done.

Assume that $|\mathrm{Cog}\,(E/F)| = 4$. Then, as in the proof of (e), two cases arise: $\mathrm{Cog}(E/F) \cong \mathbb{Z}_2 \times \mathbb{Z}_2$, or $\mathrm{Cog}(E/F) \cong \mathbb{Z}_4$. In the first case, there exist $\beta,\ \gamma \in \mathbb{Q}_+^*$ such that

$$\mathrm{Cog}(E/F) = \mathbb{Q}^*\langle \sqrt{\beta},\ \sqrt{\gamma}\,\rangle/\mathbb{Q}^* = \left\{ \widehat{1},\ \widehat{\sqrt{\beta}},\ \widehat{\sqrt{\gamma}},\ \widehat{\sqrt{\beta\gamma}}\, \right\}.$$

We claim that $\{\, 1, \sqrt{\beta},\ \sqrt{\gamma},\ \sqrt{\beta\gamma}\, \}$ is a vector space basis of E over F. Since E/F is not a radical extension, we cannot apply Corollary 2.1.10 to deduce it, as we did in the proof of (e). However we can prove the claim as follows. It is easy to show that $\sqrt{\gamma} \notin \mathbb{Q}(\sqrt{\beta})$, hence $[\mathbb{Q}(\sqrt{\beta}, \sqrt{\gamma}) : \mathbb{Q}] = 4$. This implies that $E = \mathbb{Q}(\sqrt{\beta}, \sqrt{\gamma})$, and then E/F is a radical extension, which contradicts (e).

If $\mathrm{Cog}(E/F) \cong \mathbb{Z}_4$, then there exists $\alpha \in \mathbb{Q}_+^*$ such that

$$\mathrm{Cog}(E/F) = \mathbb{Q}^* \langle \sqrt[4]{\alpha} \rangle / \mathbb{Q}^* = \{\, \widehat{1}, \widehat{\sqrt[4]{\alpha}}, \widehat{\sqrt[4]{\alpha}^2}, \widehat{\sqrt[4]{\alpha}^3} \,\}.$$

Since $\sqrt[4]{\alpha} \notin \mathbb{Q}(\sqrt{\alpha})$, it follows that $\{\, 1, \sqrt[4]{\alpha}, \sqrt[4]{\alpha}^2, \sqrt[4]{\alpha}^3 \,\}$ is a vector space basis of E over F. Thus, E/F is a radical extension, which again contradicts (e). This completes the proof of (f).

(g) Assume that $\mathrm{ord}(\widehat{\sqrt{1 + \sqrt{2}}})$ is finite, say n. Then, $\sqrt{1 + \sqrt{2}} = \sqrt[n]{a}$ for some $a \in \mathbb{Q}_+^*$, so it would follow that the extension $\mathbb{Q}(\sqrt{1 + \sqrt{2}})/\mathbb{Q}$ is a subextension of the Cogalois extension $\mathbb{Q}(\sqrt[n]{a})/\mathbb{Q}$, hence it would be itself Cogalois, which contradicts (e). $\square$

COROLLARY 3.2.7. *The property of an extension being radical, Kneser, or Cogalois is, in general, not transitive.* $\square$

3.3. The Cogalois group of a quadratic extension

For any G-radical extension E/F, the group G/F^* is a subgroup of the Cogalois group $\mathrm{Cog}(E/F) = T(E/F)/F^*$ of E/F. Hence, the effective description of the Cogalois group of a given extension is an important problem in the investigation of radical extensions, which is also of independent interest.

In general, the concrete calculation of the Cogalois group of a given extension is quite hard. The aim of this section is to provide a very explicit description of the Cogalois group of any quadratic extension of $\mathbb{Q}$.

As usually, we shall denote the imaginary unit $\sqrt{-1} \in \mathbb{C}$ by i, and $\sqrt{-d}$ by $i\sqrt{d}$ for any $d \in \mathbb{Q}_+^*$.

LEMMA 3.3.1. *Let $d \neq 1$ be a square-free rational integer, and let W denote the group of roots of unity $\mu(\mathbb{Q}(\sqrt{d}))$ in $\mathbb{Q}(\sqrt{d})$. Then*

(1) $W = \mu_2(\mathbb{C}) = \{\pm 1\}$ *if* $d \neq -1, -3$.

(2) $W = \mu_4(\mathbb{C}) = \{\pm 1, \pm i\}$ *if* $d = -1$.

(3) $W = \mu_6(\mathbb{C}) = \{\pm 1, \pm(1 + i\sqrt{3})/2, \pm(1 - i\sqrt{3})/2\}$ *if* $d = -3$.

PROOF. If the quadratic field $\mathbb{Q}(\sqrt{d})$ is real, that is, if $d > 0$, then W consists only of real roots of unity, hence $W = \{-1, 1\}$. If the quadratic field $\mathbb{Q}(\sqrt{d})$ is imaginary, that is, if $d < 0$, then W exhausts all the units of the ring of integers A of the given quadratic field, and they are precisely those elements of A having norm 1. An easy computation of the norm of an element in A establishes the result. More details can be found, e.g., in Ribenboim [**89**, p.131]. □

PROPOSITION 3.3.2. *Let* $E = \mathbb{Q}(\sqrt{d})$, *where* $d \neq 1$ *is a square-free integer. Then*

(1) $\mathrm{Cog}\,(E/\mathbb{Q}) = \langle \widehat{\sqrt{d}} \rangle \cong \mathbb{Z}_2$ *if* $d \neq -1, \, -3$.
(2) $\mathrm{Cog}\,(E/\mathbb{Q}) = \langle \widehat{1+i} \rangle \cong \mathbb{Z}_4$ *if* $d = -1$.
(3) $\mathrm{Cog}\,(E/\mathbb{Q}) = \langle \widehat{i\sqrt{3} \cdot (1+i\sqrt{3})} \rangle \cong \mathbb{Z}_6$ *if* $d = -3$.

PROOF. Let $\alpha \in T(E/\mathbb{Q})$, $\alpha = a + b\sqrt{d}$ with $a, b \in \mathbb{Q}$, and consider its coset $\widehat{\alpha}$ in $\mathrm{Cog}\,(\mathbb{Q}(\sqrt{d})/\mathbb{Q}) = T(E/\mathbb{Q})/\mathbb{Q}$. Clearly, $\alpha \in \mathbb{Q}$ if and only if $\widehat{\alpha} = \widehat{1}$.

Now suppose that $\alpha \notin \mathbb{Q}$, i.e., $b \neq 0$. Since $\alpha \in T(E/\mathbb{Q})$, there exists $n \in \mathbb{N}$, $n > 1$ such that $\alpha^n = c \in \mathbb{Q}$, hence α is a root of the polynomial $f = X^n - c \in \mathbb{Q}[X]$. The roots of this polynomial have the form $\alpha\zeta$, where $\zeta \in \mu_n(\mathbb{C})$. The minimal polynomial $\mathrm{Min}(\alpha, \mathbb{Q})$ of α over $\mathbb{Q}$ has degree 2 since $E = \mathbb{Q}(\alpha)$, and is a divisor of f in $\mathbb{Q}[X]$. Consequently, the roots of $\mathrm{Min}(\alpha, \mathbb{Q})$ are α and $\zeta\alpha$, where $\zeta \in \mu(\mathbb{C})$. Moreover, $\zeta\alpha \in E$ since the extension $E/\mathbb{Q}$ is normal. But $\zeta = (\zeta\alpha)/\alpha$, hence $\zeta \in E \cap \mu(\mathbb{C}) = \mu(E) = W$.

On the other hand, the product of the roots of $\mathrm{Min}(\alpha, \mathbb{Q})$ is a rational number, hence

(†) $\zeta\alpha^2 \in \mathbb{Q}$.

Now, we are going to investigate all the cases considered in the statement of the proposition.

Case 1: $d \neq -1, -3$. We have $\zeta \in \{-1, 1\}$ by Lemma 3.3.1, and (†) implies that $\alpha^2 \in \mathbb{Q}$, i.e., $a^2 + db^2 + 2ab\sqrt{d} \in \mathbb{Q}$. Then $ab = 0$. Since $b \neq 0$, we necessarily have $a = 0$, hence $\alpha = b\sqrt{d}$, and then, $\widehat{\alpha} = \widehat{\sqrt{d}}$. It follows that

$$\mathrm{Cog}\,(E/\mathbb{Q}) = \{ \widehat{1}, \widehat{\sqrt{d}} \} = \langle \widehat{\sqrt{d}} \rangle \cong \mathbb{Z}_2.$$

Case 2: $d = -1$. We have $\alpha = a + bi$ and $\alpha^2 - a^2 - b^2 + 2abi$. By Lemma 3.3.1, $\zeta \in \{ 1, -1, i, -i \}$.

i) If $\zeta = \pm 1$, then (†) implies that $\alpha^2 \in \mathbb{Q}$, i.e., $a^2 - b^2 + 2abi \in \mathbb{Q}$, so $ab = 0$. Then $a = 0$, hence $\alpha = bi$, and thus $\widehat{\alpha} = \widehat{i}$.

ii) If $\zeta = \pm i$, then (†) implies that $i\alpha^2 \in \mathbb{Q}$, i.e., $-2ab \pm (a^2 - b^2)i \in \mathbb{Q}$, hence $a^2 - b^2 = 0$, and so $a = \pm b$.

If $a = b$, then we have $\alpha = b(1 + i)$, hence $\widehat{\alpha} = \widehat{1 + i}$, and if $a = -b$, then we have $\alpha = -b(1 - i)$, hence $\widehat{\alpha} = \widehat{1 - i}$. Thus

$$\operatorname{Cog}(\mathbb{Q}(i)/\mathbb{Q}) = \{\widehat{1}, \widehat{i}, \widehat{1 + i}, \widehat{1 - i}\} = \langle \widehat{1 + i}\rangle \cong \mathbb{Z}_4.$$

Case 3: $d = -3$. We have $\alpha = a + bi\sqrt{3}$ and $\alpha^2 = a^2 - 3b^2 + 2abi\sqrt{3}$. By Lemma 3.3.1, $\zeta \in \{\pm 1, \pm(1 + i\sqrt{3})/2, \pm(1 - i\sqrt{3})/2\}$.

i) If $\zeta = \pm 1$, then (†) implies that $\alpha^2 \in \mathbb{Q}$, i.e., $a^2 - 3b^2 + 2abi\sqrt{3} \in \mathbb{Q}$, hence $ab = 0$. Then $a = 0$ and $\alpha = bi\sqrt{3}$, hence $\widehat{\alpha} = \widehat{i\sqrt{3}}$.

ii) If $\zeta = \pm(1 + i\sqrt{3})/2$, then we have $(1 + i\sqrt{3})\alpha^2 \in \mathbb{Q}$ by (†), i.e., $(1 + i\sqrt{3})(a^2 - 3b^2 + 2abi\sqrt{3}) \in \mathbb{Q}$. Thus

$$a^2 - 3b^2 - 6ab + (a^2 - 3b^2 + 2ab)i\sqrt{3} \in \mathbb{Q},$$

and so, $a^2 - 3b^2 + 2ab = (a - b)(a + 3b) = 0$. It follows that $a = b$ or $a = -3b$.

If $a = b$, then $\alpha = b(1 + i\sqrt{3})$, hence $\widehat{\alpha} = \widehat{1 + i\sqrt{3}}$, and if $a = -3b$, then $\alpha = b(-3 + i\sqrt{3})$, hence $\widehat{\alpha} = \widehat{-3 + i\sqrt{3}} = \widehat{i\sqrt{3} \cdot (1 + i\sqrt{3})}$.

iii) If $\zeta = \pm(1 - i\sqrt{3})/2$, then we have $(1 - i\sqrt{3})\alpha^2 \in \mathbb{Q}$ by (†), i.e., $(1 - i\sqrt{3})(a^2 - 3b^2 + 2abi\sqrt{3}) \in \mathbb{Q}$. Thus

$$a^2 - 3b^2 + 6ab + (-a^2 + 3b^2 + 2ab)i\sqrt{3} \in \mathbb{Q},$$

and so, $-a^2 + 3b^2 + 2ab = (a + b)(-a + 3b) = 0$. It follows that $a = -b$ or $a = 3b$.

If $a = -b$, then $\alpha = -b(1 - i\sqrt{3})$, hence $\widehat{\alpha} = \widehat{1 - i\sqrt{3}}$, and if $a = 3b$, then $\alpha = b(3 + i\sqrt{3})$, hence $\widehat{\alpha} = \widehat{3 + i\sqrt{3}} = \widehat{i\sqrt{3} \cdot (1 - i\sqrt{3})}$. Consequently,

$$\operatorname{Cog}(\mathbb{Q}(i\sqrt{3})/\mathbb{Q}) = \{\widehat{1}, \widehat{i\sqrt{3}}, \widehat{1 + i\sqrt{3}}, \widehat{i\sqrt{3} \cdot (1 + i\sqrt{3})}, \widehat{1 - i\sqrt{3}},$$

$$\widehat{i\sqrt{3} \cdot (1 - i\sqrt{3})}\} = \langle \widehat{i\sqrt{3} \cdot (1 + i\sqrt{3})}\rangle \cong \mathbb{Z}_6.$$

$\square$

COROLLARY 3.3.3. *The following statements are equivalent for a square-free rational integer $d \neq 1$.*

(1) $\mathbb{Q}(\sqrt{d})/\mathbb{Q}$ *is a pure extension.*
(2) $\mathbb{Q}(\sqrt{d})/\mathbb{Q}$ *is a Cogalois extension.*
(3) $d \neq -1, -3$.

PROOF. By definition, $\mathbb{Q}(\sqrt{d})/\mathbb{Q}$ is a Cogalois extension if and only if $|\mathrm{Cog}(\mathbb{Q}(\sqrt{d})/\mathbb{Q})| = 2$, i.e., by Proposition 3.3.2, if and only if $d \neq -1, -3$, so (1) $\Longleftrightarrow$ (3). The equivalence (1) $\Longleftrightarrow$ (2) follows from the Greither-Harrison Criterion. $\square$

REMARK 3.3.4. Any extension E/F of algebraic number fields has a finite Cogalois group. This will be proved in Section 6.1 using the description of the Cogalois group of a finite Galois extension by means of crossed homomorphisms (see Corollary 6.1.4). $\square$

3.4. Exercises to Chapter 3

1. (*Greither and Harrison* [63]). Prove that for any odd prime p and any $n \in \mathbb{N}^*$, the extension $\mathbb{Q}(\zeta_{p^n})/\mathbb{Q}(\zeta_p)$ is pure. (*Hint*: Show that the fields $\mathbb{Q}(\zeta_{p^n})$ and $\mathbb{Q}(\zeta_q)$ are linearly disjoint over $\mathbb{Q}$ whenever $q \neq p$ is a prime or $q = 4$.)

2. Show that the result in Exercise 1 fails for $p = 2$.

3. (*Greither and Harrison* [63]). Let $F = \mathbb{Q}(\zeta_3)$ and $E = F(\zeta_9, \sqrt[9]{5})$. Prove that the extension E/F is pure.

4. (*Greither and Harrison* [63]). Prove the following statements for the fields $F = \mathbb{Q}(\zeta_3)$ and $E = F(\zeta_9, \sqrt[9]{5})$.
   (a) E/F is a Galois extension of degree 27, with
   $$\mathrm{Gal}(E/F) = \langle \sigma, \tau \rangle \text{ and } \tau \circ \sigma = \sigma^4 \circ \tau,$$
   where $\sigma(\sqrt[9]{5}) = \zeta_9 \sqrt[9]{5}$, $\sigma(\zeta_9) = \zeta_9, \tau(\sqrt[9]{5}) = \sqrt[9]{5}$, $\sigma(\zeta_9) = \zeta_9^4$.
   (b) E/F is a Cogalois extension with
   $$\mathrm{Cog}(E/F) = \langle \widehat{\zeta_9}, \widehat{\sqrt[9]{5}} \rangle = \langle \widehat{\zeta_9} \rangle \oplus \langle \widehat{\sqrt[9]{5}} \rangle \cong \mathbb{Z}_3 \times \mathbb{Z}_9.$$

5. Let $F = \mathbb{Q}$, $E_1 = \mathbb{Q}(\sqrt{-2})$, and $E_2 = \mathbb{Q}(\sqrt{2})$. Show that E_1/F and E_2/F are Cogalois extensions, but their compositum $E_1 E_2/F$ is not pure, so, not Cogalois.

6. Let E_1/F and E_2/F be subextensions of an extension E/F, and assume that the extension $E_1 E_2/F$ is Cogalois. Prove that the groups $\mathrm{Cog}(E_1/F) \times \mathrm{Cog}(E_2/F)$ and $\mathrm{Cog}(E_1 E_2/F)$ are isomorphic if and only if the fields E_1 and E_2 are linearly disjoint over F.

7. Let E/F be a finite extension with $|\mathrm{Cog}(E/F)| = [E:F] = n$.
   (a) Show that E/F is a Cogalois extension whenever n is a prime number, or $n = 4$ and E is a subfield of $\mathbb{R}$.
   (b) Is E/F a Cogalois extension for arbitrary n?

8. Show that the extension $\mathbb{F}_4/\mathbb{F}_2$ is not Cogalois.

9. Let $n \in \mathbb{N}^*$. Prove that the extension $\mathbb{Q}(\zeta_n)/\mathbb{Q}$ is Cogalois if and only if $n = 1$ or $n = 2$.

10. Show that any quadratic extension E/F, with E a subfield of $\mathbb{R}$ is Cogalois.

11. Prove that the extension $\mathbb{Q}(\sqrt{3+\sqrt{2}})/\mathbb{Q}$ is neither radical, nor Kneser, nor Cogalois.

12. Prove that $\mathrm{Cog}(\mathbb{Q}(\sqrt{3+\sqrt{2}})/\mathbb{Q}) = \{\widehat{1}, \widehat{\sqrt{2}}\}$.

13. Let $\nu_r = \underbrace{\sqrt{1+\sqrt{1+\sqrt{1+\ldots+\sqrt{2}}}}}_{r \text{ radicals}}$, and $F_r = \mathbb{Q}(\nu_r)$, $r \geqslant 2$.

Show that $F_r/\mathbb{Q}$ is a non Galois extension of degree 2^r which is neither radical, nor Kneser, nor Cogalois.

14. With the notation of Exercise 13, determine $\mathrm{Cog}\,(F_r/\mathbb{Q})$.

15. With the notation of Exercise 13, check whether or not, for a given $r \in \mathbb{N}^*$, the quartic extension F_{r+1}/F_{r-1} is Cogalois.

16. Let $d \in \mathbb{Q} \setminus \mathbb{Q}^2$. Show that

$$
\mathrm{Cog}\,(\mathbb{Q}(\sqrt{d})/\mathbb{Q}) = \begin{cases} \{\widehat{1}, \widehat{\sqrt{d}}\} \cong \mathbb{Z}_2 & \text{if } -d, -3d \notin \mathbb{Q}^2 \\ \langle \widehat{1+i} \rangle \cong \mathbb{Z}_4 & \text{if } -d \in \mathbb{Q}^2 \\ \langle \widehat{i\sqrt{3} \cdot (1+i\sqrt{3})} \rangle \cong \mathbb{Z}_6 & \text{if } -3d \in \mathbb{Q}^2. \end{cases}
$$

3.5. Bibliographical comments to Chapter 3

Section 3.1. To the best of our knowledge, the term of "Cogalois extension" appeared for the first time in the literature in 1986 in the fundamental paper of Greither and Harrison [63], where *Cogalois extensions* were introduced and investigated. A finite extension E/F is called *conormal* (resp. *coseparable*) by Greither and Harrison if $|\mathrm{Cog}(E/F)| \leqslant [\,E : F\,]$ (resp. if E/F is radical), and it is called *Cogalois* if it is both conormal and coseparable. Their pioneering work on Cogalois extensions was continued in 1991 by Barrera-Mora, Rzedowski-Calderón, and Villa-Salvador [30].

The Cogalois extensions, as well as the Galois extensions are related to the Galois H-objects from the theory of Hopf algebras in a dual manner (see Section 10.2 for connections between Cogalois Theory and Hopf algebras).

Theorem 3.1.7 is due to Greither and Harrison [63]. The easy implication (1) $\implies$ (2) follows the original proof in [63], while the difficult one is a simplified proof of Albu and Nicolae from [19], based on Lemma 3.1.6, which, in turn, is an immediate consequence of the Kneser Criterion and Proposition 2.1.11. Note that the original proof in Greither and Harrison [63] is much longer and uses some cohomology of groups.

Theorem 3.1.9 is an extended reformulation of Theorem 1.7 in Gay and Vélez [61]. Note that Gay and Vélez proved only the implication (3) $\implies$ (1), but they omitted in the statement of their theorem the indispensable separability condition of the given field extension.

This section basically follows the outline in Albu and Nicolae [19], although there are several extensions, simplifications, and improvements.

Section 3.2. Proposition 3.2.2 and Theorem 3.2.3 are due to Greither and Harrison [63]. Proposition 3.2.4, which is an extended version of Proposition 10 in Barrera-Mora, Rzedowski-Calderón, and Villa-Salvador [30] is the finite case of Proposition 3.6 in Albu [8]. Proposition 3.2.6 is taken from Albu [7].

Section 3.3. The determination of the Cogalois group of a finite field extension is quite hard. In the literature there are several rather theoretical, but less concrete approaches to this matter, especially due to Acosta de Orozco and Vélez [2] and Gay and Vélez [61].

This section essentially follows the presentation in Albu, Nicolae, and Ţena [23] and Ţena [104].

CHAPTER 4

STRONGLY KNESER EXTENSIONS

This chapter contains the main results of Part I of the monograph. The notions of *Cogalois connection, strongly G-Kneser extension*, and *G-Cogalois extension* are introduced. The last ones are those separable G-Kneser extensions E/F for which there exists a canonical lattice isomorphism between the lattice of all subextensions of E/F and the lattice of all subgroups of the group G/F^*. A very useful characterization of G-Cogalois extensions in terms of *n-purity* is given, where n is the exponent of the finite group G/F^*. Using this characterization, we will show in Chapter 7 that the class of G-Cogalois extensions is large enough, including important classes of finite extensions.

We show that a separable G-Kneser extension E/F is G-Cogalois if and only if the group G/F^* has a prescribed structure. As a consequence, the uniqueness of the group G is deduced. This means that if the extension E/F is simultaneously G-Cogalois and H-Cogalois, then necessarily $G = H$. Consequently, it makes sense to define the *Kneser group* of a G-Cogalois extension E/F as the group G/F^*, which will be denoted by $\mathrm{Kne}(E/F)$.

Finally, we investigate *almost G-Cogalois* extensions, introduced by Lam-Estrada, Barrera-Mora, and Villa-Salvador [**79**] under the name of *pseudo G-Cogalois* extensions. These are the finite separable G-radical extensions E/F with G/F^*-Cogalois correspondence and G/F^* finite. Thus, the G-Cogalois extensions are precisely those almost G-Cogalois extensions which are also G-Kneser.

Throughout this chapter E/F will denote a fixed extension and G a group such that $F^* \leqslant G \leqslant E^*$. We shall also use the following notation:

$$\mathcal{G} := \{ \, H \mid F^* \leqslant H \leqslant G \, \},$$

$$\mathcal{E} := \underline{\mathrm{Intermediate}}\,(E/F) = \{ \, K \mid F \subseteq K, \ K \text{ subfield of } E \, \}.$$

4.1. Galois and Cogalois connections

In this section we present the dual concepts of *Galois connection* and *Cogalois connection* for arbitrary posets, and related to them, the concepts of *closed element*. Then, we illustrate them with examples occurring in Field Theory, and introduce the concepts of *field extension with Galois correspondence* and *field extension with Cogalois correspondence*.

DEFINITION 4.1.1. *A Galois connection between the posets* $(X, \leqslant)$ *and* $(Y, \leqslant)$ *is a pair of order-reversing maps*

$$\alpha : X \longrightarrow Y \ \text{ and } \ \beta : Y \longrightarrow X$$

satisfying the following conditions:

$$x \leqslant (\beta \circ \alpha)(x), \ \ \forall\, x \in X, \ \text{ and } \ y \leqslant (\alpha \circ \beta)(y), \ \ \forall\, y \in Y.$$

$\square$

Whenever we have a Galois connection as in Definition 4.1.1, we shall use the notation

$$X \ \underset{\beta}{\overset{\alpha}{\rightleftarrows}} \ Y.$$

If the maps α and β are both order-preserving instead of order-reversing, we obtain a *Cogalois connection* between X and Y. More precisely, we introduce the following definition.

DEFINITION 4.1.2. *A Cogalois connection between the posets* $(X, \leqslant)$ *and* $(Y, \leqslant)$ *is a pair of order-preserving maps*

$$\alpha : X \longrightarrow Y \ \text{ and } \ \beta : Y \longrightarrow X$$

satisfying the following conditions:

$$(\beta \circ \alpha)(x) \leqslant x, \ \ \forall\, x \in X, \ \text{ and } \ y \leqslant (\alpha \circ \beta)(y), \ \ \forall y \in Y.$$

$\square$

If we denote by X^{op} the opposite poset of X, then it is clear that

$$X \ \underset{\beta}{\overset{\alpha}{\rightleftarrows}} \ Y$$

is a Cogalois connection if and only if

$$X^{op} \ \underset{\beta}{\overset{\alpha}{\rightleftarrows}} \ Y$$

is a Galois Connection. Observe that if

$$X \xrightarrow[\beta]{\alpha} Y$$

is a Galois connection, then

$$Y \xrightarrow[\alpha]{\beta} X$$

is also a Galois connection. However, this sort of symmetry does not hold for a Cogalois connection.

Let

$$X \xrightarrow[\beta]{\alpha} Y$$

be a Galois or Cogalois connection. If $x \in X$ (resp. $y \in Y$), then we shall briefly denote by x' (resp. y') the element $\alpha(x)$ (resp. $\beta(y)$); also, we shall use the notation:

$$x'' := (x')', \ x''' := (x'')', \ y'' := (y')', \ y''' := (y'')'.$$

An element z of X or Y is said to be a *closed element* of X or Y, if $z = z''$. A closed element is also called *Galois object* (resp. *Cogalois object*) in the case of a Galois (resp. Cogalois) connection. We shall denote by $\overline{X}$ (resp. $\overline{Y}$) the set of all closed elements of X (resp. Y).

The next result collects the basic facts on Galois and Cogalois connections which will be freely used throughout this monograph.

PROPOSITION 4.1.3. *With notation above, the following assertions hold for a Galois or Cogalois connection between the posets X and Y.*

(1) $z' = z'''$ *for every element z of X or Y.*
(2) $\overline{X} = \beta(Y)$ *and* $\overline{Y} = \alpha(X)$.
(3) *The restrictions $\overline{\alpha} : \overline{X} \longrightarrow \overline{Y}$ and $\overline{\beta} : \overline{Y} \longrightarrow \overline{X}$ of α and β to the sets of closed elements of X and Y are bijections inverse to one another.*

PROOF. We will examine only the case of a Cogalois connection. Let $x \in X$ and $y \in Y$. Then, the conditions from Definition 4.1.2 can be shortly expressed as $x'' \leqslant x$ and $y \leqslant y''$. Since the priming operation is an order-preserving map, we have

$$x''' = (x'')' \leqslant x' \quad \text{and} \quad y' \leqslant (y'')' = y'''.$$

If we replace in the relation $y \leqslant y''$ the element $y \in Y$ by x', we deduce that $x' \leqslant (x')'' = x'''$, which proves that $x' = x'''$.

Similarly, if we replace in the relation $x'' \leqslant x$ the element $x \in X$ by y' we deduce that $(y')'' = y''' \leqslant y'$, which proves that $y' = y'''$. This establishes (1).

Now, let $x \in \overline{X}$. Then $x = x'' = (x')' = \beta(x') \in \beta(Y)$. Conversely, if $x \in \beta(Y)$, then $x = \beta(y) = y'$ for some $y \in Y$. But $y' = y''' = (y')''$ by (1), so $y' = x$ is a closed element of X. This shows that $\overline{X} = \beta(Y)$. In a similar manner one proves the other equality in (2).

To prove (3), let $y \in \overline{Y}$. Then $y = \alpha(x) = x'$ for some $x \in X$ by (2), hence

$$(\overline{\alpha} \circ \overline{\beta})(y) = \alpha(\beta(\alpha(x))) = x''' = x' = y.$$

Similarly, $(\overline{\beta} \circ \overline{\alpha})(x) = x$ for all $x \in \overline{X}$. $\square$

The most relevant example of a Galois connection, which actually originated the name of this concept, is that appearing in Galois Theory. Let E/F be an arbitrary field extension, and denote by Γ the Galois group $\text{Gal}(E/F)$ of E/F. Then, it is easily seen that the maps

$$\alpha : \underline{\text{Intermediate}}\,(E/F) \longrightarrow \underline{\text{Subgroups}}\,(\Gamma), \quad \alpha(K) = \text{Gal}(E/K),$$

and

$$\beta : \underline{\text{Subgroups}}\,(\Gamma) \longrightarrow \underline{\text{Intermediate}}\,(E/F), \quad \beta(\Delta) = \text{Fix}(\Delta),$$

yield a canonical Galois connection between the lattice $\underline{\text{Intermediate}}\,(E/F)$ of all intermediate fields of the extension E/F and the lattice $\underline{\text{Subgroups}}\,(\Gamma)$ of all subgroups of Γ. We will call it the *standard Galois connection associated with the extension E/F*.

PROPOSITION 4.1.4. *With the notation above, the following assertions are equivalent for a finite extension E/F with Galois group Γ.*

(1) *E/F is a Galois extension.*

(2) *Every intermediate field of the extension E/F is a closed element in the standard Galois connection associated with E/F.*

(3) *F is a closed element in the standard Galois connection associated with E/F.*

(4) *The map α is injective.*

(5) *The map β is surjective.*

(6) *The maps α and β establish anti-isomorphisms of lattices, inverse to one another, between the lattices* $\underline{\text{Intermediate}}\,(E/F)$ *and* $\underline{\text{Subgroups}}\,(\Gamma)$.

PROOF. The implications $(1) \Longrightarrow (2) \Longrightarrow (3) \Longrightarrow (1) \Longrightarrow (6)$ are well known (see 1.2.9).

The implications $(6) \Longrightarrow (5)$ and $(6) \Longrightarrow (4)$ are trivial.

$(5) \Longrightarrow (3)$: If β is surjective, then $F = \beta(\Delta) = \text{Fix}(\Delta)$ for some subgroup Δ of Γ, i.e., F is a closed element of $\underline{\text{Intermediate}}\,(E/F)$.

$(4) \Longrightarrow (3)$: We know that $F' = F''' = (F'')'$, i.e., $\alpha(F) = \alpha(F'')$. Since α is injective, we deduce that $F = F''$, hence F is a closed element of $\underline{\text{Intermediate}}\,(E/F)$. $\qquad \square$

The prototype of a Cogalois connection is that canonically associated with any radical extension. Let E/F be an arbitrary G-radical extension. Then, the maps

$$\chi : \underline{\text{Intermediate}}\,(E/F) \to \underline{\text{Subgroups}}\,(G/F^*),\ \chi(K) = (K \cap G)/F^*,$$

and

$$\omega : \underline{\text{Subgroups}}\,(G/F^*) \to \underline{\text{Intermediate}}\,(E/F),\ \omega(H/F^*) = F(H),$$

establish a Cogalois connection between the lattices $\underline{\text{Intermediate}}\,(E/F)$ and $\underline{\text{Subgroups}}\,(G/F^*)$. We will call it the *standard Cogalois connection associated with the extension E/F*. Notice that, in contrast with the standard Galois connection which is associated with any extension, the standard Cogalois connection can be associated only with radical extensions.

Clearly, the lattice $\underline{\text{Subgroups}}\,(G/F^*)$ is canonically isomorphic to the lattice

$$\mathcal{G} = \{\, H \mid F^* \leqslant H \leqslant G \,\},$$

so, if we denote by $\mathcal{E}$ the lattice of all intermediate fields of the extension E/F, then the Cogalois connection described above is essentially the same with the following one, which is also called the *standard Cogalois connection associated with E/F*:

$$\mathcal{E} \ \underset{\psi}{\overset{\varphi}{\rightleftarrows}} \ \mathcal{G},$$

where

$$\varphi : \mathcal{E} \longrightarrow \mathcal{G}, \quad \varphi(K) = K \cap G,$$
$$\psi : \mathcal{G} \longrightarrow \mathcal{E}, \quad \psi(H) = F(H).$$

The considerations above naturally lead us to define the following dual concepts.

DEFINITIONS 4.1.5. *An extension E/F with Galois group Γ is said to be an extension with Γ-Galois correspondence if the standard Galois connection associated with E/F yields a lattice anti-isomorphism between the lattices* Intermediate (E/F) *and* Subgroups (Γ).

Dually, a G-radical extension E/F is said to be an extension with G/F^*-Cogalois correspondence if the standard Cogalois connection associated with E/F gives rise to a lattice isomorphism between the lattices Intermediate (E/F) *and* Subgroups (G/F^*). □

REMARK 4.1.6. Note that any finite extension E/F with Γ-Galois correspondence is necessarily a Galois extension by Proposition 4.1.4. Consequently, the equality $[E : F] = |\mathrm{Gal}(E/F)|$ is a consequence of the fact that E/F is an extension with Γ-Galois correspondence. Conversely, if a finite extension E/F is such that $[E : F] = |\mathrm{Gal}(E/F)|$, then E/F is necessarily a Galois extension (see Exercise 21, Chapter 1).

This is not the case for finite extensions E/F with G/F^*-Cogalois correspondence; namely, for such extensions, the equality $[E : F] = |G/F^*|$, saying precisely that E/F is G-Kneser is, in general, not a consequence of the fact that E/F is an extension with G/F^*-Cogalois correspondence. We will examine this situation more closely in Section 4.5. □

4.2. Strongly G-Kneser extensions

As we noticed in Remarks 2.1.14 (1), a subextension of a Kneser extension is not necessarily a Kneser extension. In this section we introduce and investigate strongly G-Kneser extensions; namely, those G-Kneser extensions E/F such that every subextension K/F of E/F is $K^* \cap G$-Kneser. It turns out that such extensions are precisely the G-Kneser extensions with G/F^*-Cogalois correspondence.

Recall that throughout this chapter E/F will denote a fixed G-radical extension, and

$$\mathcal{G} := \{\, H \mid F^* \leqslant H \leqslant G \,\},$$
$$\mathcal{E} := \underline{\mathrm{Intermediate}}\,(E/F) = \{\, K \mid F \subseteq K, \ K \text{ subfield of } E \,\}.$$

PROPOSITION 4.2.1. *Let E/F be a finite G-Kneser extension, and let K be an intermediate field of E/F. Then, the following assertions are equivalent.*

(1) K/F *is H-Kneser for some $H \in \mathcal{G}$.*

(2) K/F *is $K^* \cap G$-Kneser.*

(3) E/K *is K^*G-Kneser.*

PROOF. (1) $\Longrightarrow$ (3): Suppose that K/F is H-Kneser for some $H \in \mathcal{G}$. Then $K = F(H)$, $[K : F] = |H/F^*|$, and $K^* \cap G = K \cap G = H$ by Proposition 2.1.11. Clearly $E = F(G) = K(G) = K(K^*G)$. But

$$(K^*G)/K^* \cong G/(K^* \cap G) = G/H,$$

and $(K^*G)/K^*$ is finite since G/F^* is finite and $F^* \leqslant H \leqslant G$. Thus, $K^* \leqslant K^*G \leqslant T(E/K)$, hence E/K is a radical K^*G-extension. Moreover,

$$[E : K] = [E : F]/[K : F] = |G/F^*|/|H/F^*|$$
$$= |G/H| = |(K^*G)/K^*|,$$

hence E/K is K^*G-Kneser.

(3) $\Longrightarrow$ (2): Suppose that E/K is K^*G-Kneser. Then

$$[E : K] = |(K^*G)/K^*| = |G/(K^* \cap G)|,$$

hence

$$[K : F] = [E : F]/[E : K] = |G/F^*|/|G/(K^* \cap G)| = |(K^* \cap G)/F^*|.$$

On the other hand, $F(K^* \cap G)/F$ is $K^* \cap G$-Kneser by Proposition 2.1.11, so $[F(K^* \cap G) : F] = |(K^* \cap G)/F^*| = [K : F]$. Since $F(K^* \cap G) \subseteq K$, we deduce that $K = F(K^* \cap G)$. Observe that $F^* \leqslant K^* \cap G \leqslant T(K/F)$ since $F^* \leqslant G \leqslant T(E/F)$ and $T(E/F) \cap K^* = T(K/F)$, hence K/F is a $K^* \cap G$-radical extension. Consequently, K/F is $K^* \cap G$-Kneser, again by Proposition 2.1.11.

(2) $\Longrightarrow$ (1) is obvious. $\qquad\qquad\qquad\qquad\qquad\qquad\qquad\square$

PROPOSITION 4.2.2. *Let $F \subseteq K \subseteq E$ be a tower of fields, and let G be a group such that $F^* \leqslant G \leqslant E^*$. If K/F is $K^* \cap G$-Kneser and E/K is K^*G-Kneser, then E/F is G-Kneser.*

PROOF. In order to prove that the extension E/F is G-Kneser, we have to check first that E/F is G-radical. To do that, observe that, by hypotheses, $K = F(K^* \cap G) \subseteq F(G)$, hence $E = K(K^*G) \subseteq F(G)$, and so, $E = F(G)$. Now, let $g \in G$. Since the extension E/K is K^*G-radical, $(K^*G)/K^*$ is a torsion group, hence $g^m \in K^*$ for some $m \in \mathbb{N}^*$. Further,

K/F is a $K^* \cap G$-radical extension, so $g^{mn} = (g^m)^n \in F^*$ for some $n \in \mathbb{N}^*$. This shows that G/F^* is a torsion group, in other words, E/F is a G-radical extension.

On the other hand, we have

$$[E : F] = [E : K] \cdot [K : F] = |(K^*G)/K^*| \cdot |(K^* \cap G)/F^*|$$
$$= |G/(K^* \cap G)| \cdot |(K^* \cap G)/F^*| = |G/F^*|,$$

Thus, E/F is a G-Kneser extension. $\square$

Propositions 4.2.1 and 4.2.2 can be reformulated together as follows.

THEOREM 4.2.3. *Let $F \subseteq K \subseteq E$ be a tower of fields, and let G be a group such that $F^* \leqslant G \leqslant E^*$. Consider the following assertions.*

(1) K/F *is $K^* \cap G$-Kneser.*
(2) E/K *is K^*G-Kneser.*
(3) E/F *is G-Kneser.*

Then, any two of the assertions (1)-(3) *imply the remaining one.* $\square$

EXAMPLE 4.2.4. Denote by $\sqrt[4]{-9}$ one of the complex roots, say $\sqrt{6}(1+i)/2$, of the irreducible polynomial $X^4 + 9 \in \mathbb{Q}[X]$. Then

$$\mathbb{Q}^*\langle\sqrt[4]{-9}\rangle/\mathbb{Q}^* = \left\{ \widehat{1}, \widehat{\sqrt[4]{-9}}, \widehat{(\sqrt[4]{-9})^2}, \widehat{(\sqrt[4]{-9})^3} \right\},$$

hence

$$|\mathbb{Q}^*\langle\sqrt[4]{-9}\rangle/\mathbb{Q}^*| = 4 = \lceil \mathbb{Q}(\sqrt[4]{-9}) : \mathbb{Q}\rceil.$$

Consequently $\mathbb{Q}(\sqrt[4]{-9})/\mathbb{Q}$ is a $\mathbb{Q}^*\langle\sqrt[4]{-9}\rangle$-Kneser extension. Since

$$(\sqrt[4]{-9})^2 = \left(\frac{\sqrt{6}}{2}(1+i)\right)^2 = 3i,$$

we deduce that $i = (\sqrt[4]{-9})^2/3 \in \mathbb{Q}(\sqrt[4]{-9})$, hence

$$\sqrt{6} = \frac{2\sqrt[4]{-9}}{1+i} \in \mathbb{Q}(\sqrt[4]{-9}).$$

It follows that we can consider the intermediate field $K = \mathbb{Q}(\sqrt{6})$ of the extension $\mathbb{Q}(\sqrt[4]{-9})/\mathbb{Q}$.

We claim that $\mathbb{Q}(\sqrt[4]{-9})/K$ is not a $K^*\mathbb{Q}^*\langle\sqrt[4]{-9}\rangle$-Kneser extension, for otherwise, by the proof of Proposition 4.2.1, it would follow that

$$2 = [\mathbb{Q}(\sqrt[4]{-9}) : K] = [\mathbb{Q}(\sqrt[4]{-9}) : \mathbb{Q}(\sqrt{6})]$$
$$= |\mathbb{Q}^*\langle\sqrt[4]{-9}\rangle/(\mathbb{Q}(\sqrt{6})^* \cap \mathbb{Q}^*\langle\sqrt[4]{-9}\rangle)| = |\mathbb{Q}^*\langle\sqrt[4]{-9}\rangle/\mathbb{Q}^*| = 4,$$

which is a contradiction.

Thus, by Proposition 4.2.1, for every H with $\mathbb{Q}^* \leqslant H \leqslant \mathbb{Q}^* \langle \sqrt[4]{-9} \rangle$, $\mathbb{Q}(\sqrt{6})/\mathbb{Q}$ is not a H-Kneser extension. However, $\mathbb{Q}(\sqrt{6})/\mathbb{Q}$ is clearly $\mathbb{Q}^* \langle \sqrt{6} \rangle$-Kneser. □

Theorem 4.2.3 and Example 4.2.4 naturally lead to the following definition.

DEFINITION 4.2.5. *A finite extension E/F is said to be strongly G-Kneser if it is a G-radical extension such that the extension E/K is K^*G-Kneser for every intermediate field K of E/F.*

The extension E/F is called strongly Kneser *if it is strongly G-Kneser for some group G.* □

The concept of strongly G-Kneser extension can be reformulated as follows.

THEOREM 4.2.6. *The following statements are equivalent for a finite G-radical extension E/F.*

(1) *E/F is strongly G-Kneser.*
(2) *K/F is $K^* \cap G$-Kneser for every intermediate field K of E/F.*
(3) *E/K is K^*G-Kneser for every intermediate field K of E/F.*
(4) *$[E : K] = |G/(K^* \cap G)|$ for every intermediate field K of E/F.*
(5) *$[K : F] = |(K^* \cap G)/F^*|$ for every intermediate field K of E/F.*

PROOF. The equivalences (1) $\Longleftrightarrow$ (2) $\Longleftrightarrow$ (3) follow immediately from Proposition 4.2.1.

The implications (2) $\Longrightarrow$ (5) and (3) $\Longrightarrow$ (4) are clear in view of the definitions of the involved concepts.

(5) $\Longrightarrow$ (2): First, observe that if we take in (5) as K the field E we obtain

$$[E : F] = |(E^* \cap G)/F^*| = |G/F^*|,$$

that is, E/F is a G-Kneser extension.

Now let K be an arbitrary intermediate field of the extension E/F. Then, the extension $F(K^* \cap G)/F$ is $K^* \cap G$-Kneser by Proposition 2.1.11, so $[F(K^* \cap G) : F] = |(K^* \cap G)/F^*|$. But $[K : F] = |(K^* \cap G)/F^*|$ by hypothesis, and $F(K^* \cap G) \subseteq K$, hence $K = F(K^* \cap G)$. Consequently K/F is $K^* \cap G$-Kneser.

(4) $\Longrightarrow$ (5): If we take in (4) as K the field F we obtain

$$[E : F] = |G/(F^* \cap G)| = |G/F^*|,$$

that is, E/F is a G-Kneser extension.

Now, for any intermediate field K of E/F, we have

$$[K : F] = [E : F]/[E : K] = |G/F^*|/|G/(K^* \cap G)| = |(K^* \cap G)/F^*|,$$

which proves the desired implication.   □

Clearly, any strongly G-Kneser extension is G-Kneser, but not conversely, as Example 4.2.4 shows.

Now consider an arbitrary G-radical extension E/F. Recall that we have previously introduced the following notation:

$$\mathcal{G} = \{\, H \mid F^* \leqslant H \leqslant G \,\},$$
$$\mathcal{E} = \{\, K \mid F \subseteq K, \ K \text{ subfield of } E \,\}.$$

Clearly, the posets $(\mathcal{G}, \subseteq)$ and $(\mathcal{E}, \subseteq)$ are lattices. We have noticed in Section 4.1 that the maps

$$\varphi : \mathcal{E} \longrightarrow \mathcal{G}, \ \varphi(K) = K \cap G,$$
$$\psi : \mathcal{G} \longrightarrow \mathcal{E}, \ \psi(H) = F(H),$$

define a Cogalois connection

$$\mathcal{E} \ \underset{\psi}{\overset{\varphi}{\rightleftarrows}} \ \mathcal{G},$$

which we called the standard Cogalois connection associated with E/F.

The next result gives a characterization of G-Kneser extensions E/F for which the standard associated Cogalois connection gives rise to a bijective correspondence between $\mathcal{E}$ and $\mathcal{G}$, and is somewhat dual to the corresponding result for Galois extensions stated in Proposition 4.1.4.

THEOREM 4.2.7. *The following assertions are equivalent for a finite G-radical extension E/F.*

(1) *E/F is strongly G-Kneser.*

(2) *E/F is G-Kneser, and the map $\psi : \mathcal{G} \longrightarrow \mathcal{E}, \ \psi(H) = F(H)$, is surjective.*

(3) *E/F is G-Kneser, and every element of $\mathcal{E}$ is a closed element in the standard Cogalois connection associated with E/F.*

(4) *E/F is G-Kneser, and the map $\varphi : \mathcal{E} \longrightarrow \mathcal{G}, \ \varphi(K) = K \cap G$, is injective.*

(5) E/F is G-Kneser, and the maps $-\cap G : \mathcal{E} \longrightarrow \mathcal{G}$, $F(-) : \mathcal{G} \longrightarrow \mathcal{E}$ are isomorphisms of lattices, inverse to one another.

(6) E/F is a G-Kneser extension with G/F^*-Cogalois correspondence.

PROOF. (1) $\Longrightarrow$ (2): Suppose that E/F is strongly G-Kneser and let $K \in \mathcal{E}$. Then, the extension K/F is $K^* \cap G$-Kneser by definition, so $K = F(H) = \psi(H)$, with $H = K \cap G = K^* \cap G$.

(2) $\Longleftrightarrow$ (3) follows from Proposition 4.1.3 (2).

(2) $\Longrightarrow$ (4): Let K_1, $K_2 \in \mathcal{E}$ be such that $\varphi(K_1) = \varphi(K_2)$, i.e., $K_1 \cap G = K_2 \cap G$. We have $K_1 = \psi(H_1) = F(H_1)$ and $K_2 = \psi(H_2) = F(H_2)$ for some H_1, $H_2 \in \mathcal{G}$. Then $K_1 \cap G = F(H_1) \cap G = H_1$ and $K_2 \cap G = F(H_2) \cap G = H_2$ by Proposition 2.1.11. Since $K_1 \cap G = K_2 \cap G$, it follows that $H_1 = H_2$, and so $K_1 = K_2$. Thus φ is injective.

(4) $\Longrightarrow$ (5): For every $H \in \mathcal{G}$ we have

$$(\varphi \circ \psi)(H) = \varphi(\psi(H)) = \varphi(F(H)) = F(H) \cap G = H$$

by Proposition 2.1.11. This shows that $\varphi \circ \psi = 1_{\mathcal{G}}$, which implies that φ is surjective. Then φ is bijective, and ψ is its inverse. Consequently, φ and ψ are isomorphisms of posets, and so, isomorphisms of lattices inverse to one another.

(5) $\Longrightarrow$ (1): Let $K \in \mathcal{E}$. We are going to show that K/F is a $K^* \cap G$-Kneser extension for every $K \in \mathcal{E}$. Since ψ is surjective, it follows that $K = \psi(H) = F(H)$ for some $H \in \mathcal{G}$, i.e., $H = \varphi(K) = K \cap G$. Since E/F is G-Kneser, the extension $F(H)/F$ is H-Kneser by Proposition 2.1.11, i.e., K/F is $K^* \cap G$-Kneser.

(5) $\Longleftrightarrow$ (6): For simplicity, denote by $\widetilde{\mathcal{G}}$ the lattice $\underline{\mathrm{Subgroups}\,(G/F^*)}$ of all subgroups of the quotient group $\widetilde{G} = G/F^*$. Since the lattices $\mathcal{G}$ and $\widetilde{\mathcal{G}}$ are canonically isomorphic, it is clear that everything about the standard Cogalois connection associated with E/F

$$\mathcal{E} \underset{\psi}{\overset{\varphi}{\rightleftarrows}} \mathcal{G}$$

can be expressed equivalently using the Cogalois connection

$$\mathcal{E} \underset{\widetilde{\psi}}{\overset{\widetilde{\varphi}}{\rightleftarrows}} \widetilde{\mathcal{G}},$$

where $\widetilde{\varphi}(K) = (K \cap G)/F^*$ and $\widetilde{\psi}(H/F^*) = F(H)$.

So, the maps $- \cap G : \mathcal{E} \longrightarrow \mathcal{G}$ and $F(-) : \mathcal{G} \longrightarrow \mathcal{E}$ are isomorphisms of lattices, inverse to one another if and only if the maps

$$\widetilde{\varphi} : \mathcal{E} \longrightarrow \widetilde{\mathcal{G}}$$

and

$$\widetilde{\psi} : \widetilde{\mathcal{G}} \longrightarrow \mathcal{E}$$

are lattice isomorphisms, inverse to one another, i.e., by Definition 4.1.5, if and only if E/F is an extension with $\widetilde{G}$-Cogalois correspondence. $\square$

The next result shows that the property of an extension being strongly Kneser behaves nicely with respect to subextensions and quotient extensions.

PROPOSITION 4.2.8. *Let E/F be a strongly G-Kneser extension. Then, for any intermediate field K, the following assertions hold.*

(1) K/F *is strongly $K^* \cap G$-Kneser.*
(2) E/K *is strongly K^*G-Kneser.*

PROOF. (1) Let L be an intermediate subfield of the extension K/F. Since $L \in \mathcal{E}$, there exists H with $F^* \leqslant H \leqslant G$ and $L = F(H)$ by Theorem 4.2.7. But $H \leqslant F(H)^* = L^* \leqslant K^*$ and $H \leqslant G$, hence $F^* \leqslant H \leqslant G \cap K^*$. This shows that K/F is strongly $K^* \cap G$ Kneser.

(2) For every subfield M of E with $K \subseteq M \subseteq E$, the extension E/M is M^*G-Kneser since E/F is strongly G-Kneser. But $M^*G = M^*(K^*G)$, so E/M is $M^*(K^*G)$-Kneser. It follows that the extension E/K is strongly K^*G-Kneser. $\square$

4.3. G-Cogalois extensions

In this section we introduce the basic concept of G-*Cogalois extension*, which plays in the Cogalois Theory the same role as that of Galois extension in Galois Theory. A G-Cogalois extension is nothing else than a separable G-Kneser extension with G/F^*-Cogalois correspondence.

We characterize G-Cogalois extensions E/F within the class of G-Kneser extensions by means of a certain sort of "local purity" we have called n-*purity*, where n is the exponent of the finite group G/F^*. Using this characterization, we will show in Chapter 7 that the class of G-Cogalois extensions is large enough, including important classes of finite extensions: Cogalois extensions, classical Kummer extensions, generalized Kummer extensions, quasi-Kummer extensions.

First, let us recall the basic notation we will use in the sequel:

$$
\begin{aligned}
\mathbb{P} &= \{\, p \mid p \in \mathbb{N}^*,\, p \text{ prime}\,\}, \\
\mathcal{P} &= (\mathbb{P} \setminus \{2\}) \cup \{4\}, \\
\mathbb{P}_n &= \{\, p \mid p \in \mathbb{P},\, p \mid n \,\}, \\
\mathbb{D}_n &= \{\, m \mid m \in \mathbb{N},\, m \mid n \,\}, \\
\mathcal{P}_n &= \mathcal{P} \cap \mathbb{D}_n,
\end{aligned}
$$

where $n \in \mathbb{N}^*$.

Recall from Section 3.1 that an extension E/F is called *pure* when $\mu_p(E) \subseteq F$ for all $p \in \mathcal{P}$.

The next definition introduces a concept which can be interpreted as a sort of "local purity".

DEFINITION 4.3.1. *Let E/F be an arbitrary extension and let $n \in \mathbb{N}^*$. The extension E/F is called n-pure if $\mu_p(E) \subseteq F$ for every $p \in \mathcal{P}_n$.* □

Clearly, an extension E/F is pure if and only if it is n-pure for every $n \in \mathbb{N}^*$. Also, note that an n-pure extension is not necessarily pure; e.g., $\mathbb{Q}(\sqrt{-3})/\mathbb{Q}$ is 2-pure but not pure.

Recall that for an arbitrary G-radical extension E/F, which is not necessarily finite, we have defined the standard Cogalois connection

$$
\mathcal{E} \xrightarrow[\psi]{\varphi} \mathcal{G}
$$

between the lattices

$$
\mathcal{E} = \{\, K \mid F \subseteq K,\ K \text{ subfield of } E \,\},
$$

$$
\mathcal{G} = \{\, H \mid F^* \leqslant H \leqslant G \,\}
$$

as follows:

$$
\varphi : \mathcal{E} \longrightarrow \mathcal{G}, \ \ \varphi(K) = K \cap G,
$$

$$
\psi : \mathcal{G} \longrightarrow \mathcal{E}, \ \ \psi(H) = F(H).
$$

The characterizations of strongly G-Kneser in terms of this standard Cogalois connection given in Theorem 4.2.7 are strengthened for separable strongly G-Kneser extensions by the next result.

THEOREM 4.3.2 (THE n-PURITY CRITERION). *The following assertions are equivalent for a finite separable G-radical extension E/F with G/F^* finite and $n = \exp(G/F^*)$.*

(1) *E/F is strongly G-Kneser.*
(2) *E/F is G-Kneser, and the map $\psi : \mathcal{G} \longrightarrow \mathcal{E}$, $\psi(H) = F(H)$, is surjective.*
(3) *E/F is G-Kneser, and every element of $\mathcal{E}$ is a closed element in the standard Cogalois connection associated with E/F.*
(4) *E/F is G-Kneser, and the map $\varphi : \mathcal{E} \longrightarrow \mathcal{G}$, $\varphi(K) = K \cap G$, is injective.*
(5) *E/F is G-Kneser, and the maps $- \cap G : \mathcal{E} \longrightarrow \mathcal{G}$, $F(-) : \mathcal{G} \longrightarrow \mathcal{E}$ are isomorphisms of lattices, inverse to one another.*
(6) *E/F is a G-Kneser extension with G/F^*-Cogalois correspondence.*
(7) *E/F is n-pure.*

PROOF. The equivalences (1) through (6) hold for extensions which are not necessarily separable, and are contained in Theorem 4.2.7.

(7) $\Longrightarrow$ (1): Assume that the extension E/F is n-pure, and let $K \in \mathcal{E}$. We have to prove that K/F is $K^* \cap G$ Kneser. For this, we shall use the Kneser Criterion (Theorem 2.2.1).

Let p be an odd prime such that $\zeta_p \in K^* \cap G$. If $p \,|\, n$, then $\zeta_p \in F$ by n-purity. If $\gcd(p, n) = 1$, then there exist $a, b \in \mathbb{Z}$ with $1 = ap + bn$, hence

$$\zeta_p = \zeta_p^{ap+bn} = \zeta_p^{bn} \in F^*$$

since $G^n \subseteq F^*$.

Now suppose that $1 + \zeta_4 \in K^* \cap G$. Then necessarily $\mathrm{Char}(F) \neq 2$, for otherwise, it would follow that $\zeta_4 = 1$, hence $1 + \zeta_4 = 0 \in K^* \cap G$, which is a contradiction.

Denote $m = \mathrm{ord}(\widehat{1 + \zeta_4})$. Since $(1 + \zeta_4)^4 = -4 \in F^*$ one has $m \in \{1, 2, 4\}$. If $m = 1$ then $1 + \zeta_4 \in F^*$, hence $\zeta_4 \in F$. If $m = 2$ then $(1 + \zeta_4)^2 = 2\zeta_4 \in F^*$, hence $\zeta_4 \in F$ since $\mathrm{Char}(F) \neq 2$. If $m = 4$ then $4 \,|\, n$. But $1 + \zeta_4 \in K^* \cap G \subseteq E$ implies that $\zeta_4 \in E$, hence $\zeta_4 \in F$ by n-purity. Thus, E/F is strongly G-Kneser. Observe that we did not use here the separability condition of the extension E/F.

(2) $\implies$ (7) Suppose that the extension E/F is G-Kneser and ψ is surjective. Let p be an odd prime with $p \,|\, n$. There exists then a $g \in G$ such that $\mathrm{ord}(\widehat{g}) = p$ by Proposition 1.4.8. By Proposition 2.1.10, the extension $F(F^*\langle g \rangle)/F$ is $F^*\langle g \rangle$-Kneser, hence

$$[F(g) : F] = [F(F^*\langle g \rangle) : F] = |F^*\langle g \rangle / F^*| = |\langle \widehat{g} \rangle| = p.$$

Assume that $\zeta_p \in E$. Since $X^p - g^p \in F[X]$ and $[F(g) : F] = p$, it follows that $\mathrm{Min}(g, F) = X^p - g^p$. Then $\mathrm{Min}(\zeta_p g, F) = X^p - g^p$, and so, $[F(\zeta_p g) : F] = p$. Since $F(\zeta_p g) \in \mathcal{E}$ and ψ is surjective, there exists an $h \in G$ such that $\mathrm{ord}(\widehat{h}) = p$ and $F(\zeta_p g) = F(h) = F(F^*\langle h \rangle)$, hence $\zeta_p g \in F(h)$, and thus $\zeta_p \in F(g, h)$. Since the subgroups $\langle \widehat{g} \rangle$ and $\langle \widehat{h} \rangle$ of G/F^* have both the same order p, they are either equal, or independent (i.e., $\langle \widehat{g} \rangle \cap \langle \widehat{h} \rangle = \{\widehat{1}\}$). Observe that if $\langle \widehat{g} \rangle$ and $\langle \widehat{h} \rangle$ are independent subgroups of G/F^*, then $\langle \widehat{g}, \widehat{h} \rangle = \langle \widehat{g} \rangle \oplus \langle \widehat{h} \rangle$. On the other hand, by Proposition 2.1.11, $F(F^*\langle g, h \rangle)/F$ is $F^*\langle g, h \rangle$-Kneser, hence

$$[F(g, h) : F] = [F(F^*\langle g, h \rangle) : F] = |\langle \widehat{g}, \widehat{h} \rangle| \in \{p, p^2\}.$$

Since $F \subseteq F(\zeta_p) \subseteq F(g, h)$, $[F(\zeta_p) : F] \leqslant p - 1$, and $[F(g, h) : F] \in \{p, p^2\}$, we deduce that $[F(\zeta_p) : F] = 1$, i.e., $\zeta_p \in F$.

Suppose that $4 \,|\, n$ and $\zeta_4 \in E \setminus F$. Then necessarily $\mathrm{Char}(F) \neq 2$ and G/F^* contains an element of order 4, say $\widehat{g}$. Since $F(\zeta_4) \subseteq E$ and $[F(\zeta_4) : F] = 2$, there exists $h \in G$ such that $\mathrm{ord}(\widehat{h}) = 2$ and $F(\zeta_4) = F(F^*\langle h \rangle) = F(h)$. It follows that $\zeta_4 = \lambda + \mu h$, for some $\lambda, \mu \in F$, hence $-1 = \lambda^2 + 2\lambda\mu h + \mu^2 h^2$ and $h^2 \in F$. Then necessarily $2\lambda\mu = 0$. Since $\zeta_4 \notin F$ we deduce that $\mu \neq 0$, hence $\lambda = 0$. It follows that $\zeta_4 = \mu h$, hence $\zeta_4 \in G$.

Set $K = F((1 + \zeta_4)g)$. Then, using the implication (2) $\implies$ (1), we deduce that E/K is K^*G-Kneser. Since $1 + \zeta_4 = (1 + \zeta_4)g g^{-1} \in K^*G$, it follows that $\zeta_4 \in K$ by the Kneser Criterion.

But $[(1 + \zeta_4)g]^4 = -4g^4 \in F^*$ implies that $[K : F] \leqslant 4$, and so,

$$\zeta_4 = \lambda_0 + \lambda_1(1 + \zeta_4)g + \lambda_2(1 + \zeta_4)^2 g^2 + \lambda_3(1 + \zeta_4)^3 g^3$$
$$= \lambda_0 + \lambda_1 g + \lambda_1 \zeta_4 g + 2\lambda_2 \zeta_4 g^2 + 2\lambda_3 \zeta_4 g^3 - 2\lambda_3 g^3$$

for some $\lambda_0, \lambda_1, \lambda_2, \lambda_3 \in F$.

Since $\zeta_4 \in G$, we deduce by Corollary 2.1.10 that ζ_4 must be congruent modulo F^* with one of the following elements

$$1, \ g, \ \zeta_4 g, \ \zeta_4 g^2, \ \zeta_4 g^3, \ g^3.$$

But $\mathrm{ord}(\widehat{\zeta_4}) = 2$ and $\mathrm{ord}(\widehat{g}) = 4$, hence necessarily $\widehat{\zeta_4} = \widehat{1}$, i.e., $\zeta_4 \in F^*$, which contradicts our initial assumption that $\zeta_4 \notin F$. This completes the proof of the theorem. $\square$

In the class of finite extensions E/F with G/F^*-Cogalois correspondence the most interesting are those extensions which additionally are separable. In view of the equivalence (1) $\iff$ (6) in Theorem 4.3.2, these are precisely the separable strongly G-Kneser extensions. They deserve a special name.

DEFINITION 4.3.3. *An extension E/F is called G-Cogalois if it is a separable strongly G-Kneser extension.* $\square$

REMARKS 4.3.4. (1) The condition "E/F is G-Kneser" cannot be removed from the statement of Theorem 4.3.2 (see Exercise 2).

(2) The example in Exercise 2 is an extension with $G/\mathbb{Q}^*$-Cogalois correspondence, which is not G-Cogalois, but, by Exercise 3, it is $\widetilde{G}$-Cogalois for another group $\widetilde{G}$. We shall examine such situations more closely in Section 4.5.

(3) A strongly G-Kneser extension is not necessarily separable, as the example in Remark 3.1.10 shows. $\square$

The next result shows that the property of an extension being G-Cogalois behaves nicely with respect to subextensions and quotient extensions.

PROPOSITION 4.3.5. *Let E/F be a G-Cogalois extension. Then, for every intermediate field K of E/F, the following assertions hold.*
(1) *K/F is $K^* \cap G$-Cogalois.*
(2) *E/K is K^*G-Cogalois.*

PROOF. The result follows immediately from Proposition 4.2.8. $\square$

4.4. The Kneser group of a G-Cogalois extension

This section essentially shows that a separable G-Kneser extension is G-Cogalois if and only if the group G has a prescribed structure. This implies that the group G/F^* of any G-Cogalois extension E/F is uniquely determined; it is called the *Kneser group* of E/F and denoted by $\mathrm{Kne}(E/F)$.

Recall that if A is a multiplicative group with identity element e, then for any $p \in \mathbb{P}$ we denote by

$$t_p(A) = \{\, x \in A \,|\, x^{p^n} = e \text{ for some } n \in \mathbb{N} \,\}$$

the *p-primary component* of the group A. Also recall that for any extension E/F we have denoted by $\mathrm{Cog}(E/F)$ the torsion subgroup $t(E^*/F^*)$ of the quotient group E^*/F^*. By $\mathrm{Cog}_2(E/F)$ we shall denote the subgroup of $\mathrm{Cog}(E/F)$ consisting of all its elements of order $\leqslant 2$.

THEOREM 4.4.1. *Let E/F be a separable G-Kneser extension with $n = \exp(G/F^*)$.*

(1) *Suppose that $n \not\equiv 2 \pmod 4$. Then E/F is a G-Cogalois extension if and only if*

$$G/F^* = \bigoplus_{p \in \mathbb{P}_n} t_p(\mathrm{Cog}(E/F)).$$

(2) *Suppose that $n \equiv 2 \pmod 4$. Then E/F is a G-Cogalois extension if and only if*

$$G/F^* = \Big(\bigoplus_{p \in \mathbb{P}_n \setminus \{2\}} t_p(\mathrm{Cog}(E/F)) \Big) \bigoplus \mathrm{Cog}_2(E/F).$$

PROOF. (1) "$\Longrightarrow$": Since $\exp(G/F^*) = n$, one has $t_p(G/F^*) = 0$ for all $p \in \mathbb{P} \setminus \mathbb{P}_n$ by Proposition 1.4.7. Therefore, decomposing the finite Abelian group G/F^* into the direct sum of its p-primary components, we obtain

$$G/F^* = \bigoplus_{p \in \mathbb{P}_n} t_p(G/F^*).$$

So, it is sufficient to show that

$$t_p(G/F^*) = t_p(\mathrm{Cog}(E/F))$$

for every $p \in \mathbb{P}_n$. The inclusion " $\subseteq$ " is clear since $G/F^* \subseteq \mathrm{Cog}(E/F)$.

For the opposite inclusion, let $p \in \mathbb{P}_n$, and let $g \in T(E/F)$ with $\widehat{g} \in t_p(\mathrm{Cog}(E/F))$. Set $G' = G\langle g \rangle$ and $n' = \exp(G'/F^*)$. Now observe that $\mathbb{P}_{n'} = \mathbb{P}_n$, and $4 \,|\, n$ if and only if $4 \,|\, n'$, hence E/F is n'-pure since it is n-pure by Theorem 4.3.2. But E/F is a separable G'-radical extension with G'/F^* finite, hence E/F is G'-Cogalois. Since clearly $F^* \leqslant G \leqslant G'$ and $E = F(G) = F(G')$, one deduces that $G = G'$, again by Theorem 4.3.2. Consequently, $g \in G$, as desired.

"$\Longleftarrow$": If $G/F^* = \bigoplus_{p \in \mathbb{P}_n} t_p(\mathrm{Cog}(E/F))$, then we have to prove that E/F is G-Cogalois, that is, E/F is n-pure. Let p be an odd prime, $p \,|\, n$.

If $\zeta_p \in E$, then clearly $\widehat{\zeta_p} \in t_p(\mathrm{Cog}(E/F))$, hence $\widehat{\zeta_p} \in G/F^*$, i.e., $\zeta_p \in G$. Using the Kneser Criterion we deduce that $\zeta_p \in F$.

If $4 \mid n$ and $\zeta_4 \in E$, then $1 + \zeta_4 \in E$. If $\mathrm{Char}(K) = 2$, then $\zeta_4 = 1 \in F$. So, we may assume that $\mathrm{Char}(F) \neq 2$. Then $\widehat{1 + \zeta_4} \in t_2(\mathrm{Cog}(E/F)) \subseteq G/F^*$ since $(1 + \zeta_4)^4 = -4$. Thus $1 + \zeta_4 \in G$, so, by the Kneser Criterion we deduce that $\zeta_4 \in F$. This proves that E/F is n-pure, as desired.

(2) "$\Longrightarrow$": Suppose that E/F is a G-Cogalois extension. As in (1) one shows that

$$t_p(G/F^*) = t_p(\mathrm{Cog}(E/F))$$

holds for every odd prime $p \in \mathbb{P}_n$. Observe that since $n \equiv 2 \ (\mathrm{mod}\,4)$, G/F^* contains no element of order 4, hence $t_2(G/F^*)$ is exactly the set of all elements of G/F^* of order $\leqslant 2$, and so, $t_2(G/F^*) \subseteq \mathrm{Cog}_2(E/F)$.

To finish the proof of this implication we only have to check the opposite inclusion

$$\mathrm{Cog}_2(E/F) \subseteq t_2(G/F^*).$$

So, let $\widehat{h} \in \mathrm{Cog}_2(E/F)$. Set $G' = G\langle h \rangle$ and $n' = \exp(G'/F^*)$. Then $n' \not\equiv 0 \ (\mathrm{mod}\,4)$, hence E/F is n'-pure. Now continue as in (1).

"$\Longleftarrow$": If $G/F^* = \left(\bigoplus_{p \in \mathbb{P}_n \setminus \{2\}} t_p(\mathrm{Cog}(E/F)) \right) \bigoplus \mathrm{Cog}_2(E/F)$, we have to prove that E/F is G-Cogalois, that is, E/F is n-pure. Let p be an odd prime in $\mathbb{P}_n$. As above, one shows that $\mu_p(E) = \mu_p(F)$. But $n \equiv 2 \ (\mathrm{mod}\,4)$, hence we cannot have $4 \mid n$. It follows that the extension E/F is n-pure. $\square$

COROLLARY 4.4.2. *Let E/F be an extension which is simultaneously G-Cogalois and H-Cogalois. Then $G = H$.*

PROOF. Set $m = \exp(G/F^*)$, $n = \exp(H/F^*)$, and $k = [E : F]$. Then

$$|G/F^*| = |H/F^*| = [E : F] = k.$$

On the other hand, by Proposition 1.4.10, the order and the exponent of a finite Abelian group have the same prime divisors. Consequently, we have $\mathbb{P}_m = \mathbb{P}_n = \mathbb{P}_k$, hence by Theorem 4.4.1, it is sufficient to prove that $4 \mid m \Longleftrightarrow 4 \mid n$.

Assume that $4 \mid m$. Then, by Proposition 1.4.8, G/F^* contains an element of order 4, say $\widehat{g}$. Set $G_1 = F^*\langle g \rangle$ and $E_1 = F(G_1)$. By Theorem 4.3.2, there exists H_1 such that $F^* \leqslant H_1 \leqslant H$, $E_1 = F(H_1)$ and $|H_1/F^*| = 4$. By Proposition 4.3.5, E_1/K is an $E_1^* \cap G$ Cogalois extension. But $E_1^* \cap G = E_1 \cap G = F(G_1) \cap G = G_1$ by Proposition 2.1.1, hence E_1/F

is a G_1-Cogalois extension. Then, using the bijective Cogalois correspondence (provided by Theorem 4.3.2) between the lattice of all intermediate fields of E_1/F and the lattice of all subgroups of the cyclic group G_1/F^* of order 4, one deduces that the extension E_1/F has only one proper intermediate field. On the other hand, one shows similarly that the extension E_1/F is H_1-Cogalois. Now, using the bijective Cogalois correspondence between the lattice of all intermediate fields of E_1/F and the lattice of all subgroups of H_1/F^*, one deduces that the group H_1/F^* of order 4 is necessarily cyclic, and then, $4 \mid n$. $\square$

The next corollary provides a more precise result than that given by Lemma 3.1.6.

COROLLARY 4.4.3. *Let E/F be a finite G-radical pure separable extension. Then, the extension E/F is simultaneously Cogalois and G-Cogalois, and $\mathrm{Cog}\,(E/F) = G/F^*$.*

PROOF. By the Greither-Harrison Criterion, the extension E/F is Cogalois, hence, by Theorem 3.2.3 and Theorem 4.3.2, if follows that E/F is $T(E/F)$-Cogalois. In particular, it follows that G/F^* is finite.

On the other, if $n = \exp(G/F^*)$, then E/F is also n-pure; so, by Theorem 4.3.2, E/F is G-Cogalois. By Corollary 4.4.2, we have $G = T(E/F)$. Hence $\mathrm{Cog}\,(E/F) = T(E/F)/F^* = G/F^*$. Note that the equality $G = T(E/F)$ also follows from Lemma 3.1.6. $\square$

REMARKS 4.4.4. (1) The condition "E/F is a separable G-Kneser extension" in Theorem 4.4.1 cannot be replaced by the weaker one "E/F is a finite separable G-radical extension with G/F^* finite". Indeed, the quadratic G-radical extension $\mathbb{Q}(\zeta_3)/\mathbb{Q}$, with $G = \mathbb{Q}^*\langle\zeta_3\rangle$, considered in Exercise 2, is not G-Cogalois, but $G/\mathbb{Q}^* = t_3(\mathrm{Cog}(\mathbb{Q}(\zeta_3)/\mathbb{Q}))$ and $|G/\mathbb{Q}^*| = \exp(G/\mathbb{Q}^*) = 3$.

(2) If E/F is a G-Cogalois extension with $\exp(G/F^*) \equiv 2 \,(\mathrm{mod}\,4)$, then we may have $t_2(G/F^*) \neq t_2(\mathrm{Cog}(E/F))$ (see Exercise 5). $\square$

By Corollary 4.4.2, for any G-Cogalois extension, the group G is uniquely determined. So, it makes sense to introduce the following concept.

DEFINITION 4.4.5. *If E/F is a G-Cogalois extension, then the group G/F^* is called the Kneser group of the extension E/F and is denoted by $\mathrm{Kne}(E/F)$.* $\square$

Note that for any extension E/F, $\mathrm{Kne}(E/F)$ is a subgroup, in general proper, of the Cogalois group $\mathrm{Cog}(E/F)$ of E/F.

4.5. Almost G-Cogalois extensions

The aim of this section is to investigate the G-radical extensions with G/F^*-Cogalois correspondence. First, we present some general results about such extensions which are not necessarily finite. Then, we focus on finite separable G-radical extensions E/F with G/F^*-Cogalois correspondence. We call these extensions *almost G-Cogalois*. Finally, we characterize those almost G-Cogalois extensions which are not G-Cogalois, but are $\widetilde{G}$-Cogalois for a suitable group $\widetilde{G}$.

In Section 4.1 we have introduced the concept of extension with G/F^*-Cogalois correspondence as follows. Let E/F be an arbitrary G-radical extension which is not necessarily finite, and denote

$$\mathcal{G} = \{ H \mid F^* \leqslant H \leqslant G \},$$
$$\mathcal{E} = \{ K \mid F \subseteq K, \ K \text{ subfield of } E \}.$$

We say that E/F is an extension with G/F^*-Galois correspondence if the canonical maps

$$\varphi : \mathcal{E} \longrightarrow \mathcal{G}, \ \varphi(K) = K \cap G,$$
$$\psi : \mathcal{G} \longrightarrow \mathcal{E}, \ \psi(H) = F(H),$$

are bijections, inverse to one another. In that case, they are actually isomorphisms of lattices.

Such extensions, which additionally are separable and have the group G/F^* finite, deserve a special name.

DEFINITION 4.5.1. *An extension E/F is said to be* almost G-Cogalois *if it is a separable G-radical extension with G/F^*-Cogalois correspondence, and with G/F^* finite. A* strictly almost G-Cogalois extension *is an almost G-Cogalois extension which is not G-Cogalois.*   □

Observe that, by Theorem 4.3.2, a G-Cogalois extension is precisely an almost G-Cogalois extension which is also G-Kneser.

REMARK 4.5.2. The condition "G/F^* is a finite group" in Definition 4.5.1 can be replaced by the condition "E/F is a finite extension". Indeed, since E/F is finite and separable, then $\mathcal{E}$ is a finite lattice, hence the lattice $\mathcal{G}$, which is isomorphic to $\mathcal{E}$, also has to be finite. This clearly implies that G/F^* is a finite group, since any infinite group has infinitely many subgroups. $\qquad\qquad\square$

The next result shows that the property of a radical extension being with Cogalois correspondence is inherited by subextensions and quotient extensions.

PROPOSITION 4.5.3. *Let E/F be a G-radical extension with G/F^*- Cogalois correspondence. Then, for every $K \in \mathcal{E}$, K/F is an extension with $(K \cap G)/F^*$-Cogalois correspondence, and E/K is an extension with $(K^*G)/K^*$-Cogalois correspondence.*

PROOF. Let $H = K \cap G$, and set

$$\begin{aligned}
\mathcal{H}' &= \{ H' \mid F^* \leqslant H' \leqslant H \}, \\
\mathcal{H}'' &= \{ H'' \mid K^* \leqslant H'' \leqslant K^*G \}, \\
\mathcal{S} &= \{ S \mid H \leqslant S \leqslant G \}, \\
\mathcal{K}' &= \{ L' \mid F \subseteq L', \ L' \text{ subfield of } K \}, \\
\mathcal{K}'' &= \{ L'' \mid K \subseteq L'', \ L'' \text{ subfield of } E \}.
\end{aligned}$$

Now, consider the following order-preserving maps:

$$\begin{aligned}
\alpha' &: \mathcal{H}' \longrightarrow \mathcal{K}', \ H' \mapsto F(H'), \\
\beta' &: \mathcal{K}' \longrightarrow \mathcal{H}', \ L' \mapsto L' \cap H, \\
\alpha'' &: \mathcal{H}'' \longrightarrow \mathcal{K}'', \ H'' \mapsto K(H''), \\
\beta'' &: \mathcal{K}'' \longrightarrow \mathcal{H}'', \ L'' \mapsto L'' \cap K^*G.
\end{aligned}$$

Since E/F is an extension with G/F^*-Cogalois correspondence, then, by definition, the maps

$$\begin{aligned}
\psi &: \mathcal{G} \longrightarrow \mathcal{E}, \ S \mapsto F(S), \\
\varphi &: \mathcal{E} \longrightarrow \mathcal{G}, \ L \mapsto L \cap G,
\end{aligned}$$

are isomorphisms of lattices, inverse to one another, hence so are also their restrictions to the sublattices $\mathcal{H}'$ and $\mathcal{K}'$, respectively, since $H = \varphi(K)$ and $K = \psi(H)$. This shows that the extension K/F is an extension with H/F^*-Cogalois correspondence. A similar argument shows that the maps

$$\begin{aligned}
\psi'' &: \mathcal{S} \longrightarrow \mathcal{K}'', \ S \mapsto K(S), \\
\varphi'' &: \mathcal{K}'' \longrightarrow \mathcal{S}, \ L'' \mapsto L'' \cap G,
\end{aligned}$$

are isomorphisms of lattices, inverse to one another.

Since

$$G/H = G/(K^* \cap G) \cong (K^*G)/K^*,$$

the maps

$$\lambda : \mathcal{S} \longrightarrow \mathcal{H}'', \ S \mapsto K^*S,$$
$$\mu : \mathcal{H}'' \longrightarrow \mathcal{S}, \ H'' \mapsto H'' \cap G,$$

are isomorphisms of lattices, inverse to one another.
For every $H'' \in \mathcal{H}''$ and $L'' \in \mathcal{K}''$ we have

$$\alpha''(H'') = K(H'') = K((\lambda \circ \mu)(H'')) = K(K^*(H'' \cap G))$$
$$= K(H'' \cap G) = (\psi'' \circ \mu)(H'')$$

and

$$\beta''(L'') = L'' \cap K^*G = (\lambda \circ \mu)(L'' \cap K^*G) = K^*((L'' \cap K^*G) \cap G)$$
$$= K^*(L'' \cap G) = (\lambda \circ \varphi'')(L'').$$

Thus, $\alpha'' = \psi'' \circ \mu$ and $\beta'' = \lambda \circ \varphi''$. Hence α'' and β'' are isomorphisms of lattices, inverse to one another, i.e., E/K is an extension with $(K^*G)/K^*$-Cogalois correspondence. $\qquad\square$

From Proposition 4.5.3 we deduce at once the next result.

COROLLARY 4.5.4. *Let E/F be an almost G-Cogalois extension. Then, for every $K \in \mathcal{E}$, K/F is an almost $K \cap G$-Cogalois extension, and E/K is an almost K^*G-Cogalois extension.* $\qquad\square$

LEMMA 4.5.5. *Let E/F be a finite extension, and let G be a group such that $F^* \leqslant G \leqslant E^*$ and G/F^* is finite of exponent n. Then, for every $p \in \mathcal{P}_n$, one has either $\zeta_p \in G$ or $p \mid [E : F]$.*

PROOF. Let $p \in \mathcal{P}_n$. By Proposition 1.4.8, there exists $g \in G$ such that $\mathrm{ord}(\hat{g}) = p$. Then $g^p = a$ for some $a \in F^*$.

First, assume that $p = 4$. If $\mathrm{Char}(F) = 2$, then $\zeta_4 = 1 \in G$, as desired. So, we can suppose that $\mathrm{Char}(F) \neq 2$.

If $a \in -4F^4$, then there exists $b \in F^*$ such that $a = -4b^4$. Then $g^4 = -4b^4 = 4\zeta_4^2 b^4$. Hence $0 = g^4 - 4\zeta_4^2 b^4 = (g^2 - 2\zeta_4 b^2)(g^2 + 2\zeta_4 b^2)$. This implies that $g^2 = \pm 2\zeta_4 b^2$, and so, $\zeta_4 \in G$ since $\mathrm{Char}(F) \neq 2$.

If $a \notin -4F^4$, then $X^4 - a$ is irreducible in $F[X]$ by the Vahlen-Capelli Criterion (Theorem 1.3.5) since $a \notin F^2$. Then $[F(g) : F] = 4$, hence $4 \mid [E : F]$.

Now, assume that $p \in \mathcal{P}_n$, $p \neq 4$. We have $g^p = a \in F^*$. If $a \in F^p$, then there exists $c \in F$ such that $c^p = a = g^p$. Observe that $\mathrm{Char}(F) \neq p$, for otherwise we would have $g = c \in F$, i.e., $\operatorname{ord}(\widehat{g}) = 1$, which is a contradiction. Then, $g = \zeta_p^j c \in G$ for some j, $0 \leqslant j \leqslant p-1$. We cannot have $j = 0$ because $\operatorname{ord}(\widehat{g}) = p > 1$, hence $\zeta_p \in \langle \zeta_p^j \rangle \leqslant G$, as desired.

If $a \notin F^p$, then $X^p - a$ is irreducible in $F[X]$ by Lemma 1.3.2, so $[F(g) : F] = p$, hence $p \,|\, [E : F]$.　　　　□

LEMMA 4.5.6. *If E/F is an extension with G/F^*-Cogalois correspondence, then $\zeta_4 \in F$ whenever $1 + \zeta_4 \in G$.*

PROOF. Assume that $\zeta_4 \notin F$. Observe that $\zeta_4 = 1 \in F$ if $\operatorname{Char}(F) = 2$. So, necessarily $\operatorname{Char}(F) \neq 2$. By the proof of implication $(1) \Longrightarrow (2)$ in Theorem 2.2.1, we have

$$\operatorname{ord}(\widehat{1 + \zeta_4}) = |\, F^* \langle 1 + \zeta_4 \rangle / F^* \,| = 4.$$

On the other hand,

$$F(F^* \langle 1 + \zeta_4 \rangle) = F(\zeta_4) = F(F^* \langle \zeta_4 \rangle).$$

Since $(1 + \zeta_4)^2 = 2\zeta_4$ and $\operatorname{Char}(F) \neq 2$, we deduce that $\zeta_4 \in \langle 1 + \zeta_4 \rangle \leqslant G$. Thus $F^* \langle \zeta_4 \rangle$, $F^* \langle 1 + \zeta_4 \rangle \in \mathcal{G}$, and then, since E/F is an extension with G/F^*-Cogalois correspondence, we deduce that $F^* \langle \zeta_4 \rangle = F^* \langle 1 + \zeta_4 \rangle$. This implies that

$$4 = |\, F^* \langle 1 + \zeta_4 \rangle / F^* \,| = |\, F^* \langle \zeta_4 \rangle / F^* \,| = \operatorname{ord}(\widehat{\zeta_4}) = 2,$$

which is a contradiction.　　　　□

We are now going to introduce the notation that will be used in the next theorem.

Let E/F be a strictly almost G-Cogalois extension, with $\exp(G/F^*) = n$. Then, the finite separable extension E/F is not G-Kneser, hence, by the Kneser Criterion and Lemma 4.5.6, there exists at least an odd prime p with $\zeta_p \in G \setminus F$. Then $\operatorname{ord}(\widehat{\zeta_p}) = p$, and so, $p \,|\, n$. Let $p_1, \ldots, p_r$ be all distinct odd primes in $\mathcal{P}_n$ such that $\zeta_{p_i} \in G \setminus F$, and denote $N_i = F^* \langle \zeta_{p_i} \rangle$ for every i, $1 \leqslant i \leqslant r$. Also denote

$$N = N_1 \cdot \ldots \cdot N_r = F^* \langle \zeta_{p_1}, \ldots, \zeta_{p_r} \rangle.$$

Since $|N_i / F^*| = p_i$ for every i, $1 \leqslant i \leqslant r$, we have

$$N / F^* = \bigoplus_{i=1}^{r} (N_i / F^*) \quad \text{and} \quad |N / F^*| = p_1 \cdot \ldots \cdot p_r.$$

The symbol "$\bigoplus$" stands for "internal direct sum of subgroups" (see 1.1.2).

THEOREM 4.5.7. *Let E/F be a strictly almost G-Cogalois extension. Then, with the notation above, there exists a group $\widetilde{G}$ such that E/F is $\widetilde{G}$-Cogalois if and only if the following three conditions are satisfied.*

(1) *For every i, $1 \leqslant i \leqslant r$, there exists $\alpha_i \in E^*$ such that $F(\zeta_{p_i}) = F(\alpha_i)$, $\mathrm{ord}(\widehat{\alpha_i}) = l_i \in \mathbb{P}$, and $\zeta_{l_i} \in F$, in other words, $F(\zeta_{p_i})/F$ is a classical Kummer extension of prime exponent l_i.*

(2) *There exists $M \in \mathcal{G}$ such that*

$$G/F^* = (M/F^*) \oplus (N/F^*),$$

$$\gcd(|M/F^*|, |N/F^*|) = 1, \quad and \quad l_1 \cdot \ldots \cdot l_r \cdot |M/F^*| = [E:F].$$

(3) *$F(M)/M$ is M-Cogalois, and $\gcd(|M/F^*|, l_1 \cdot \ldots \cdot l_r) = 1$.*

If the conditions above hold, then

$$\widetilde{G}/F^* = (M/F^*) \oplus \left(\bigoplus_{i=1}^{r} (F^*\langle \alpha_i \rangle / F^*) \right).$$

PROOF. First, we assume that there exists $\widetilde{G}$ such that E/F is $\widetilde{G}$-Cogalois. Then $F^* \leqslant \widetilde{G} \leqslant T(E/F)$. Denote $\widetilde{\mathcal{G}} = \{\, S \mid F^* \leqslant S \leqslant \widetilde{G} \,\}$. Since E/F is an extension with both G/F^*-correspondence and $\widetilde{G}/F^*$-correspondence, it follows that for every $H \in \mathcal{G}$ there exists a unique $\widetilde{H} \in \widetilde{\mathcal{G}}$ such that $F(H) = F(\widetilde{H})$, and the map

$$\tau : \mathcal{G} \longrightarrow \widetilde{\mathcal{G}}, \quad H \mapsto \widetilde{H},$$

is a lattice isomorphism.

Let $n_0 = \exp(\widetilde{G}/F^*)$. By Theorem 4.3.2, the extension E/F is n_0-pure. We claim that $\gcd(n_0, p_i) = 1$ for every i, $1 \leqslant i \leqslant r$. Indeed, if we would have $p_j \mid n_0$, i.e., $p_j \in \mathcal{P}_{n_0}$ for some j, then $\zeta_{p_j} \in G \subseteq E$, which would imply that $\zeta_{p_j} \in F$ by n_0-purity. But, this would contradict the choice of the primes p_i, $1 \leqslant i \leqslant r$, being such that $\zeta_{p_i} \in G \setminus F$. Thus, by Proposition 1.4.10, we also have $\gcd(|\widetilde{G}/F^*|, p_i) = 1$ for every i, $1 \leqslant i \leqslant r$.

By the definition of the lattice isomorphism τ considered above, we have

$$F(\widetilde{N_i}) = F(N_i) = F(\zeta_{p_i})$$

for every i, $1 \leqslant i \leqslant r$. Since E/F is an almost G-Cogalois extension, the map $\psi : \mathcal{G} \longrightarrow \mathcal{E}$, $\psi(H) = F(H)$, is a lattice isomorphism, so, for every i, $1 \leqslant i \leqslant r$, its restriction to the lattice $\{\, S \mid F^* \leqslant S \leqslant N_i \,\}$ yields a lattice isomorphism Subgroups(N_i/F^*) $\longrightarrow$ Intermediate($F(N_i)/F$). Since N_i/F^* is a cyclic group of prime order, it follows that the Abelian extension

$F(\zeta_{p_i})/F$ does not have any proper intermediate field, hence its degree is a prime number, say l_i, with $l_i \mid p_i - 1$ for all i, $1 \leqslant i \leqslant r$.

On the other hand, since E/F is $\widetilde{G}$-Cogalois, the extension $F(\widetilde{N_i})$ is $\widetilde{N_i}$-Kneser, so

$$[F(\widetilde{N_i}) : F] = [F(\zeta_{p_i}) : F] = |\widetilde{N_i}/F^*| = l_i$$

for all i, $1 \leqslant i \leqslant r$.

Let $\alpha_i \in \widetilde{G}$ be such that $\widetilde{N_i} = F^*\langle \alpha_i \rangle$, $1 \leqslant i \leqslant r$. Then $F(\widetilde{N_i}) = F(\alpha_i)$, and so $[F(\alpha_i) : F] = [F(\widetilde{N_i}) : F] = l_i$. This implies that $X^{l_i} - \alpha_i^{l_i}$ is an irreducible polynomial in $F[X]$. Since $F(\zeta_{p_i})/F$ is a Galois extension of degree l_i, every conjugate of α_i over F, in particular $\zeta_{l_i}\alpha_i$ must belong to $F(\zeta_{p_i}) = F(\alpha_i)$, hence $\zeta_{l_i} \in F(\zeta_{p_i})$. Then $F \subseteq F(\zeta_{l_i}) \subseteq F(\zeta_{p_i})$. Note that $F(\zeta_{l_i}) \neq F(\zeta_{p_i})$ since $[F(\zeta_{l_i}) : F] \leqslant l_i - 1 < [F(\zeta_{p_i}) : F] = l_i$. But l_i is a prime number, hence $F(\zeta_{l_i}) = F$, i.e., $\zeta_{l_i} \in F$. This proves condition (1).

Since τ is a lattice isomorphism, we have

$$\widetilde{N} = \tau(N) = \tau(N_1 \cdot \ldots \cdot N_r) = \tau(N_1) \cdot \ldots \cdot \tau(N_r)$$

$$= \widetilde{N_1} \cdot \ldots \cdot \widetilde{N_r} = F^* \langle \alpha_1, \ldots, \alpha_r \rangle.$$

If we set $K = F(\zeta_{p_1}, \ldots, \zeta_{p_r})$, then we have

$$K = F(N) = F(\widetilde{N}) = F(\alpha_1, \ldots, \alpha_r).$$

Since N/F^* is the internal direct sum $\bigoplus_{i=1}^r (N_i/F)$ of its finite family $(N_i/F^*)_{1 \leqslant i \leqslant r}$ of subgroups, it follows that

$$\left(N_1 \cdot \ldots \cdot N_{i-1} \cdot N_{i+1} \cdot \ldots \cdot N_r \right) \cap N_i = F$$

for every $i = 1, \ldots, r$. If we transfer this equality into $\widetilde{G}$ via the lattice isomorphism $\tau : \mathcal{G} \longrightarrow \widetilde{\mathcal{G}}$, we deduce that

$$\widetilde{N}/F^* = \bigoplus_{i=1}^r (\widetilde{N_i}/F).$$

Since E/F is $\widetilde{G}$-Cogalois, it follows that K/F is $\widetilde{N}$-Cogalois and E/K is $K^*\widetilde{G}$-Cogalois by Proposition 4.3.5, so, in particular K/F is $\widetilde{N}$-Kneser. Thus,

$$[K : F] = [F(\widetilde{N}) : F] = |\widetilde{N}/F^*| = \prod_{i=1}^r |\widetilde{N_i}/F^*| = l_1 \cdot \ldots \cdot l_r.$$

We are now going to prove that E/K is K^*G-Kneser. To do that, we will apply the Kneser Criterion. By Corollary 4.5.4, E/K is an almost K^*G-Cogalois extension, so, in view of Lemma 4.5.6, it suffices to prove that if q is any odd prime with $\zeta_q \in K^*G$, i.e., if $\zeta_q = a \cdot y$ for some $a \in K^*$ and $g \in G$, then $\zeta_q \in K$.

Let $m = \exp((K^*G)/K^*)$. If $q \nmid m$, then $1 = uq + vm$ for some $u, v \in \mathbb{Z}$, so

$$\zeta_q = a \cdot g^{uq+vm} = a \cdot (a^{-1}\zeta_q)^{uq} \cdot g^{vm} = a^{1-uq} \cdot g^{vm} \in K.$$

If $q \mid m$, then $q \mid \exp(G/F^*)$, hence, by Lemma 4.5.5, we have either $\zeta_q \in G$ or $q \mid [E : F]$. If $\zeta_q \in G \setminus F$, then $q = p_i$ for some i, $1 \leqslant i \leqslant r$. Thus $\zeta_q = \zeta_{p_i} \in K$. If $q \mid [E : F]$, then $q \mid |\widetilde{G}/F^*|$ since $[E : F] = |\widetilde{G}/F^*|$, hence $q \mid \exp(\widetilde{G}/F^*)$. It follows that $\zeta_q \in F \subseteq K$ since E/F is n_0-pure. Consequently, we have in any case that $\zeta_q \in K$.

Therefore, E/K is K^*G-Kneser, so also K^*G-Cogalois, since it is almost K^*G-Cogalois by Corollary 4.5.4. But, we have already seen that E/K is $K^*\widetilde{G}$-Cogalois. Then, by Corollary 4.4.2, we have $K^*G = K^*\widetilde{G}$. Now, using the fact that $K = F(N) = F(\widetilde{N})$ and the fact that E/F is an extension with both G/F^*-correspondence and $\widetilde{G}/F^*$-correspondence, we deduce that

$$N = K^* \cap G \quad \text{and} \quad \widetilde{N} = K^* \cap \widetilde{G}.$$

Hence

$$
\begin{aligned}
|G/F^*|/(p_1 \cdot \ldots \cdot p_r) &= |G/F^*|/|N/F^*| = |G/N| \\
&= |(K^*G)/K^*| = [E : K] = |(K^*\widetilde{G})/K^*| \\
&= |\widetilde{G}/\widetilde{N}| = |\widetilde{G}/F^*|/|\widetilde{N}/F^*| \\
&= |\widetilde{G}/F^*|/(l_1 \cdot \ldots \cdot l_r),
\end{aligned}
$$

and so,

$$[E : F] = |\widetilde{G}/F^*| = (l_1 \cdot \ldots \cdot l_r) \cdot |G/F^*|/(p_1 \cdot \ldots \cdot p_r).$$

Note that

$$\gcd(|\widetilde{G}/F^*|, p_1 \cdot \ldots \cdot p_r) = 1$$

since we have already seen at the beginning of the proof of this theorem that $\gcd(\exp(\widetilde{G}/F^*), p_i) = 1$ for every i, $1 \leqslant i \leqslant r$.

Consequently, $t_{p_i}(G/F^*) = N_i/F^*$ for every i, $1 \leqslant i \leqslant r$. Since any finite Abelian group is the internal direct sum of its p-primary components, we deduce that there exists $M \in \mathcal{G}$ such that

$$G/F^* = (M/F^*) \oplus (N/F^*),$$

$$\gcd(|M/F^*|, |N/F^*|) = 1, \quad \text{and} \quad l_1 \cdot \ldots \cdot l_r \cdot |M/F^*| = [E : F].$$

This proves condition (2).

We are now going to prove (3). Set $L = F(M)$. Using the fact that the extension E/F is an extension with G/F^*-Cogalois correspondence, we obtain that

$$L \cap K = F(M) \cap F(N) = F(M \cap N) = F(F^*) = F.$$

To prove that the extension $F(M)/F$ is M-Cogalois, we will proceed as above when we showed that the extension E/K is K^*G-Kneser. So, it will be sufficient to check that $\zeta_q \in F$ for every odd prime q with $\zeta_q \in M$. Let q be such a prime. If $q = p_i$ for some i, $1 \leqslant i \leqslant r$, then $\zeta_q = \zeta_{p_i} \in M \cap K \subseteq L \cap K = K$, as desired. If $q \notin \{p_1, \ldots, p_r\}$, then $\zeta_q \in F$ since $\zeta_{p_i} \in G \setminus F$ for every i, $1 \leqslant i \leqslant r$.

Consider $\widetilde{M} = \tau(M) \in \widetilde{\mathcal{G}}$. Then, the extension L/F is simultaneously M-Cogalois and $\widetilde{M}$-Cogalois, hence $M = \widetilde{M}$ by Corollary 4.4.2. The internal direct sum decomposition

$$G/F^* = (M/F^*) \oplus (N/F^*)$$

yields via the lattice isomorphism

$$\tau : \mathcal{G} \longrightarrow \widetilde{\mathcal{G}}, \ H \mapsto \widetilde{H},$$

the internal direct sum decomposition

$$\widetilde{G}/F^* = (M/F^*) \oplus (\widetilde{N}/F^*).$$

Finally, we shall prove that $|M/F^*|$ is not divisible by any l_i, $1 \leqslant i \leqslant r$. Assume that this is not the case, hence $l_i \,|\, |M/F^*|$ for some i. Choose $\beta \in M$ such that $\text{ord}(\widehat{\beta}) = l_i$. Since the extension $F(M)/F$ is M-Cogalois and $F(\beta) = F(F^*\langle \beta \rangle)$, we have $[F(\beta) : F] = |F^*\langle \beta \rangle / F^*| = \text{ord}(\widehat{\beta}) = l_i$. Let $L_i = F(\beta, \alpha_i) = F(\beta, \zeta_{p_i})$. Observe that

$$F^* \subseteq F^*\langle \beta \rangle \cap F^*\langle \zeta_{p_i} \rangle \subseteq M \cap N = F^*,$$

i.e.,

$$F^*\langle \beta \rangle \cap F^*\langle \zeta_{p_i} \rangle = F^*.$$

Thus,

$$F^*\langle \beta, \zeta_{p_i} \rangle / F^* = (F^*\langle \beta \rangle / F^*) \oplus (F^*\langle \zeta_{p_i} \rangle / F^*) \cong \mathbb{Z}_{l_i} \times \mathbb{Z}_{p_i} \cong \mathbb{Z}_{l_i p_i},$$

hence the group $F^*\langle \beta, \zeta_{p_i}\rangle/F^*$ has exactly four subgroups. Using the lattice isomorphism

$$\psi : \mathcal{G} \longrightarrow \mathcal{E}, \quad \psi(S) = F(S),$$

we deduce that the corresponding extension $F(F^*\langle \beta, \zeta_{p_i}\rangle)/F$ has exactly four intermediate fields. Now, observe that $L_i = F(F^*\langle \beta, \zeta_{p_i}\rangle)$, hence the extension L_i/F has exactly four intermediate fields.

On the other hand, if we transfer the equality $F^*\langle \beta\rangle \cap F^*\langle \zeta_{p_i}\rangle = F^*$ from the lattice $\mathcal{G}$ into the lattice $\mathcal{E}$ via the same lattice isomorphism $\psi : \mathcal{G} \longrightarrow \mathcal{E}$, we obtain that $F(\beta) \cap F(\alpha_i) = F$. But, L_i/F is a Galois extension, so (see 1.2.9),

$$\mathrm{Gal}(L_i/F) \cong \mathrm{Gal}(F(\beta)/F) \times \mathrm{Gal}(F(\alpha_i)/F) \cong \mathbb{Z}_{l_i} \times \mathbb{Z}_{l_i}.$$

By the Fundamental Theorem of Galois Theory we deduce that the extension L_i/F has more than four intermediate fields, which contradicts our previous statement. This completes the proof of condition (3).

Conversely, assume that conditions (1)-(3) are satisfied, and let $\widetilde{G}$ be the subgroup of E^* containing F^* such that

$$\widetilde{G}/F^* = (M/F^*) \oplus \left(\bigoplus_{i=1}^{r} (F^*\langle \alpha_i\rangle/F^*) \right).$$

Observe that we have indeed an internal direct sum decomposition in the equality above since $G/F^* = (M/F^*) \oplus (N/F^*)$, $N/F^* = \bigoplus_{i=1}^{r} (N_i/F^*)$, $F(N_i) = F(\zeta_{p_i}) = F(\alpha_i)$ for all i, $1 \leqslant i \leqslant r$, and since $\psi : \mathcal{G} \longrightarrow \mathcal{E}$ is a lattice isomorphism.

We are going to prove that the extension E/F is $\widetilde{G}$-Cogalois. Let $n_0 = \exp(\widetilde{G}/F^*)$. By Theorem 4.3.2, we have only to prove that the extension E/F is n_0-pure.

Let $p \in \mathcal{P}_{n_0}$. If $p = l_i$ for some i, $1 \leqslant i \leqslant r$, then $\zeta_p = \zeta_{l_i} \in F$ by condition (1). So, we may assume that $p \notin \{l_1, \dots, l_r\}$. Then, by (3), we have $p\,|\,|M/F^*|$, i.e., $p\,|\,\exp(M/F^*)$. Since the extension $F(M)/F$ is M-Cogalois by condition (3), we deduce that $\zeta_p \in F$ whenever $\zeta_p \in F(M)$. Next, we assume that $\zeta_p \in E \setminus F(M)$ and we show that we shall obtain a contradiction.

If $p = 4$, then there exists $H \in \mathcal{G}$ such that $F(\zeta_4) = F(H)$ since E/F is an extension with G/F^*-Cogalois correspondence. Since $\zeta_4 \notin F(M)$, we have $[F(\zeta_4) : F] = 2$, hence the extension $F(\zeta_4)/F$ has no proper intermediate field. Using the lattice isomorphism $\varphi : \mathcal{E} \longrightarrow \mathcal{G}$, $K \mapsto K \cap G$, we deduce that H/F^* is a cyclic group of prime order, say q. We claim that

$H = F^* \langle \zeta_{p_i} \rangle$ for some i, $1 \leqslant i \leqslant r$. Indeed, since $\zeta_4 \in E \setminus F(M)$, we have $F(M) \cap F(H) = F(M) \cap F(\zeta_4) = F$, hence using the lattice isomorphism $\psi : \mathcal{G} \longrightarrow \mathcal{E}$, $S \mapsto F(S)$, one finds that $(M/F^*) \cap (H/F^*) = \{\widehat{1}\}$. Now, the direct sum decomposition in (2) implies that

$$H/F^* \leqslant t_q(G/F^*) = t_q(M/F^*) \oplus t_q(N/F^*),$$

hence $H/F^* \leqslant t_q(M/F^*) \leqslant M/F^*$ in case $q \notin \{p_1, \dots, p_r\}$, which is a contradiction. Thus $q = p_i$ for some i, $1 \leqslant i \leqslant r$, and consequently $H/F^* \leqslant t_{p_i}(N/F^*) = F^* \langle \zeta_{p_i} \rangle / F^*$. Then $H = F^* \langle \zeta_{p_i} \rangle$, hence $F(\zeta_4) = F(\zeta_{p_i}) = F(\alpha_i)$. Observe that condition (1) implies that $[F(\alpha_i) : F] = l_i$. Thus $2 = [F(\zeta_4) : F] = [F(\zeta_{p_i}) : F] = l_i$. Since $|N/F^*|$ is odd, this implies that $l_i \,||\, |M/F^*|$, which is a contradiction.

If p is an odd prime, choose $\beta \in M$ such that $\mathrm{ord}(\widehat{\beta}) = p$. Since the extension $F(M)/M$ is M-Cogalois, we have

$$[F(\beta) : F] = [F(F^* \langle \beta \rangle) : F] = |F^* \langle \beta \rangle / F^*| = \mathrm{ord}(\widehat{\beta}) = p,$$

hence $\mathrm{Min}(\beta, F) = X^p - \beta^p$. Then, we have also $\mathrm{Min}(\beta\zeta_p, F) = X^p - \beta^p$, and so, $[F(\beta\zeta_p) : F] = p$. Since $F(\beta\zeta_p) \in \mathcal{E}$ and E/F is an extension with G/F^*-Cogalois correspondence, there exists an $H \in \mathcal{G}$ such that $F(H) = F(\beta\zeta_p)$. Since $\zeta_p \in E \setminus F(M)$ and $\beta \in M$, it follows that $F(M) \cap F(\beta\zeta_p) = F$. As in the proof of the case $p = 4$, one deduces that $H = F^* \langle \zeta_{p_j} \rangle$ for some j, $1 \leqslant j \leqslant r$. Therefore, $F(H) = F(\beta\zeta_p) = F(\zeta_{p_j})$, and then, $p = [F(\beta\zeta_p) : F] = [F(\zeta_{p_j}) : F] = l_j$, which contradict our initial assumption that $p \notin \{l_1, \dots, l_r\}$. Therefore E/F is a $\widetilde{G}$-Cogalois extension, and we are done. $\qquad\square$

Now, we will apply Theorem 4.5.7 to number fields. Recall that a number field is nothing else than a subfield K of the field $\mathbb{C}$ of complex numbers such that the extension $K/\mathbb{Q}$ is finite.

COROLLARY 4.5.8. *Let E be a number field such that the extension $E/\mathbb{Q}$ is strictly almost G-Cogalois. Then $E/\mathbb{Q}$ is $\widetilde{G}$-Cogalois for some group $\widetilde{G}$ if and only if there exists an intermediate field K of $E/\mathbb{Q}$ such that $K/\mathbb{Q}$ is Cogalois, with $\gcd(|\mathrm{Cog}(K/\mathbb{Q})|, 6) = 1$, $E = K(\zeta_3)$, and*

$$G/\mathbb{Q}^* = \mathrm{Cog}(K/\mathbb{Q}) \oplus (\mathbb{Q}^* \langle \zeta_3 \rangle / \mathbb{Q}^*).$$

In this case, we have

$$\widetilde{G}/\mathbb{Q}^* = \mathrm{Cog}(K/\mathbb{Q}) \oplus (\mathbb{Q}^* \langle \sqrt{-3} \rangle / \mathbb{Q}^*).$$

PROOF. Suppose that the extension $E/\mathbb{Q}$ is $\widetilde{G}$-Cogalois for some group $\widetilde{G}$. Then, the extension $E/\mathbb{Q}$ satisfies the conditions (1)-(3) of Theorem 4.5.7. Observe that if p is an odd prime, then $\mathbb{Q}(\zeta_p)/\mathbb{Q}$ is a classical Kummer extension of prime exponent if and only if $p = 3$. Hence, retaining the notation of Theorem 4.5.7, we have

$$G/\mathbb{Q}^* = M/\mathbb{Q}^* \oplus (\mathbb{Q}^*\langle \zeta_3 \rangle/\mathbb{Q}^*) \text{ and } \widetilde{G}/\mathbb{Q}^* = M/\mathbb{Q}^* \oplus (\mathbb{Q}^*\langle \sqrt{-3} \rangle/\mathbb{Q}^*),$$

where $\mathbb{Q}(M)/\mathbb{Q}$ is M-Cogalois, $\gcd(|M/\mathbb{Q}^*|,6) = 1$, and $[E : \mathbb{Q}] = 2 \cdot |M/\mathbb{Q}^*|$.

For simplicity, denote $K = \mathbb{Q}(M)$. We shall prove that $\zeta_p \notin K$ for every $p \in \mathcal{P}$. Assume that $\zeta_p \in K$. Then $\mathbb{Q} \subseteq \mathbb{Q}(\zeta_p) \subseteq K$. Let $H \leqslant M$ with $\mathbb{Q}^* \leqslant H$ be such that $\mathbb{Q}(\zeta_p) = \mathbb{Q}(H)$. Since the extension $\mathbb{Q}(\zeta_p)/\mathbb{Q}$ is H-Kneser, we have $|H/\mathbb{Q}^*| = [\mathbb{Q}(\zeta_p) : \mathbb{Q}]$. If p is an odd prime, then $[\mathbb{Q}(\zeta_p) : \mathbb{Q}] = p-1$ is an even number, and so is also $|M/\mathbb{Q}^*|$ (since $|H/\mathbb{Q}^*|$ divides $|M/\mathbb{Q}^*|$), which is a contradiction. If $p = 4$, then again $|M/\mathbb{Q}^*|$ is an even number since $[\mathbb{Q}(\zeta_4) : \mathbb{Q}] = 2$, which is a contradiction. Thus, we proved that $\zeta_p \notin K$ for every $p \in \mathcal{P}$.

Therefore, the separable M-radical extension $K/\mathbb{Q}$ is pure, hence it is Cogalois by the Greither-Harrison Criterion. Then, by Corollary 4.4.3, we necessarily have

$$M/\mathbb{Q}^* = \mathrm{Cog}(K/\mathbb{Q}).$$

Finally, observe that $E = \mathbb{Q}(G) = \mathbb{Q}(M)(\zeta_3) = K(\zeta_3)$.

Conversely, let K be an intermediate field of the extension $E/\mathbb{Q}$ which satisfies the conditions in the statement of Corollary 4.5.8.

We are going to check that the conditions of Theorem 4.5.7, with $F = \mathbb{Q}$, are fulfilled. First, let p be an odd prime such that $\zeta_p \in G = T(K/\mathbb{Q})\langle \zeta_3 \rangle$. Then $\zeta_p = \gamma\zeta_3^k$, with $\gamma \in T(K/\mathbb{Q})$ and $k = 0,1,2$, hence $\gamma = \zeta_p\zeta_3^{-k}$. Thus, $(\gamma^3)^p = 1$. If $\gamma^3 = 1$, then $\gamma \in \langle \zeta_3 \rangle \cap T(K/\mathbb{Q}) = \mathbb{Q}^*$, hence $\zeta_p \in \langle \zeta_3 \rangle$, and then, necessarily $p = 3$. If $\gamma^3 \neq 1$, then we must have $\gamma^3 \in \langle \zeta_p \rangle \setminus \{1\}$, which implies that $\zeta_p \in \langle \gamma^3 \rangle \in T(K/\mathbb{Q}) \subseteq K$. Being a Cogalois extension, $K/\mathbb{Q}$ is pure, hence $\zeta_p \in \mathbb{Q}$, which is a contradiction.

Consequently we have shown that 3 is the only odd prime in $\mathcal{P}_n$ such that $\zeta_p \in G \setminus \mathbb{Q}$, where $n = \exp(G/\mathbb{Q}^*)$, hence $N = \mathbb{Q}^*\langle \zeta_3 \rangle$. Now, observe that $\mathbb{Q}(\zeta_3)/\mathbb{Q}$ a classical Kummer extension of exponent 2 since $\mathbb{Q}(\zeta_3) = \mathbb{Q}(\sqrt{-3})$, so condition (1) of Theorem 4.6.7 is satisfied.

Since $G/\mathbb{Q}^* = \mathrm{Cog}(K/\mathbb{Q}) \oplus (\mathbb{Q}^*\langle \zeta_3 \rangle/\mathbb{Q}^*)$, we have $\zeta_3 \notin K$, hence $[E : \mathbb{Q}] = [K(\zeta_3) : K] \cdot [K : \mathbb{Q}] = 2 \cdot [K : \mathbb{Q}] = 2 \cdot |\mathrm{Cog}(K/\mathbb{Q})|$. Therefore, if we put $M = T(K/\mathbb{Q})$, then conditions (2) and (3) also hold since $K/\mathbb{Q}$ is Cogalois and $\gcd(|\operatorname{Cog}(K/\mathbb{Q})|, 6) = 1$.

Finally, observe that $\mathbb{Q}(\zeta_3) = \mathbb{Q}(\sqrt{-3})$, with $\operatorname{ord}(\widehat{\sqrt{-3}}) = 2$, hence the group $\widetilde{G}$ in Theorem 4.6.7 is precisely the subgroup of E^* defined by

$$\widetilde{G}/\mathbb{Q}^* = \operatorname{Cog}(K/\mathbb{Q}) \oplus (\mathbb{Q}^* \langle \sqrt{-3} \rangle / \mathbb{Q}^*).$$

Therefore, the extension $E/\mathbb{Q}$ is $\widetilde{G}$-Cogalois by Theorem 4.6.7, and we are done. $\qquad\square$

EXAMPLES 4.5.9. (1) Let p a positive prime number, and let $q = p^r \geqslant 3$, $r \in \mathbb{N}^*$. We can choose positive prime numbers t and l such that

$$l \mid q^t - 1 \quad \text{and} \quad \gcd(l, q - 1) = \gcd(l, q - 1) = 1.$$

Indeed, let t be any positive prime number which does not divide $q - 1$, and let

$$d = \gcd(1 + q + \cdots + q^{t-1}, q - 1).$$

Then $q \equiv 1 \pmod{d}$, so $1 + q + \cdots + q^{t-1} \equiv 1 + \cdots + 1 \equiv t \pmod{d}$. Thus, d divides the relatively prime numbers $q - 1$ and t, hence $d = 1$.

Since $q^t - 1 = (q - 1) \cdot (1 + q + \cdots + q^{t-1})$, surely there exists a prime number l such that $l \mid q^t - 1$ and $\gcd(l, q - 1) = 1$.

We claim that the extension $\mathbb{F}_{q^t}/\mathbb{F}_q$ is strictly almost G-Cogalois, where $G = \mathbb{F}_q^* \langle \zeta_l \rangle$, but is not $\widetilde{G}$-Cogalois for any group $\widetilde{G}$.

Indeed, by Fermat's Little Theorem for the field $\mathbb{F}_{q^t}$, the field $\mathbb{F}_{q^t}$ contains all the $(q^t - 1)$-th roots of unity of the algebraic closure $\overline{\mathbb{F}_q}$ of $\mathbb{F}_q$. In particular, since $l \mid q^t - 1$, we have $\zeta_l \in \mathbb{F}_{q^t}$. If we would have $\zeta_l \in \mathbb{F}_q$, then $\zeta_l^{q-1} = 1$, so $l \mid q-1$, which is a contradiction. Therefore, $\zeta_l \in \mathbb{F}_{q^t} \setminus \mathbb{F}_q$, which implies that $\mathbb{F}_q(\zeta_l) = \mathbb{F}_{q^t}$ since the degree $[\mathbb{F}_{q^t} : \mathbb{F}_q] = t$ is a prime number. But t is a divisor of $\varphi(l) = l - 1$, hence $t \neq l$.

Let $G = \mathbb{F}_q^* \langle \zeta_l \rangle$. Then clearly $\mathbb{F}_q^* \leqslant G \leqslant \mathbb{F}_{q^t}^*$. If $m = \operatorname{ord}(\widehat{\zeta_l})$, then $m \leqslant l$, and $\zeta_l^m \in \mathbb{F}_q^*$. Hence $\zeta_l^{m(q-1)} = 1$. Thus $l \mid m(q-1)$, which implies that $l \mid m$. Consequently, $l = m$, i.e., $|G/\mathbb{F}_q^*| = l \neq t = [\mathbb{F}_{q^t} : \mathbb{F}_q]$, hence the extension $\mathbb{F}_{q^t}/\mathbb{F}_q$ is not G-Kneser.

Since l and t are both prime numbers, the lattices $\underline{\operatorname{Subgroups}}(G/\mathbb{F}_q^*)$ and $\underline{\operatorname{Intermediate}}(\mathbb{F}_{q^t}/\mathbb{F}_q)$ have both only two elements, so the extension $\mathbb{F}_{q^t}/\mathbb{F}_q$ is almost G-Cogalois. But we have just shown that it is not G-Kneser, so $\mathbb{F}_{q^t}/\mathbb{F}_q$ is a strictly almost G-Cogalois extension. We claim that for any group $\widetilde{G}$ with $\mathbb{F}_q^* \leqslant \widetilde{G} \leqslant \mathbb{F}_{q^t}^*$, the extension $\mathbb{F}_{q^t}/\mathbb{F}_q$ is not $\widetilde{G}$-Cogalois, for otherwise, we would have $[\mathbb{F}_{q^t} : \mathbb{F}_q] = |\widetilde{G}/\mathbb{F}_q^*| = t$, with

$t \mid (\mathbb{F}_{q^t}^* : \mathbb{F}_q^*) = (q^t - 1)/(q - 1)$. Then $t \mid (q^t - 1)$, i.e., $q^t \equiv 1 \pmod{t}$. From Fermat's Little Theorem, we have $q^t \equiv q \pmod{t}$. Hence $q \equiv 1 \pmod{t}$, i.e., $t \mid (q - 1)$. Since $\gcd(t, q - 1) = 1$, it follows that $t = 1$, which is a contradiction.

(2) Let $E = \mathbb{Q}(\zeta_3, \sqrt[3]{2})$, $F = \mathbb{Q}$, and $G = \mathbb{Q}^* \langle \zeta_3, \sqrt[3]{2} \rangle$. Then E/F is a strictly almost G-Cogalois extension by Exercise 25. We claim that there exists no group $\widetilde{G}$ such that E/F is $\widetilde{G}$-Cogalois. Indeed, as in the proof of Corollary 4.5.8, with the notation of Theorem 4.5.7, we have $|N/F^*| = 3$. Clearly, $G/F^* = (\mathbb{Q}^* \langle \sqrt[3]{2} \rangle / \mathbb{Q}^*) \oplus (\mathbb{Q}^* \langle \zeta_3 \rangle / \mathbb{Q}^*)$, hence $|G/F^*| = 9$. Thus, for any proper subgroup R/F^* of G/F^*, we must have $|R/F^*| = 3$, and so, condition (2) of Theorem 4.5.7 does not hold.

(3) By Exercise 26, there exist infinitely many number fields satisfying the conditions of Corollary 4.5.8. $\qquad \square$

4.6. Exercises to Chapter 4

1. Prove that Proposition 4.1.4 fails for infinite extensions. (*Hint:* Show that for any field F, the extension $F(X)/F$, where X is an indeterminate, satisfies condition (3) in Proposition 4.1.4, but it is not Galois. Also, any infinite Galois extension satisfies of course condition (1) but not necessarily condition (6) in Proposition 4.1.4.)

2. Show that the G-radical extension $\mathbb{Q}(\zeta_3)/\mathbb{Q}$, where $G = \mathbb{Q}^* \langle \zeta_3 \rangle$, is an extension with $G/\mathbb{Q}^*$-Cogalois correspondence which is not G-Cogalois.

3. Show that the extension $\mathbb{Q}(\zeta_3)/\mathbb{Q}$ is $\mathbb{Q}^* \langle \sqrt{-3} \rangle$-Cogalois, but is not Cogalois.

4. Show that any finite separable G-radical extension, with $\exp(G) = 2$ is G-Cogalois.

5. Prove that the extension $\mathbb{Q}(i, \sqrt{6})/\mathbb{Q}$ is $\mathbb{Q}^* \langle i, \sqrt{6} \rangle$-Cogalois, but $\sqrt{6}(1 + i)/2 \in t_2(\mathrm{Cog}(\mathbb{Q}(i, \sqrt{6})/\mathbb{Q})) \setminus t_2(\mathbb{Q}^* \langle i, \sqrt{6} \rangle / \mathbb{Q}^*)$.

6. Let E/F be a G-Cogalois extension with $\exp(G/F^*) \equiv 2 \pmod{4}$. Prove that if $\mu_4(E) \subseteq F$, then $t_2(G/F^*) = t_2(\mathrm{Cog}(E/F))$. (*Hint:* Proceed as in the proof of Theorem 4.4.1 (1).)

7. Let E/F be a finite separable G-radical extension with $m = |G/F^*|$, satisfying the condition

(†) $\mu_p(E) \subseteq F$ for every odd $p \in \mathbb{P}_m$,   and
           $\mu_4(E) \subseteq F$ whenever m is even.
Prove that E/F is G-Cogalois and $G/F^* = \bigoplus_{p \in \mathbb{P}_m} t_p(\mathrm{Cog}(E/F))$.
(*Hint*: Use Theorem 4.4.1 and Exercise 6.)

8. Prove that condition (†) in Exercise 7 is equivalent to condition
   (††) $\zeta_{2p} \notin E \setminus F$ for every $p \in \mathbb{P}_m$.
   (*Hint*: Use the Gay-Vélez Criterion.)

9. Show that a G-Cogalois extension E/F may not satisfy condition
   (††) in Exercise 8.

10. Let E/F be a G-Cogalois extension. Prove that for any interme-
    diate fields K, L of E/F such that $L \subseteq K$, the extension K/L is
    $L^*G \cap K^*$-Cogalois.

11. Let E/F be an arbitrary extension such that $\mathrm{Cog}(E/F)$ is a finite
    group. If $n = \exp(\mathrm{Cog}(E/F))$, then show that the extension E/F is
    pure if and only if it is n-pure.

12. Give an example of a finite separable radical extension which is not
    G-Cogalois for any group G.

13. (*Acosta de Orozco and Vélez* [**61**]). A finite extension E/F is said
    to have the *unique subfield property*, abbreviated USP, if for every
    divisor m of $[E:F]$ there exists a unique intermediate field K of
    E/F such that $[K:F] = m$.
        Let F be any field, and let $u \in \Omega$ be a root of an irreducible
    binomial $X^n - a \in F[X]$ with $\gcd(n, e(F)) = 1$, where Ω in an
    algebraically closed overfield of F. Prove that the extension $F(u)/F$
    has the USP if and only if the following two conditions are satisfied.
    (a) $\zeta_p \notin F(u) \setminus F$ for every odd prime divisor p of n.
    (b) If $4 \mid n$, then $\zeta_4 \notin F(u) \setminus F$.

14. Show that a finite extension E/F of degree n has the USP if and
    only if the canonical map

    $$\underline{\mathrm{Intermediate}}(E/F) \longrightarrow \mathbb{D}_n, \ K \mapsto [K:F],$$

    is a lattice isomorphism.

15. Prove that the following assertions are equivalent for a finite G-
    Cogalois extension E/F of degree n.
    (a) E/F has the USP.
    (b) The Kneser group G/F^* of E/F is cyclic.

(c) $G/F^* \cong \mathbb{Z}_n$.

16. Prove that the following assertions are equivalent for a finite Cogalois extension E/F of degree n.

    (a) E/F has the USP.

    (b) The Cogalois group $\mathrm{Cog}(E/F)$ of E/F is cyclic.

    (c) $\mathrm{Cog}(E/F) \cong \mathbb{Z}_n$.

17. Let F be any field, and let $u \in \Omega$ be a root of an irreducible binomial $X^n - a \in F[X]$ with $\gcd(n, e(F)) = 1$, where Ω in an algebraically closed overfield of F. Prove that the following assertions are equivalent.

    (a) The extension $F(u)/F$ has the USP.

    (b) The extension $F(u)/F$ is n-pure.

    (c) The extension $F(u)/F$ is $F^*\langle u \rangle$-Cogalois.

    (d) The extension $F(u)/F$ is G-Cogalois for some group G, and G/F^* is a cyclic group.

18. Prove the following statements.

    (a) Any extension of degree a prime number has the USP.

    (b) A finite G-radical extension having the USP is not necessarily G-Cogalois.

    (c) A finite G-Cogalois extension may fail to have the USP.

19. Show that the condition "G/F^* is a cyclic group" in the statement (d) of Exercise 17 is essential. (*Hint*: Take $F = \mathbb{Q}$ and $u = \sqrt{2}\,(1+i)$ $\in \mathbb{C}$. Then u is a root of the irreducible binomial $X^4 + 16 \in F[X]$, and the extension $F(u)/F$ is $\mathbb{Q}^*\langle i, \sqrt{2}\,\rangle$-Cogalois, but it does not have the USP.)

20. Let $u \in \Omega$ be a root of an irreducible binomial $X^n - a \in F[X]$. If $n \not\equiv 0 \pmod 4$ and $\gcd(n, e(F)) = 1$, then prove that the following statements are equivalent.

    (a) The extension $F(u)/F$ has the USP.

    (b) The extension $F(u)/F$ is $F^*\langle u \rangle$-Cogalois.

    (c) The extension $F(u)/F$ is G-Cogalois for some group G.

21. Let E/F be a G-Cogalois extension, with F a field of characteristic $\neq 2$, $E = F(u)$, and u a root in Ω of an irreducible binomial $X^{2^s} - a \in F[X]$. Assume that the Kneser group G/F^* of E/F is noncyclic. Prove the following statements.

    (a) $\exp(G/F^*) = 2$.

    (b) E/F is a classical 2-Kummer extension.

(c) $\zeta_{2^s} \in E$ and $\zeta_4 \notin F$.

(d) There exist $s - 1$ elements $a_2, \ldots, a_s \in F^*$ such that $E = F(\zeta_4, \sqrt{a_2}, \ldots, \sqrt{a_s})$ and $G/F^* = F^*\langle \zeta_4, \sqrt{a_2}, \ldots, \sqrt{a_s}\rangle/F^* \cong (\mathbb{Z}_2)^s$.

22. Show that for any field F of characteristic $\neq 2$ such that $\zeta_4 \notin F$, and for a root $\sqrt{a} \in \Omega$ of a polynomial $X^2 - a \in F[X]$ such that $\sqrt{a} \notin F^*\langle\zeta_4\rangle$, the extension $F(\zeta_4, \sqrt{a})/F$ is a simple radical quartic $F^*\langle\zeta_4, \sqrt{a}\rangle$-Cogalois extension with a noncyclic Kneser group.

23. Prove that any cyclic Galois extension $E/\mathbb{Q}$ of degree > 2 is not G-Cogalois, but has the USP.

24. Let K be an intermediate field of an extension E/F such that the extension K/F is almost $K \cap G$-Cogalois and the extension E/K is almost K^*G-Cogalois. Does it follow that the extension E/F is almost G-Cogalois?

25. (*Lam-Estrada, Barrera-Mora, and Villa-Salvador* [**79**]). Show that the extension $\mathbb{Q}(\zeta_3, \sqrt[3]{2})/\mathbb{Q}$ is strictly almost G-Cogalois, where $G = \mathbb{Q}^*\langle\zeta_3, \sqrt[3]{2}\rangle$.

26. (*Lam-Estrada, Barrera-Mora, and Villa-Salvador* [**79**]). Let $a, p \in \mathbb{P}$, $p \geqslant 5$, and let $F = \mathbb{Q}$, $E = \mathbb{Q}(\zeta_3, \sqrt[p]{a})$, and $G = \mathbb{Q}^*\langle\zeta_3, \sqrt[p]{a}\rangle$. Show that the extension $E/\mathbb{Q}$ is strictly almost G-Cogalois and $\widetilde{G}$-Cogalois, where $\widetilde{G}/\mathbb{Q}^* = (\mathbb{Q}^*\langle\sqrt{-3}\rangle/\mathbb{Q}^*) \oplus (\mathbb{Q}^*\langle\sqrt[p]{a}\rangle/\mathbb{Q}^*)$.

4.7. Bibliographical comments to Chapter 4

Section 4.1. The concept of Galois connection is discussed in Stenström [**100**, Chapter III, Section 8] for complete lattices, and in Albu and Năstăsescu [**18**, Section 3] for arbitrary posets. The dual concept of Cogalois connection was introduced by Albu and Nicolae [**19**]. The nice idea to use the "prime" notation in the standard Galois connection associated with a field extensions goes back to Kaplansky [**74**, Section 3].

Proposition 4.1.4, inspired by the nice approach of the Fundamental Theorem of Galois Theory in Kaplansky [**74**, Theorem 10], seems to be new.

This section basically follows the outline in Albu and Năstăsescu [**18**, Section 3] and Albu and Nicolae [**19**], although there are several extensions and improvements.

Section 4.2. The concept of strongly G-Kneser extension is due to Albu and Nicolae [**19**]. The results of this section are taken from Albu and Nicolae [**19**] and Albu and Ţena [**25**].

Section 4.3. The concept of G-Cogalois extension, as well as its basic properties including the n-Purity Criterion (Theorem 4.3.2) are due to Albu and Nicolae [**19**]. This criterion is the main tool in the whole Cogalois Theory. It permits to reobtain in a simple and unifying manner, even in a more general setting, a series of results of Albu [**3**], Barrera-Mora, Rzedowski-Calderón, and Villa-Salvador [**30**], Greither and Harrison [**63**], as well as a considerable part of the classical Finite Kummer Theory (see Chapter 7). A very elementary approach, at the undergraduate level, to the basic concepts and properties of Kneser and G-Cogalois extensions is available in Albu [**4**], [**5**].

This section basically follows the presentation in Albu and Nicolae [**19**] and Albu and Ţena [**25**], although there are several extensions and improvements.

Section 4.4. The structure of the group G/F^* of a G-Cogalois extension E/F and its uniqueness were established by Albu and Nicolae [**19**]. The term of Kneser group of a G-Cogalois extension appeared first in Albu and Nicolae [**22**].

Section 4.5. An analogue of Theorem 5.3 of Greither and Pareigis [**64**] prompted Albu and Nicolae to raise in [**19**] the following problem: *if E/F is a separable G-radical extension which is not G-Cogalois, but is an extension with G/F^*-Cogalois correspondence, then does there exist another group $\widetilde{G}$ such that E/F is $\widetilde{G}$-Cogalois?* This problem was solved in *negative* by Lam-Estrada, Barrera-Mora, and Villa-Salvador in [**79**, Section 4]. They introduced the concept of a *pseudo G-Cogalois extension*, which is precisely our concept of finite separable G-radical extension with G/F^*-Cogalois correspondence.

This section mainly presents with some improvements, simplifications and corrections of several inaccuracies, the negative answer of Lam-Estrada, Barrera-Mora, and Villa-Salvador [**79**] to the question raised by Albu and Nicolae [**19**, Problem 3.10] mentioned above.

CHAPTER 5

GALOIS G-COGALOIS EXTENSIONS

The purpose of this chapter is to study finite extensions which are simultaneously Galois and G-Cogalois. In Section 5.1 we characterize G-radical extensions, not necessarily finite, which are separable or Galois. In the next section we prove that the Kneser group and the Galois group of any finite Abelian G-Cogalois extension are isomorphic, but not in a canonical way. Section 5.3 contains some applications of Section 5.1 and Section 5.2 to elementary Field Arithmetic. Further applications, involving results we will prove in Section 6.1, Chapter 7, and Section 8.1 will be given in Section 8.2.

5.1. Galois G-radical extensions

In this section we characterize G-radical extensions, not necessarily finite, which are separable or Galois.

Recall that for any torsion group T with identity element e we introduced in Section 1.4 the notation:

$$\mathcal{O}_T = \{\, \mathrm{ord}(x) \,|\, x \in T \,\}.$$

When the subset $\mathcal{O}_T$ of $\mathbb{N}$ is a bounded set, or equivalently, a finite set, one says that the torsion group T is a *group of bounded order*, and the least number $n \in \mathbb{N}^*$ with the property that $T^n = \{e\}$ is the *exponent* $\exp(T)$ of T. The group T is said to be *n-bounded* if T is a group of bounded order and $\exp(T) = n$.

Observe that for any G-radical extension E/F, which is not necessarily finite, the group G/F^* is a torsion Abelian group, so it makes sense to consider the set $\mathcal{O}_{G/F^*}$ of natural numbers.

The next concept is important for infinite extensions; we will deal with it much more deeply in Chapters 12 and 13.

DEFINITION 5.1.1. *A G-radical extension E/F, which is not necessarily finite, is said to be a* bounded extension *if G/F^* is a group of bounded order; in this case, if $\exp(G/F^*) = n$, we say that E/F is an n-bounded extension.* □

If E/F is an n-bounded extension, then, in view of Remark 1.4.9 we have

$$\mathcal{O}_{G/F^*} = \mathbb{D}_n.$$

LEMMA 5.1.2. *Let E/F be a G-radical extension which is not necessarily finite. Then E/F is separable if and only if $\gcd(m, e(F)) = 1$ for all $m \in \mathcal{O}_{G/F^*}$.*

PROOF. We can suppose that $\mathrm{Char}(F) = p > 0$, so $e(F) = p$. Assume that E/F is separable, and let $m \in \mathcal{O}_{G/F^*}$. Then, there exists $u \in G$ such that $m = \mathrm{ord}(\widehat{u})$. If $p \mid m$ then $m = ps$ for some $s \in \mathbb{N}^*$. It follows that u^s is simultaneously separable and purely inseparable over F, so $u^s \in F$. This shows that $\widehat{u}^s = \widehat{1}$, contrary to $\mathrm{ord}(\widehat{u}) = m$.

Conversely, let $u \in G$ and set $m = \mathrm{ord}(\widehat{u})$. By hypothesis $(p, m) = 1$, hence the derivative of the polynomial $f_u = X^m - u^m \in F[X]$ is nonzero. This means that f_u has only simple roots in Ω. But, the minimal polynomial of u over F is a divisor of f_u, hence u is separable over F. Since $E = F(G)$ we deduce that E/F is a separable extension. □

COROLLARY 5.1.3. *Let E/F be an n-bounded G-radical extension, which is not necessarily finite. Then E/F is a separable extension if and only if $\gcd(n, e(F)) = 1$.*

PROOF. Apply Lemma 5.1.2, and use the fact that $\mathcal{O}_{G/F^*} = \mathbb{D}_n$ for any n-bounded G-radical extension E/F. □

An immediate consequence of Corollary 5.1.3 is the following result.

COROLLARY 5.1.4. *Let E/F be a finite G-radical extension with G/F^* finite, and let $n = \exp(G/F^*)$. Then E/F is separable if and only if $\gcd(n, e(F)) = 1$.* □

PROPOSITION 5.1.5. *Let E/F be a G-radical extension, which is not necessarily finite. Then E/F is a Galois extension if and only if $\gcd(m, e(F)) = 1$ and $\zeta_m \in E$ for all $m \in \mathcal{O}_{G/F^*}$.*

PROOF. " $\Longrightarrow$ " Assume that E/F is a Galois extension, that is, E/F is a separable and normal extension, and let $m \in \mathcal{O}_{G/F^*}$. We can assume that $m \geqslant 2$. By Lemma 5.1.2 we have $(m, e(F)) = 1$. Let $m = p_1^{k_1} \cdot \ldots \cdot p_r^{k_r}$, where $p_1, \ldots, p_r$ are mutually distinct positive primes and $k_1, \ldots, k_r \in \mathbb{N}^*$. If we set $q_i = p_i^{k_i}$, $i = 1, \ldots, r$, then clearly $q_i \in \mathcal{O}_{G/F^*}$ for all i, $1 \leqslant i \leqslant r$, hence $q_i = \mathrm{ord}(\widehat{u_i})$ for some $u_i \in G$, $i = 1, \ldots, r$.

We claim that $\zeta_{q_i} \in E$ for all i, $1 \leqslant i \leqslant r$. Indeed, set $a_i = u_i^{q_i}$ and observe that u_i is a root of the polynomial $f_i = X^{q_i} - a_i \in F[X]$. Since E/F is normal, E contains all the conjugates of u_i over F. Any conjugate of u_i over F is necessarily a root of f_i, hence all these conjugates are $u_i\xi_1, \ldots, u_i\xi_s$, where $\xi_1, \ldots, \xi_s$ are certain q_i-th roots of unity. Suppose that no ξ_j is a primitive q_i-th roots of unity. Then every ξ_j, $1 \leqslant j \leqslant s$, will be necessarily a q_i/p_i-th root of unity. Now, observe that all the conjugates of the element $u_i^{q_i/p_i} \in E$ over F are

$$(u_i\xi_1)^{q_i/p_i} = u_i^{q_i/p_i}, \ldots, (u_i\xi_s)^{q_i/p_i} = u_i^{q_i/p_i}.$$

Thus, $u_i^{q_i/p_i}$ coincides with all its conjugates, hence $u_i^{q_i/p_i} \in F$ since E/F is a Galois extension. But this is a contradiction since q_i is minimum with the property that $u_i^{q_i} \in F^*$. This proves that there exists j, $1 \leqslant j \leqslant s$, such that ξ_j is a primitive q_i-th root of unity. Since $u_i, u_i\xi_j \in E$, we deduce that $\xi_j \in E$, and so, $\zeta_{q_i} \in E$. This proves our claim.

We are now going to show that $\zeta_m \in E$. For every i, $1 \leqslant i \leqslant r$, set $b_i = m/p_i^{k_i}$. Since $\gcd(b_1, \ldots, b_r) = 1$, we have $1 = \sum_{1 \leqslant i \leqslant r} b_i c_i$ for some $c_1, \ldots, c_r \in \mathbb{Z}$. It follows that

$$\zeta_m = \zeta_m^{b_1 c_1 + \ldots + b_r c_r} = (\zeta_m^{b_1})^{c_1} \cdot \ldots \cdot (\zeta_m^{b_r})^{c_r} \in E,$$

since $\zeta_{q_i} \in E$ and $\zeta_m^{b_i}$ is a primitive q_i-th root of unity for all i, $1 \leqslant i \leqslant r$.

" $\Longleftarrow$ " Assume that $(m, e(F)) = 1$ and $\zeta_m \in E$ for all $m \in \mathcal{O}_{G/F^*}$. By Lemma 5.1.2, E/F is a separable extension. To prove that E/F is a normal extension, let $u \in G$ and set $m = \mathrm{ord}(u)$. As above, every conjugate of u over F is a product of u by a power of ζ_m, so it belongs to E. Consequently E/F is a normal extension, and we are done. $\square$

COROLLARY 5.1.6. *Let E/F be an n-bounded G-radical extension. Then E/F is a Galois extension if and only if $\gcd(n, e(F)) = 1$ and $\zeta_n \in E$.*

PROOF. Apply Proposition 5.1.5, and use the fact that $\mathcal{O}_{G/F^*} = \mathbb{D}_n$ for any n-bounded G-radical extension E/F. $\square$

COROLLARY 5.1.7. *Let E/F be a finite G-radical extension with G/F^* a finite group of exponent n. Then E/F is a Galois extension if and only if $\gcd(n, e(F)) = 1$ and $\zeta_n \in E$.* $\square$

5.2. Abelian G-Cogalois extensions

The aim of this section is to prove that the Kneser group and the Galois group of any finite Abelian G-Cogalois extension are isomorphic.

LEMMA 5.2.1. *Let E/F be a G-Cogalois extension and let E_1/F, E_2/F be subextensions of E/F such that $E_1 \cap E_2 = F$. Then, there exists a canonical group isomorphism*

$$(G \cap (E_1 E_2)^*)/F^* \cong ((G \cap E_1^*)/F^*) \times ((G \cap E_2^*)/F^*).$$

PROOF. By Proposition 4.3.5, the extension E_i/F is $G \cap E_i^*$-Cogalois, $i = 1, 2$, and the extension $E_1 E_2$ is $G \cap (E_1 E_2)^*$-Cogalois. So, we have

$$|(G \cap (E_1 E_2)^*)/F^*| = [E_1 E_2 : F] \leqslant [E_1 : F] \cdot [E_2 : F] =$$
$$= |(G \cap E_1^*)/F^*| \cdot |(G \cap E_2^*)/F^*|.$$

Now consider the group morphism

$$\varphi : ((G \cap E_1^*)/F^*)) \times ((G \cap E_2^*)/F^*) \longrightarrow (G \cap (E_1 E_2)^*)/F^*,$$

defined by $\varphi(\hat{x}, \hat{y}) = \widehat{xy}$, $x \in G \cap E_1^*$, $y \in G \cap E_2^*$. We are going to show that φ is injective. If $(\hat{x}, \hat{y}) \in \mathrm{Ker}(\varphi)$, then $\widehat{xy} = \hat{1}$, hence $xy \in F^*$. Since $x \in E_1$ it follows that $y \in E_1$. But $y \in E_2$, so $y \in E_1 \cap E_2 = F$. This implies that $x \in F^*$. Consequently, $(\hat{x}, \hat{y}) = (\hat{1}, \hat{1})$, hence $\mathrm{Ker}(\varphi) = \{(\hat{1}, \hat{1})\}$. This shows that φ is injective, and then

$$|(G \cap E_1^*)/F^*| \cdot |(G \cap E_2^*)/F^*| \leqslant |(G \cap (E_1 E_2)^*)/F^*|.$$

Therefore, we have

$$|(G \cap (E_1 E_2)^*)/F^*| = |((G \cap E_1^*)/F^*) \times ((G \cap E_2^*)/F^*)|.$$

This equality shows that the monomorphism φ is surjective, which finishes the proof.   □

THEOREM 5.2.2. *For any finite Abelian G-Cogalois extension E/F, the groups $\mathrm{Gal}(E/F)$ and $\mathrm{Kne}(E/F)$ are isomorphic.*

PROOF. First we will examine the case when G/F^* is a cyclic group. Then, for every divisor m of $[E : F] = |G/F^*|$ there exists a unique subgroup of G/F^* of order m. Using the bijective Cogalois correspondence given by Theorem 4.3.2, we deduce that there exists a unique intermediate field E_m of the extension E/F with $[E_m : F] = m$. Now, using the bijective Galois correspondence given by the Fundamental Theorem of Galois Theory, it follows that the Galois group $\mathrm{Gal}(E/F)$ of E/F has a unique subgroup of order m for every divisor m of $|\mathrm{Gal}(E/F)| = |G/F^*| = [E : F]$.

This implies that $\mathrm{Gal}(E/F)$ is a cyclic group (see Exercise 18, Chapter 1), which is isomorphic to the cyclic group $G/F^* = \mathrm{Kne}\,(E/F)$.

Now, assume that the group G/F^* is not necessarily cyclic. We will proceed by induction on $n = |G/F^*| = [E:F]$. For $n = 2$ and $n = 3$ the group G/F^* is cyclic, and this case was already considered above.

Let $n \geqslant 4$. Assume that the property is true for all extensions of degree $< n$, and prove it for extensions of degree n. If G/F^* is cyclic, then this case was already settled. If the group G/F^* is not cyclic, then we can write $G/F^* = AB$ with

$$A = G_1/F^*,\ B = G_2/F^*,\ G_1, G_2 \in \mathcal{G},\ G_1 \cap G_2 = F^*,\ G_1 G_2 = G,$$

and $|A| < n$, $|B| < n$, where $\mathcal{G} = \{H \,|\, F^* \leqslant H \leqslant G\}$. Let $E_1 = F(G_1)$ and $E_2 = F(G_2)$. The extensions E_1/F and E_2/F are $G \cap E_1^*$-Cogalois and $G \cap E_2^*$-Cogalois, respectively. Obviously, they are also Abelian extensions, since the corresponding Galois groups are factor groups of the Abelian group $\mathrm{Gal}(E/F)$. But $|G_1/F^*| < n$ and $|G_2/F^*| < n$, hence, by the inductive hypothesis, we have

$$G_1/F^* \cong \mathrm{Gal}\,(E_1/F)\ \text{ and }\ G_2/F^* \cong \mathrm{Gal}\,(E_2/F).$$

Since E/F is a G-Cogalois extension, the map $H \mapsto F(H)$ is a lattice isomorphism between the lattice $\mathcal{G} = \{H \,|\, F^* \leqslant H \leqslant G\}$ and the lattice $\underline{\mathrm{Intermediate}}(E/F)$, so

$$E_1 \cap E_2 = F(G_1) \cap F(G_2) = F(G_1 \cap G_2) = F$$

and

$$E_1 E_2 = F(G_1)F(G_2) = F(G_1 G_2) = F(G) = E.$$

Observe that $G \cap E_i^* = G \cap E_i = G \cap F(G_i) = G_i$, $i = 1, 2$, by Proposition 2.1.11, and $\mathrm{Gal}\,(E_1 E_2/F) \cong \mathrm{Gal}\,(E_1/F) \times \mathrm{Gal}\,(E_2/F)$ since $E_1 \cap E_2 = F$ (see 1.2.9). Now, using Lemma 5.2.1, we obtain

$$
\begin{aligned}
G/F^* &= (G_1 G_2)/F^* = (G \cap (E_1 E_2)^*)/F^* \\
&\cong ((G \cap E_1^*)/F^*) \times ((G \cap E_2^*)/F^*) \\
&= (G_1/F^*) \times (G_2/F^*) \\
&\cong \mathrm{Gal}\,(E_1/F) \times \mathrm{Gal}\,(E_2/F) \\
&\cong \mathrm{Gal}\,(E_1 E_2/F) = \mathrm{Gal}\,(E/F).
\end{aligned}
$$

The proof is complete. $\qquad\qquad\qquad\qquad\qquad\qquad\qquad\qquad\qquad$ $\square$

COROLLARY 5.2.3. *For any finite Abelian Cogalois extension E/F, the groups $\mathrm{Gal}(E/F)$ and $\mathrm{Cog}(E/F)$ are isomorphic.*

PROOF. The extension E/F is $T(E/F)$-Kneser by Examples 2.1.15 (2), hence it is $T(E/F)$-Cogalois by Theorem 3.2.3 and Theorem 4.3.2. Thus, $\mathrm{Gal}(E/F)$ is isomorphic to $\mathrm{Kne}(E/F) = T(E/F)/F^* = \mathrm{Cog}(E/F)$ by Theorem 5.2.2.   □

REMARK 5.2.4. An alternative proof of Theorem 5.2.2, involves the concept of lattice-isomorphism of groups (see Section 15.2). What follows is a brief description of this approach which will be presented in full detail in Section 15.3 for arbitrary (i.e., not necessarily finite) extensions. If E/F is a finite Abelian G-Cogalois extension with Galois group Γ, then the lattice anti-isomorphism from the lattice Intermediate (E/F) onto the lattice Subgroups(Γ) given by the Fundamental Theorem of Finite Galois Theory, produces, by taking the characters, a lattice isomorphism from the lattice Intermediate (E/F) onto the lattice Subgroups$(\mathrm{Ch}\,(\Gamma))$ of all subgroups of the character group $\mathrm{Ch}\,(\Gamma) = \mathrm{Hom}(\Gamma, \mathbb{C}^*)$ of Γ.

On the other hand, since E/F is a G-Cogalois extension, the map $H/F^* \mapsto F(H)$ is an isomorphism from the lattice Subgroups $(\mathrm{Kne}\,(E/F))$ onto the lattice Intermediate (E/F). If we compose these two lattice isomorphisms, we obtain a lattice isomorphism from Subgroups $(\mathrm{Kne}\,(E/F))$ onto Subgroups $(\mathrm{Ch}\,(\Gamma))$. According to the terminology of Section 15.2, this means that the groups $\mathrm{Kne}\,(E/F)$ and $\mathrm{Ch}\,(\Gamma)$ are *lattice-isomorphic*. In general, two lattice-isomorphic groups A and B are not necessarily isomorphic, excepting few cases, when the groups A and B satisfy various restrictive conditions, in view of a series of quite technical results due to Baer [27] (see Section 15.2). By chance, some of these conditions are satisfied in our case, and so, one deduces that the finite Abelian groups $\mathrm{Kne}\,(E/F)$ and $\mathrm{Ch}\,(\Gamma)$ are isomorphic. Since any finite Abelian group is isomorphic, but not in a canonical way, to its character group, it follows that the groups $\mathrm{Gal}\,(E/F)$ and $\mathrm{Kne}\,(E/F)$ are isomorphic.

This approach, which is unavoidable for infinite extensions, is more natural than the inductive one provided in the proof of Theorem 5.2.2, but it involves a certain amount of technical facts related to lattice-isomorphisms of groups.   □

5.3. Applications to elementary Field Arithmetic (I)

The aim of this section is to apply the results of Sections 5.1 and 5.2 to elementary arithmetic of fields. Further applications of Cogalois Theory to

Field Arithmetic, based on results we will establish in subsequent chapters, will be presented in Section 8.2.

In the sequel, the algebraically closed field Ω containing any algebraic number field K is the field $\mathbb{C}$ of complex numbers, and, for any $n \in \mathbb{N}^*$, the primitive n-th root of unity is $\zeta_n = \cos(2\pi/n) + i\sin(2\pi/n)$.

The next result is very similar to Proposition 3.2.6, but in contrast to that, it provides a quartic Galois extension.

PROPOSITION 5.3.1. *The following assertions hold.*

(a) $\mathbb{Q}(\sqrt{2})$ *is a subfield of the field* $\mathbb{Q}(\sqrt{2+\sqrt{2}})$.

(b) $[\mathbb{Q}(\sqrt{2+\sqrt{2}}) : \mathbb{Q}(\sqrt{2})] = 2$ *and* $[\mathbb{Q}(\sqrt{2+\sqrt{2}}) : \mathbb{Q}] = 4$.

(c) $\mathbb{Q}(\sqrt{2+\sqrt{2}})/\mathbb{Q}$ *is an Abelian extension.*

(d) $\mathbb{Q}(\sqrt{2+\sqrt{2}})/\mathbb{Q}(\sqrt{2})$ *and* $\mathbb{Q}(\sqrt{2})/\mathbb{Q}$ *are both Cogalois extensions.*

(e) $\mathbb{Q}(\sqrt{2+\sqrt{2}})/\mathbb{Q}$ *is neither a radical extension, nor a Kneser extension, nor a Cogalois extension.*

(f) $\mathrm{Cog}\,(\mathbb{Q}(\sqrt{2+\sqrt{2}})/\mathbb{Q}) = \{\widehat{1},\, \widehat{\sqrt{2}}\,\}$.

(g) *The element* $\widehat{\sqrt{2+\sqrt{2}}}$ *of the group* $\mathbb{Q}(\sqrt{2+\sqrt{2}})^*/\mathbb{Q}^*$ *has infinite order.*

PROOF. For simplicity, set

$$F = \mathbb{Q},\ E = \mathbb{Q}\left(\sqrt{2+\sqrt{2}}\right),\ \text{and}\ \theta = \sqrt{2+\sqrt{2}}.$$

As in the proof of Proposition 3.2.6, the assertions (a),(b) and (d) are easily shown.

(c) Observe that $\mathrm{Min}(\theta, \mathbb{Q}) = X^4 - 4X^2 + 2$ hence the conjugates of θ over $\mathbb{Q}$ are

$$\sqrt{2+\sqrt{2}},\ -\sqrt{2+\sqrt{2}},\ \sqrt{2-\sqrt{2}},\ -\sqrt{2-\sqrt{2}},$$

and all of them belong to E, since

$$\sqrt{2-\sqrt{2}} = \frac{\sqrt{2}}{\sqrt{2+\sqrt{2}}} \in E.$$

This shows that E/F is a quartic Galois extension. It is easily shown that E/F has a cyclic Galois group generated by the automorphism sending $\sqrt{2+\sqrt{2}}$ to $\sqrt{2-\sqrt{2}}$.

(e) As in the proof of Proposition 3.2.6, it suffices to show that E/F is not a Cogalois extension. Assume that this is not the case, and look for a contradiction. By Corollary 5.2.3, the assumption that E/F is a Cogalois extension implies that $\mathrm{Cog}\,(E/F) \cong \mathrm{Gal}\,(E/F) \simeq \mathbb{Z}_4$. Then clearly $\exp(\mathrm{Cog}\,(E/F) = 4$, so $i = \zeta_4 \in E$ by Corollary 5.1.7, which is a contradiction. This proves that E/F is not a Cogalois extension.

For (f) and (g) proceed as in the proof of Proposition 3.2.6. $\qquad\square$

REMARK 5.3.2. More generally, consider the extension E_r/F, where $F = \mathbb{Q}$, $E_r = \mathbb{Q}(\theta_r)$, and

$$\theta_r = \underbrace{\sqrt{2 + \sqrt{2 + \sqrt{2 + \ldots + \sqrt{2}}}}}_{r \text{ radicals}}.$$

We claim that for every $r \in \mathbb{N}$, $r \geqslant 2$, the extension E_r/F is not a radical extension. To do that, observe that

$$\theta_r = 2\cos(\pi/2^{r+1}) = \zeta_{2^{r+2}} + \zeta_{2^{r+2}}^{-1},$$

and E_r/F is a Galois extension with Galois group the cyclic group of order 2^r generated by the automorphism sending $\cos(\pi/2^{r+1})$ to $\cos(5\pi/2^{r+1})$ (see Exercises 19 and 20).

Similar arguments to those used above prove that if E_r/F would be a radical extension, then it would be a Cogalois extension, and

$$\mathrm{Cog}\,(E_r/F) \simeq \mathrm{Gal}\,(E_r/F) \simeq \mathbb{Z}_{2^r}.$$

This would imply that $\zeta_{2^r} \in E_r \subseteq \mathbb{R}$ by Corollary 5.1.7, which is a contradiction. This proves our claim. $\qquad\square$

The examples discussed in Proposition 3.2.6 and Proposition 5.3.1 have the form $\mathbb{Q}\left(\sqrt{r + \sqrt{d}}\right)$. For simplicity, denote

$$\mathbb{Q}_{r,d} = \mathbb{Q}\left(\sqrt{r + \sqrt{d}}\right)$$

for any $r, d \in \mathbb{Q}$.

Next, we are interested to investigate similar questions for the extension $\mathbb{Q}_{r,d}/\mathbb{Q}$, namely the following ones:

- When is $\mathbb{Q}_{r,d}/\mathbb{Q}$ a quartic extension?
- When is $\mathbb{Q}_{r,d}/\mathbb{Q}$ a Galois extension?
- When is $\mathbb{Q}_{r,d}/\mathbb{Q}$ a Cogalois, or Kneser, or radical extension?

In the sequel, the notation $x \pm y = z$ (resp. $x \pm y \in M$), where $x, y, z \in \mathbb{C}$ and $M \subseteq \mathbb{C}$, means that either $x + y = z$ or $x - y = z$ (resp. either $x + y \in M$ or $x - y \in M$). Also, the notation $x \pm y \notin M$ means that $x + y \notin M$ and $x - y \notin M$.

PROPOSITION 5.3.3. *The following statements are equivalent for* $d \in \mathbb{Q}_+ \setminus \mathbb{Q}^2$ *and* $r \in \mathbb{Q}_+^*$.

(1) *The polynomial* $X^4 - 2rX^2 + r^2 - d$ *is reducible in* $\mathbb{Q}[X]$.

(2) *There exist* $c, k \in \mathbb{Q}_+^*$ *such that* $r^2 - d = c^2$ *and* $r \pm c = k^2/2$.

(3) *There exists* $c \in \mathbb{Q}_+^*$ *such that* $r^2 - d = c^2$ *and*

$$\mathbb{Q}_{r,d} = \mathbb{Q}\left(\sqrt{(r \pm c)/2} \right).$$

(4) $[\mathbb{Q}_{r,d} : \mathbb{Q}] = 2$.

PROOF. Set $f = X^4 - 2rX^2 + r^2 - d$, and observe that the complex roots of f are

$$\sqrt{r + \sqrt{d}}, \quad -\sqrt{r + \sqrt{d}}, \quad \sqrt{r - \sqrt{d}}, \quad -\sqrt{r - \sqrt{d}},$$

and none of them is rational. Hence (1) $\Longleftrightarrow$ (4).

(1) $\Longrightarrow$ (2): Assume that f is reducible in $\mathbb{Q}[X]$. Since $r \pm \sqrt{d} \in \mathbb{R} \setminus \mathbb{Q}$, it follows that f has necessarily a monic divisor of degree 2 of type

$$\left(X \pm \sqrt{r + \sqrt{d}} \right) \cdot \left(X \pm \sqrt{r - \sqrt{d}} \right)$$

$$= X^2 \pm \left(\sqrt{r + \sqrt{d}} \pm \sqrt{r - \sqrt{d}} \right) X \pm \sqrt{r^2 - d} \in \mathbb{Q}[X].$$

Then

$$\sqrt{r^2 - d} \in \mathbb{Q} \quad \text{and} \quad \sqrt{r + \sqrt{d}} \pm \sqrt{r - \sqrt{d}} \in \mathbb{Q},$$

hence

$$r^2 - d = c^2 \quad \text{for some} \quad c \in \mathbb{Q}_+^*.$$

On the other hand,

$$\sqrt{r + \sqrt{d}} = \sqrt{(r + c)/2} + \sqrt{(r - c)/2},$$

$$\sqrt{r - \sqrt{d}} = \sqrt{(r + c)/2} - \sqrt{(r - c)/2}.$$

Notice that we have $r > c > 0$ and $r > \sqrt{d} > 0$, hence all quadratic radicals appearing in the two formulas just above are positive real numbers. Thus,

$$\sqrt{r + \sqrt{d}} \pm \sqrt{r - \sqrt{d}} = 2\sqrt{(r \pm c)/2}.$$

This implies that $r \pm c = k^2/2$ for some $k \in \mathbb{Q}_+^*$, which proves the implication (1) $\Longrightarrow$ (2).

(2) $\Longrightarrow$ (3): If $r^2 - d = c^2$, then, by the proof of the implication (1) $\Longrightarrow$ (2), we deduce that

$$\mathbb{Q}_{r,d} = \mathbb{Q}(\sqrt{(r+c)/2} + \sqrt{(r-c)/2}),$$

hence

$$\mathbb{Q}_{r,d} = \begin{cases} \mathbb{Q}(\sqrt{(r+c)/2}) & \text{if } r - c = k^2/2, \\ \mathbb{Q}(\sqrt{(r-c)/2}) & \text{if } r + c = k^2/2. \end{cases}$$

(3) $\Longrightarrow$ (4): Assume that $r^2 - d = c^2$ and

$$\mathbb{Q}_{r,d} = \mathbb{Q}(\sqrt{(r-c)/2}).$$

To prove (4), it is sufficient to show that $\sqrt{(r-c)/2} \notin \mathbb{Q}$. Suppose otherwise. Then $r - c = 2s^2$ for some $s \in \mathbb{Q}_+$, hence

$$\sqrt{r + \sqrt{d}} = \sqrt{(r+c)/2} + \sqrt{(r-c)/2} = \sqrt{(r+c)/2} + s,$$

and so,

$$\mathbb{Q} = \mathbb{Q}(\sqrt{(r-c)/2}) = \mathbb{Q}_{r,d} = \mathbb{Q}(\sqrt{(r+c)/2}).$$

Thus, there exists $t \in \mathbb{Q}_+$ such that $r + c = 2t^2$. Now, using the equality $r - c = 2s^2$ we deduce that

$$r = s^2 + t^2 \text{ and } c = t^2 - s^2,$$

hence

$$d = r^2 - c^2 = (s^2 + t^2)^2 - (t^2 - s^2)^2 = (2st)^2 \in \mathbb{Q}^{*2},$$

which is a contradiction. $\square$

REMARKS 5.3.4. (1) It is easily seen that for $r = 0$ and for any $d \in \mathbb{Q} \setminus \mathbb{Q}^2$, the polynomial $X^4 - 2rX^2 + r^2 - d$ is irreducible in $\mathbb{Q}[X]$.

(2) If we do not assume that the rational numbers r and d in Proposition 5.3.3 are positive, then we have to be careful with the meaning of the nonreal radicals $\sqrt{r \pm \sqrt{d}}$ and $\sqrt{(r \pm c)/2}$ appearing in its proof.

We are going to show that Proposition 5.3.3 holds for any $r \in \mathbb{Q}$ and any $d \in \mathbb{Q} \setminus \mathbb{Q}^2$. Indeed, for any complex number $a \in \mathbb{C} \setminus \mathbb{R}_+^*$ we denote, as usually, by $\sqrt{a}$ one of the not specified complex roots of the equation $x^2 - a = 0$. Then, instead of the equalities

$$\sqrt{r + \sqrt{d}} = \sqrt{(r+c)/2} + \sqrt{(r-c)/2},$$

This implies that there exist a_0, a_1, a_2, $a_3 \in \mathbb{Q}$ such that

(6) $$\theta := \sqrt{r + \sqrt{d}} = a_0 + a_1\sqrt{d} + a_2\sqrt{k} + a_3\sqrt{d}\sqrt{k}.$$

We can write (6) as

(7) $$\theta - (a_0 + a_1\sqrt{d}) = \sqrt{k}\,(a_2 + a_3\sqrt{d}).$$

Squaring (7) we obtain

(8) $$r + \sqrt{d} - 2\theta(a_0 + a_1\sqrt{d}) + (a_0 + a_1\sqrt{d})^2 = k(a_2 + a_3\sqrt{d})^2.$$

Then necessarily $a_0 + a_1\sqrt{d} = 0$, for otherwise, (8) would imply that $\theta \in \mathbb{Q}(\sqrt{d})$, which is a contradiction. Consequently, (7) becomes

(9) $$\theta = \sqrt{k}\,(a_2 + a_3\sqrt{d}).$$

From (9) we deduce that

$$\left(\frac{a_2 + a_3\sqrt{d}}{\theta}\right)^2 = \frac{1}{k} \in \mathbb{Q},$$

i.e.,

$$u := \frac{\left(a_2 + a_3\sqrt{d}\right)^2}{r + \sqrt{d}} \in \mathbb{Q}.$$

Thus, u coincides with its conjugate in the quadratic extension $\mathbb{Q}(\sqrt{d})/\mathbb{Q}$, i.e.,

$$\frac{\left(a_2 + a_3\sqrt{d}\right)^2}{r + \sqrt{d}} = \frac{\left(a_2 - a_3\sqrt{d}\right)^2}{r - \sqrt{d}}.$$

After easy calculations we obtain

(10) $$a_2^2 - 2a_2a_3r + da_3^2 = 0.$$

We have $a_3 \neq 0$, for otherwise, (10) would imply $a_2 = 0$, and then, (9) would become $\theta = 0$, which is impossible. Hence, we can divide (10) by $a_3^2 \neq 0$. If we set $t := a_2/a_3$, then we see that $t \in \mathbb{Q}$ is a root of the quadratic equation

(11) $$t^2 - 2rt + d = 0.$$

Since the roots of (11) are

$$r \pm \sqrt{r^2 - d},$$

we deduce that $r^2 - d \in \mathbb{Q}^2$, and we are done.

(b) By negation, statement (a) can be obviously reformulated as follows:

$\mathbb{Q}_{r,d}$ *has at most one quadratic subfield if and only if* $r^2 - d \notin \mathbb{Q}^2$.

But $\mathbb{Q}_{r,d}$ has at least one quadratic subfield, namely $\mathbb{Q}(\sqrt{d})$. This clearly implies assertion (b). $\square$

Next, we are going to study when is $\mathbb{Q}_{r,d}/\mathbb{Q}$, $r \in \mathbb{Q}$, $d \in \mathbb{Q} \setminus \mathbb{Q}^2$, a Galois extension.

PROPOSITION 5.3.7. *The following statements hold for* $r \in \mathbb{Q}$ *and* $d \in \mathbb{Q} \setminus \mathbb{Q}^2$.
  (1) *If the polynomial* $X^4 - 2rX^2 + r^2 - d$ *is reducible in* $\mathbb{Q}[X]$, *then* $\mathbb{Q}_{r,d}/\mathbb{Q}$ *is a Galois extension, and* $\mathrm{Gal}\,(\mathbb{Q}_{r,d}/\mathbb{Q}) \cong \mathbb{Z}_2$.
  (2) *If the polynomial* $X^4 - 2rX^2 + r^2 - d$ *is irreducible in* $\mathbb{Q}[X]$, *then* $\mathbb{Q}_{r,d}/\mathbb{Q}$ *is a Galois extension if and only if* $\sqrt{r^2 - d} \in \mathbb{Q}(\sqrt{d})$. *In this case, we have*
    (a) $\mathrm{Gal}\,(\mathbb{Q}_{r,d}/\mathbb{Q}) \cong \mathbb{Z}_2 \times \mathbb{Z}_2 \Longleftrightarrow \sqrt{r^2 - d} \in \mathbb{Q}^*$,
    (b) $\mathrm{Gal}\,(\mathbb{Q}_{r,d}/\mathbb{Q}) \cong \mathbb{Z}_4 \Longleftrightarrow \sqrt{r^2 - d} = s\sqrt{d}$ *for some* $s \in \mathbb{Q}^*$.

PROOF. (1) follows immediately from Proposition 5.3.3, since any quadratic extension is a Galois extension.

(2) Set $f = X^4 - 2rX^2 + r^2 - d$ and observe that the complex roots of f are
$$\sqrt{r + \sqrt{d}},\ -\sqrt{r + \sqrt{d}},\ \sqrt{r - \sqrt{d}},\ -\sqrt{r - \sqrt{d}},$$
hence $f = \mathrm{Min}\,\left(\sqrt{r + \sqrt{d}}, \mathbb{Q}\right)$.
  Since
$$\sqrt{r - \sqrt{d}} = \pm \frac{\sqrt{r^2 - d}}{\sqrt{r + \sqrt{d}}},$$
we have
$$\mathbb{Q}_{r,d}/\mathbb{Q} \text{ is Galois} \iff \sqrt{r - \sqrt{d}} \in \mathbb{Q}_{r,d} \iff \sqrt{r^2 - d} \in \mathbb{Q}_{r,d}.$$
The following cases arise:

Case 1: $\sqrt{r^2 - d} \in \mathbb{Q}$. Then, the proof of Proposition 5.3.6 (a) shows that
$$\mathbb{Q}_{r,d} = \mathbb{Q}(\sqrt{(r + c)/2}, \sqrt{(r - c)/2}),$$
where $c \in \mathbb{Q}_+^*$ is such that $r^2 - d = c^2$, and $\mathbb{Q}_{r,d}/\mathbb{Q}$ is a Galois extension with $\mathrm{Gal}\,(\mathbb{Q}_{r,d}/\mathbb{Q}) \cong \mathbb{Z}_2 \times \mathbb{Z}_2$.

Case 2: $\sqrt{r^2 - d} \in \mathbb{Q}(\sqrt{d}) \setminus \mathbb{Q}$. Then $\mathbb{Q}_{r,d}$ has precisely one quadratic subfield by Proposition 5.3.6 (b), hence, by Galois Theory, $\mathrm{Gal}\,(\mathbb{Q}_{r,d}/\mathbb{Q})$ is a group of order 4 having only one subgroup of order 2. Therefore, this group is necessarily a cyclic group of order 4, and so,

$$\mathrm{Gal}\,(\mathbb{Q}_{r,d}/\mathbb{Q}) \cong \mathbb{Z}_4.$$

Now observe that $\sqrt{r^2 - d} \in \mathbb{Q}(\sqrt{d}) \setminus \mathbb{Q} \iff \sqrt{r^2 - d} = s\sqrt{d}$ for some $s \in \mathbb{Q}^*$.

Case 3: $\sqrt{r^2 - d} \in \mathbb{Q}_{r,d} \setminus \mathbb{Q}(\sqrt{d})$. Then $\mathbb{Q}(\sqrt{d})$ is a proper subfield of the field $\mathbb{Q}(\sqrt{d})(\sqrt{r^2 - d})$, which in turn, is a subfield of $\mathbb{Q}_{r,d}$. Thus

$$[\,\mathbb{Q}(\sqrt{d}, \sqrt{r^2 - d}) : \mathbb{Q}] = [\mathbb{Q}_{r,d} : \mathbb{Q}] = 4.$$

This clearly implies that

$$\mathbb{Q}_{r,d} = \mathbb{Q}(\sqrt{d}, \sqrt{r^2 - d}),$$

hence $\mathbb{Q}_{r,d}$ has at least two quadratic subfields, and then, by Proposition 5.3.6 (a) we must have $\sqrt{r^2 - d} \in \mathbb{Q}$, which contradicts our assumption $\sqrt{r^2 - d} \in \mathbb{Q}_{r,d} \setminus \mathbb{Q}(\sqrt{d})$. This shows that Case (3) cannot occur, and proves the proposition. $\qquad\square$

COROLLARY 5.3.8. *The following assertions are equivalent for an* $n \in \mathbb{Z}$ *and a square-free integer* $d \in \mathbb{Z} \setminus \{1\}$.

(1) $\mathbb{Q}_{n,d}/\mathbb{Q}$ *is a Galois extension.*

(2) *There exists* $k \in \mathbb{Z}^*$ *such that either* $\sqrt{n^2 - d} = k$ *or* $\sqrt{n^2 - d} = k\sqrt{d}$.

(3) $n^2 - d \in \mathbb{N}^{*2}$, *or there exist* $x, y \in \mathbb{Z}^*$ *such that* $x^2 - dy^2 = -1$ *and* $n = dy$.

PROOF. By Corollary 5.3.5, the polynomial $f = X^4 - 2nX^2 + n^2 - d$ is irreducible. Then, by Proposition 5.3.7 (2), $\mathbb{Q}_{n,d}/\mathbb{Q}$ is a Galois extension if and only if $\sqrt{n^2 - d} \in \mathbb{Q}(\sqrt{d})$. There are two cases.

Case 1: $\sqrt{n^2 - d} \in \mathbb{Q}^*$. Then we have $\sqrt{n^2 - d} \in \mathbb{Z}^*$.

Case 2: $\sqrt{n^2 - d} \in \mathbb{Q}(\sqrt{d}) \setminus \mathbb{Q}$. Then $\sqrt{n^2 - d} = z + x\sqrt{d}$ for some $x, z \in \mathbb{Q}$. This implies that

$$n^2 - d = z^2 + x^2 d + 2xz\sqrt{d}.$$

Since $\{1, \sqrt{d}\}$ is linearly independent over $\mathbb{Q}$, we deduce that $xz = 0$. But $x \neq 0$ since $\sqrt{n^2 - d} \notin \mathbb{Q}$, hence necessarily $z = 0$, and then $\sqrt{n^2 - d} =$

$x\sqrt{d}$, or equivalently

$$d(n^2 - d) = (xd)^2.$$

We deduce that $c := xd \in \mathbb{Z}$ and $d \mid c^2$. Then $d \mid c$ since d is square-free, hence $x \in \mathbb{Z}$. Thus

$$n^2 - d = dx^2.$$

Then $d \mid n^2$, and consequently $d \mid n$ since d is a square-free integer. So $n = dy$ for some $y \in \mathbb{Z}^*$ verifying the Pell-Fermat equation

$$x^2 - dy^2 = -1.$$

Conversely, if $(x, y) \in \mathbb{Z}^* \times \mathbb{Z}^*$ is any solution of the Pell-Fermat equation above, and $n = dy$, then $\sqrt{n^2 - d} = \pm x\sqrt{d}$, and we are done. $\square$

We are now going to find conditions on the rational numbers r, d for which $\mathbb{Q}_{r,d}/\mathbb{Q}$ is a Cogalois extension.

LEMMA 5.3.9. *Let $d \in \mathbb{Q}\backslash\mathbb{Q}^2$. Then, $\mathbb{Q}(\sqrt{d})/\mathbb{Q}$ is a Cogalois extension if and only if $-d \notin \mathbb{Q}^2$ and $-3d \notin \mathbb{Q}^2$.*

PROOF. For the given $d \in \mathbb{Q} \backslash \mathbb{Q}^2$ we can find a unique square-free rational integer δ such that $\mathbb{Q}(\sqrt{d}) = \mathbb{Q}(\sqrt{\delta})$. By Corollary 3.3.3, $\mathbb{Q}(\sqrt{\delta})/\mathbb{Q}$ is a Cogalois extension if and only if $\delta \neq -1, -3$. Since $d/\delta \in \mathbb{Q}^2$, the result follows at once. $\square$

LEMMA 5.3.10. *Let $d_1, d_2 \in \mathbb{Q}\backslash\mathbb{Q}^2$. Then $\mathbb{Q}(\sqrt{d_1}, \sqrt{d_2})/\mathbb{Q}$ is a Cogalois extension if and only if $-d_1, -d_2, -3d_1, -3d_2, -d_1d_2, -3d_1d_2 \notin \mathbb{Q}^2$.*

PROOF. Set $E = \mathbb{Q}(\sqrt{d_1}, \sqrt{d_2})/\mathbb{Q}$. If $E/\mathbb{Q}$ is a Cogalois extension, then its subextensions $\mathbb{Q}(\sqrt{d_1})/\mathbb{Q}$, $\mathbb{Q}(\sqrt{d_2})/\mathbb{Q}$, and $\mathbb{Q}(\sqrt{d_1d_2})/\mathbb{Q}$ are also Cogalois by Proposition 3.2.2 (2), and so,

$$-d_1, -d_2, -3d_1, -3d_2, -d_1d_2, -3d_1d_2 \notin \mathbb{Q}^2$$

by Lemma 5.3.9.

Conversely, assume that $-d_1, -d_2, -3d_1, -3d_2, -d_1d_2, -3d_1d_2 \notin \mathbb{Q}^2$. Then $\mathbb{Q}(\sqrt{d_1})/\mathbb{Q}$, $\mathbb{Q}(\sqrt{d_2})/\mathbb{Q}$, and $\mathbb{Q}(\sqrt{d_1d_2})/\mathbb{Q}$ are all Cogalois extensions by Lemma 5.3.9. If $\mathbb{Q}(\sqrt{d_1}) = \mathbb{Q}(\sqrt{d_2})$, then $E = \mathbb{Q}(\sqrt{d_1}, \sqrt{d_2}) = \mathbb{Q}(\sqrt{d_1})$, and we are done.

So, we may assume that $[E : \mathbb{Q}] = 4$, and then, $\mathbb{Q}(\sqrt{d_1}, \sqrt{d_2})/\mathbb{Q}$ is an Abelian extension with Galois group isomorphic to $\mathbb{Z}_2 \times \mathbb{Z}_2$, hence it has precisely three proper subextensions, namely $\mathbb{Q}(\sqrt{d_1})/\mathbb{Q}$, $\mathbb{Q}(\sqrt{d_2})/\mathbb{Q}$, and $\mathbb{Q}(\sqrt{d_1d_2})/\mathbb{Q}$.

In view of the Greither-Harrison Criterion (Theorem 3.1.7), we have to prove that $\mathbb{Q}(\sqrt{d_1}, \sqrt{d_2})/\mathbb{Q}$ is a pure extension. So, let $p \in \mathcal{P}$ be such that $\zeta_p \in E$. Then $\mathbb{Q}(\zeta_p) \subseteq E$, hence $p - 1 = [\mathbb{Q}(\zeta_p) : \mathbb{Q}] \leqslant 4$, and so, $p \in \{3, 4, 5\}$.

We cannot have $p = 5$, for otherwise, it would follow that $E = \mathbb{Q}(\zeta_5)$, hence $\mathbb{Z}_4 \cong \mathrm{Gal}(\mathbb{Q}(\zeta_5)/\mathbb{Q}) = \mathrm{Gal}(E/\mathbb{Q}) \cong \mathbb{Z}_2 \times \mathbb{Z}_2$, which is a contradiction.

If $p = 4$, then $\mathbb{Q}(\zeta_4) = \mathbb{Q}(i)$ has to be one of the three quadratic subfields $\mathbb{Q}(\sqrt{d_1})$, $\mathbb{Q}(\sqrt{d_2})$, and $\mathbb{Q}(\sqrt{d_1 d_2})$ of E. For instance, if we would have $\mathbb{Q}(i) = \mathbb{Q}(\sqrt{d_1})$, this would imply that $i = a\sqrt{d_1}$ for some $a \in \mathbb{Q}^*$, i.e., $-d_1 \in \mathbb{Q}^2$, which contradicts our assumption.

Similarly, if $p = 3$, then $\mathbb{Q}(\zeta_3) = \mathbb{Q}(\sqrt{-3})$ has to be one of the three quadratic subfields $\mathbb{Q}(\sqrt{d_1})$, $\mathbb{Q}(\sqrt{d_2})$, and $\mathbb{Q}(\sqrt{d_1 d_2})$ of E, and this cannot happen either.

Consequently $\zeta_p \notin E$ for every $p \in \mathcal{P}$. This shows that $E/\mathbb{Q}$ is a pure extension, hence a Cogalois extension.   □

THEOREM 5.3.11. *The following statements hold for $r \in \mathbb{Q}$ and $d \in \mathbb{Q} \setminus \mathbb{Q}^2$.*

(1) $\mathbb{Q}_{r,d}/\mathbb{Q}$ *is a quadratic Cogalois extension if and only if* $\sqrt{r^2 - d} \in \mathbb{Q}_+^*$ *and, either* $2(r - \sqrt{r^2 - d}) \in \mathbb{Q}^2$, *or* $2(r + \sqrt{r^2 - d}) \in \mathbb{Q}^2$ *and* $2(\sqrt{r^2 - d} - r), 6(\sqrt{r^2 - d} - r) \notin \mathbb{Q}^2$.

(2) *If* $[\mathbb{Q}_{r,d} : \mathbb{Q}] = 4$ *and* $\sqrt{r^2 - d} \in \mathbb{Q}$, *then* $\mathbb{Q}_{r,d}/\mathbb{Q}$ *is a Cogalois extension if and only if* $-d, -3d, 2(-r \pm \sqrt{r^2 - d}), 6(-r \pm \sqrt{r^2 - d}) \notin \mathbb{Q}^2$.

(3) *If* $[\mathbb{Q}_{r,d} : \mathbb{Q}] = 4$, $\sqrt{r^2 - d} \in \mathbb{Q}(\sqrt{d}) \setminus \mathbb{Q}$, *and* $-d \notin \mathbb{Q}^2$, *then* $\mathbb{Q}_{r,d}/\mathbb{Q}$ *is not a Cogalois extension.*

(4) *If* $[\mathbb{Q}_{r,d} : \mathbb{Q}] = 4$, $\sqrt{r^2 - d} \notin \mathbb{Q}(\sqrt{d})$, *and either* $\sqrt{d - r^2} \notin \mathbb{Q}(\sqrt{d})$ *or* $d^2 - dr^2 \notin \mathbb{Q}^2$, *then* $\mathbb{Q}_{r,d}/\mathbb{Q}$ *is not a Cogalois extension.*

PROOF. (1) By Proposition 5.3.3 and Remarks 5.3.4 (2), $[\mathbb{Q}_{r,d} : \mathbb{Q}] = 2$ if and only if there exist $c, k \in \mathbb{Q}_+^*$ such that $r^2 - d = c^2$ and $r \pm c = k^2/2$. By Remarks 5.3.4 (2) and the proof of the implication $(2) \Longrightarrow (3)$ in Proposition 5.3.3, we have

$$\mathbb{Q}_{r,d} = \begin{cases} \mathbb{Q}(\sqrt{(r + c)/2}) & \text{if } r - c = k^2/2, \\ \mathbb{Q}(\sqrt{(r - c)/2}) & \text{if } r + c = k^2/2. \end{cases}$$

If $r - c = k^2/2$, then $(r + c)/2 = (k^2 + 4c)/4 > 0$, hence the extension $\mathbb{Q}_{r,d}/\mathbb{Q}$ is Cogalois in view of Lemma 5.3.9.

Now, if $r + c = k^2/2$, then $(r - c)/2 = (k^2 - 4c)/4$, hence the extension $\mathbb{Q}_{r,d}/\mathbb{Q}$ is Cogalois if and only if $4c - k^2 \notin \mathbb{Q}^2$ and $3(4c - k^2) \notin \mathbb{Q}^2$, again by Lemma 5.3.9. Since $4c - k^2 = 2(c - r) = 2(\sqrt{r^2 - d} - r)$, we deduce that if $\mathbb{Q}_{r,d}/\mathbb{Q}$ is a Cogalois extension, then $2(\sqrt{r^2 - d} - r)$, $6(\sqrt{r^2 - d} - r) \notin \mathbb{Q}^2$.

The conditions in the statement of (1) are also sufficient to ensure that $\mathbb{Q}_{r,d}/\mathbb{Q}$ is a quadratic Cogalois extension.

(2) and (3): Assume that $[\mathbb{Q}_{r,d} : \mathbb{Q}] = 4$ and $\sqrt{r^2 - d} \in \mathbb{Q}(\sqrt{d})$. Then $\mathbb{Q}_{r,d}/\mathbb{Q}$ is a Galois extension by Proposition 5.3.7 (2). Two cases arise:

Case (i): $\sqrt{r^2 - d} \in \mathbb{Q}$. Then, as in the proof of Proposition 5.3.6 (a) we deduce that

$$\mathbb{Q}_{r,d} = \mathbb{Q}(\sqrt{(r + c)/2}, \sqrt{(r - c)/2}),$$

where $c = \sqrt{r^2 - d}$. Now apply Lemma 5.3.10 to obtain the desired necessary and sufficient conditions for $\mathbb{Q}_{r,d}/\mathbb{Q}$ being Cogalois.

Case (ii): $\sqrt{r^2 - d} \in \mathbb{Q}(\sqrt{d}) \setminus \mathbb{Q}$. Then $\mathrm{Gal}(\mathbb{Q}_{r,d}/\mathbb{Q}) \cong \mathbb{Z}_4$ by Proposition 5.3.7 (2). Assume that $\mathbb{Q}_{r,d}/\mathbb{Q}$ were a Cogalois extension. Then, by Corollary 5.2.3, we would have

$$\mathrm{Cog}(\mathbb{Q}_{r,d}/\mathbb{Q}) \cong \mathrm{Gal}(\mathbb{Q}_{r,d}/\mathbb{Q}) \cong \mathbb{Z}_4.$$

Then clearly $\exp(\mathrm{Cog}(\mathbb{Q}_{r,d}/\mathbb{Q})) = 4$, so $i = \zeta_4 \in \mathbb{Q}_{r,d}$ by Corollary 5.1.7. Thus, $\mathbb{Q}(i)$ and $\mathbb{Q}(\sqrt{d})$ are quadratic subfields of $\mathbb{Q}_{r,d}$, and by Galois Theory, it follows that they are equal, so, in particular, we deduce that $\sqrt{d} = bi$ for some $b \in \mathbb{Q}^*$, i.e., $-d \in \mathbb{Q}^2$, which contradicts our assumption. This proves that the extension $\mathbb{Q}_{r,d}/\mathbb{Q}$ is not Cogalois.

(4) In case $[\mathbb{Q}_{r,d} : \mathbb{Q}] = 4$ and $\sqrt{r^2 - d} \notin \mathbb{Q}(\sqrt{d})$, then $\mathbb{Q}_{r,d}/\mathbb{Q}$ is not a Galois extension by Proposition 5.3.7 (2). We shall adapt the idea of the proof of Proposition 3.2.6 (e) to this case.

Assume that $\mathbb{Q}_{r,d}/\mathbb{Q}$ is a Cogalois extension. Since $[\mathbb{Q}_{r,d} : \mathbb{Q}] = 4$, then, by the definition of the concept of Cogalois extension, $\mathrm{Cog}(\mathbb{Q}_{r,d}/\mathbb{Q})$ is a group of order 4, hence it is isomorphic either to $\mathbb{Z}_2 \times \mathbb{Z}_2$ or to $\mathbb{Z}_4$.

If $\mathrm{Cog}(\mathbb{Q}_{r,d}/\mathbb{Q}) \simeq \mathbb{Z}_2 \times \mathbb{Z}_2$, then there exist $\beta, \gamma \in \mathbb{Q}^*$ such that

$$\mathrm{Cog}(\mathbb{Q}_{r,d}/\mathbb{Q}) = \mathbb{Q}^* \langle \sqrt{\beta}, \sqrt{\gamma} \rangle / \mathbb{Q}^* = \{\widehat{1}, \widehat{\sqrt{\beta}}, \widehat{\sqrt{\gamma}}, \widehat{\sqrt{\beta\gamma}}\}.$$

Then $\{1, \sqrt{\beta}, \sqrt{\gamma}, \sqrt{\beta\gamma}\}$ is a vector space basis of $\mathbb{Q}_{r,d}$ over $\mathbb{Q}$ by Corollary 2.1.10. In particular, it follows that

$$\mathbb{Q}_{r,d} = \mathbb{Q}(\sqrt{\beta}, \sqrt{\gamma}).$$

Thus, the extension $\mathbb{Q}_{r,d}/\mathbb{Q}$ would be a Galois extension, which contradicts our assumption.

Consequently, we must have $\mathrm{Cog}\,(\mathbb{Q}_{r,d}/\mathbb{Q}) \simeq \mathbb{Z}_4$, hence there exists $\alpha \in \mathbb{Q}^*$ such that

$$\mathrm{Cog}\,(\mathbb{Q}_{r,d}/\mathbb{Q}) = \mathbb{Q}^*\langle\, \sqrt[4]{\alpha}\,\rangle/\mathbb{Q}^* = \{\, \widehat{1}, \widehat{\sqrt[4]{\alpha}}, \widehat{\sqrt[4]{\alpha}}^{\,2}, \widehat{\sqrt[4]{\alpha}}^{\,3} \,\}.$$

Then, again by Corollary 2.1.10, we deduce that $\{\, 1, \sqrt[4]{\alpha}, \sqrt[4]{\alpha}^{\,2}, \sqrt[4]{\alpha}^{\,3} \,\}$ is a vector space basis of $\mathbb{Q}_{r,d}$ over $\mathbb{Q}$. In particular, it follows that

$$\mathbb{Q}_{r,d} = \mathbb{Q}(\sqrt[4]{\alpha}).$$

Since $[\mathbb{Q}_{r,d} : \mathbb{Q}] = 4$, we deduce that $\sqrt{\alpha} \notin \mathbb{Q}$.

Clearly, $\mathbb{Q}(\sqrt{\alpha})$ is a proper subfield of $\mathbb{Q}(\sqrt[4]{\alpha}) = \mathbb{Q}_{r,d}$. Note that the cyclic group $\mathrm{Cog}\,(\mathbb{Q}_{r,d}/\mathbb{Q})$ of order 4 has a unique proper subgroup, hence, by Theorem 3.2.3, it follows that $\mathbb{Q}_{r,d}/\mathbb{Q}$ has a unique proper intermediate subfield. Since $\mathbb{Q}(\sqrt{d})$ is also a proper subfield of $\mathbb{Q}_{r,d}$, this implies necessarily that $\mathbb{Q}(\sqrt{\alpha}) = \mathbb{Q}(\sqrt{d})$, hence $\sqrt{d} - k\sqrt{\alpha}$ for some $k \in \mathbb{Q}^*$.

Since $\mathbb{Q}_{r,d} = \mathbb{Q}(\sqrt[4]{\alpha})$ and $\sqrt[4]{\alpha}^{\,2} = \pm\sqrt{\alpha}$, there exist $a, b, e, f \in \mathbb{Q}$ such that

(12) $$\theta := \sqrt{r + \sqrt{d}} = a + b\sqrt[4]{\alpha} + e\sqrt{\alpha} + f\sqrt{\alpha}\,\sqrt[4]{\alpha}.$$

We can also write (12) as

(13) $$\theta - (a + e\sqrt{\alpha}) = \sqrt[4]{\alpha}\,(b + f\sqrt{\alpha}).$$

Squaring (13), we obtain

$$\theta^2 - 2\theta(a + e\sqrt{\alpha}) + (a + e\sqrt{\alpha})^2 = \pm\sqrt{\alpha}\,(b + f\sqrt{\alpha})^2.$$

Since $\theta^2 \in \mathbb{Q}(\sqrt{d}) = \mathbb{Q}(\sqrt{\alpha})$, we deduce that we must have $a + e\sqrt{\alpha} = 0$, for otherwise, it would follow that $\theta \in \mathbb{Q}(\sqrt{\alpha}) = \mathbb{Q}(\sqrt{d})$, which is impossible. Thus, (13) becomes

(14) $$\theta = \sqrt[4]{\alpha}\,(b + f\sqrt{\alpha}),$$

which can be also written as

$$\frac{b + f\sqrt{\alpha}}{\theta} = \frac{1}{\sqrt[4]{\alpha}},$$

or

(15) $$\frac{bk + f\sqrt{d}}{\theta} = \frac{k}{\sqrt[4]{\alpha}},$$

since $\sqrt{\alpha} = \sqrt{d}/k$.

From (15) we deduce that

$$\left(\frac{bk + f\sqrt{d}}{\theta}\right)^4 \in \mathbb{Q},$$

i.e.,

$$u := \frac{\left(bk + f\sqrt{d}\right)^4}{(r + \sqrt{d})^2} \in \mathbb{Q}.$$

Then, u coincides with its conjugate in the quadratic extension $\mathbb{Q}(\sqrt{d})/\mathbb{Q}$, i.e.,

$$\frac{\left(bk + f\sqrt{d}\right)^4}{(r + \sqrt{d})^2} = \frac{\left(bk - f\sqrt{d}\right)^4}{(r - \sqrt{d})^2}.$$

This implies that

$$\frac{\left(bk + f\sqrt{d}\right)^2}{r + \sqrt{d}} = \pm\frac{\left(bk - f\sqrt{d}\right)^2}{r - \sqrt{d}},$$

hence

(16) $$\left(bk + f\sqrt{d}\right)^2(r - \sqrt{d}) = \pm\left(bk - f\sqrt{d}\right)^2(r + \sqrt{d}).$$

After easy calculations we obtain

(17) $$b^2k^2 - 2brfk + f^2d = 0,$$

if the sign in the right-hand side of (16) is " + ", or

(18) $$b^2rk^2 - 2bdfk + f^2dr = 0$$

if the sign in the right-hand side of (16) is " − ".

Note that (14) implies that $b + f\sqrt{\alpha} \neq 0$, hence (16) can be also written as

$$\left(\frac{bk - f\sqrt{d}}{bk + f\sqrt{d}}\right)^2 = \pm\frac{r - \sqrt{d}}{r + \sqrt{d}} = \pm\frac{r^2 - d}{(r + \sqrt{d})^2}.$$

Consequently, in case $\mathbb{Q}_{r,d}/\mathbb{Q}$ were a Cogalois extension, then necessarily we would have $\pm(r^2 - d) \in \mathbb{Q}(\sqrt{d})^2$, i.e., $\sqrt{r^2 - d} \in \mathbb{Q}(\sqrt{d})$ or $\sqrt{d - r^2} \in \mathbb{Q}(\sqrt{d})$. Since $\sqrt{r^2 - d} \notin \mathbb{Q}(\sqrt{d})$ by hypothesis, we deduce that the extension $\mathbb{Q}_{r,d}/\mathbb{Q}$ is not Cogalois whenever $\sqrt{d - r^2} \notin \mathbb{Q}(\sqrt{d})$.

Now observe that the coefficient b^2 in equation (17) is nonzero, for otherwise (17) would imply $f = 0$, and then (14) would become $\theta = 0$, which is impossible. So, the roots of (17) are

$$k_{1,2} = \frac{bfr \pm bf\sqrt{r^2 - d}}{b^2} \in \mathbb{Q},$$

hence we must have $\sqrt{r^2 - d} \in \mathbb{Q}$, which contradicts our assumption.

The roots of the quadratic equation (18) are

$$k_{3,4} = \frac{df \pm f\sqrt{d^2 - r^2 d}}{br} \in \mathbb{Q},$$

hence we must have $\sqrt{d^2 - r^2 d} \in \mathbb{Q}$, which also contradicts our assumption.

This completes the proof of the fact that $\mathbb{Q}_{r,d}/\mathbb{Q}$ is not a Cogalois extension. $\square$

COROLLARY 5.3.12. *Let $d \in \mathbb{N}$, $d \geqslant 2$ be a square-free integer, and let $n \in \mathbb{Z}^*$ be such that $\sqrt{n^2 - d} \notin \mathbb{Q}(\sqrt{d})$. Then $\mathbb{Q}_{n,d}/\mathbb{Q}$ is not a Cogalois extension.*

PROOF. By Corollary 5.3.5, $[\mathbb{Q}_{n,d} : \mathbb{Q}] = 4$. According to Theorem 5.3.11 (4) it is sufficient to prove that $d^2 - dn^2 \notin \mathbb{Q}^2$, or equivalently, that $d^2 - dn^2 \notin \mathbb{N}^2$.

First, observe that $d^2 - dn^2 \neq 0$. Now, assume that $d^2 - dn^2 \in \mathbb{N}^{*2}$. Then, for any prime divisor p of d we must have $p \mid d - n^2$ since $d^2 - dn^2 = d(d - n^2) \in \mathbb{N}^{*2}$ and d is square-free, hence $p \mid n^2$. It follows that $p \mid n$, which implies that $d \mid n$ since d is square-free. Thus $n = ds$ for some $s \in \mathbb{Z}^*$, hence

$$\sqrt{d^2 - dn^2} = d\sqrt{1 - ds^2} \in \mathbb{N},$$

which implies that $\sqrt{1 - ds^2} \in \mathbb{N}$, i.e., $1 - ds^2 = u^2$ for some $u \in \mathbb{N}^*$. But $d \geqslant 2$ by hypothesis, hence $1 = u^2 + ds^2 > 2$, which is a contradiction. $\square$

REMARK 5.3.13. The next example shows that we may have $d^2 - r^2 d \in \mathbb{Q}^2$ for $d \in \mathbb{Q} \setminus \mathbb{Q}^2$ and $r \in \mathbb{Q}^*$. If $d \in \mathbb{Q} \setminus \{-1\}$ is arbitrary, take $r := 2d/(d+1)$. Then $d^2 - r^2 d = d^2(d-1)^2/(d+1)^2 \in \mathbb{Q}^2$. If $d = -1$, take $r := 3/4$. Then $d^2 - r^2 d = (5/4)^2$. $\square$

Next, we investigate when is $\mathbb{Q}_{r,d}/\mathbb{Q}$ a radical extension. We will mainly discuss those $\mathbb{Q}_{r,d}$ which are subfields of $\mathbb{R}$. Since any extension E/F with E a subfield of $\mathbb{R}$ is clearly pure and separable, by the Greither-Harrison Criterion it follows that E/F is radical if and only if it is Cogalois.

Thus, the radical extensions of type $\mathbb{Q}_{r,d}/\mathbb{Q}$, with $r + \sqrt{d} > 0$, are precisely the Cogalois ones.

PROPOSITION 5.3.14. *The following statements hold for* $r \in \mathbb{Q}$ *and* $d \in \mathbb{Q} \setminus \mathbb{Q}^2$.

(1) *If* $\sqrt{r^2 - d} \in \mathbb{Q}$ *then* $\mathbb{Q}_{r,d}/\mathbb{Q}$ *is a radical Galois extension.*

(2) *If* $\sqrt{r^2 - d} \in \mathbb{Q}(\sqrt{d}) \setminus \mathbb{Q}$, $d > 0$, *and* $r + \sqrt{d} > 0$, *then* $\mathbb{Q}_{r,d}/\mathbb{Q}$ *is a nonradical Galois extension.*

(3) *If* $d \in \mathbb{N}$, $d \geqslant 2$ *is square-free*, $r \in \mathbb{Z}^*$, $r + \sqrt{d} > 0$, *and* $\sqrt{r^2 - d} \notin \mathbb{Q}(\sqrt{d})$, *then the extension* $\mathbb{Q}_{r,d}/\mathbb{Q}$ *is neither Galois nor radical.*

PROOF. If $[\mathbb{Q}_{r,d} : \mathbb{Q}] = 2$, then $\mathbb{Q}_{r,d}$ is a quadratic field, hence $\mathbb{Q}_{r,d}/\mathbb{Q}$ is obviously a radical Galois extension.

Next, we assume that $[\mathbb{Q}_{r,d} : \mathbb{Q}] \neq 2$. Then $[\mathbb{Q}_{r,d} : \mathbb{Q}] = 4$ by Remark 5.3.4 (2). Two cases arise:

Case 1: $\mathbb{Q}_{r,d}/\mathbb{Q}$ is a Galois extension. Then, by Proposition 5.3.7 (2) this is equivalent to $\sqrt{r^2 - d} \in \mathbb{Q}(\sqrt{d})$. There are two subcases:

Subcase (i): $\sqrt{r^2 - d} \in \mathbb{Q}$. Then, as in the proof of Proposition 5.3.6 (a) we deduce that

$$\mathbb{Q}_{r,d} = \mathbb{Q}(\sqrt{(r + c)/2}, \sqrt{(r - c)/2}),$$

where $c = \sqrt{r^2 - d}$, so $\mathbb{Q}_{r,d}/\mathbb{Q}$ is a radical extension.

Subcase (ii): $\sqrt{r^2 - d} \in \mathbb{Q}(\sqrt{d}) \setminus \mathbb{Q}$. Then, the extension $\mathbb{Q}_{r,d}/\mathbb{Q}$ is not Cogalois extension by Theorem 5.3.11 (3), so it is also not radical.

Case 2: $\mathbb{Q}_{r,d}/\mathbb{Q}$ is not a Galois extension. By Proposition 5.3.7 (2), this is equivalent to $\sqrt{r^2 - d} \notin \mathbb{Q}(\sqrt{d})$. By Corollary 5.3.12, the extension $\mathbb{Q}_{r,d}/\mathbb{Q}$ is not Cogalois, hence it is not radical too. $\square$

We end this section by showing that the abstract property of the extension $\mathbb{Q}(\sqrt{2 + \sqrt{2}})/\mathbb{Q}$ being not Cogalois, proved in Proposition 5.3.1 (e), can be equivalently expressed more attractively and elementarily as the impossibility to write $\sqrt{2 + \sqrt{2}}$ as a finite sum of real numbers of type $\pm \sqrt[n_i]{a_i}$, where $r, n_1, \ldots, n_r, a_1, \ldots, a_r \in \mathbb{N}^*$. The same problem will be discussed for any algebraic number $\alpha \in \mathbb{R}_+^*$.

PROPOSITION 5.3.15. *The following statements are equivalent for a real algebraic number field* K.

(1) *There exist* $r \in \mathbb{N}^*$ *and* $n_1, \ldots, n_r, a_1, \ldots, a_r \in \mathbb{N}^*$ *such that* $K = \mathbb{Q}(\sqrt[n_1]{a_1}, \ldots, \sqrt[n_r]{a_r})$.
(2) *The extension* $K/\mathbb{Q}$ *is radical.*
(3) *The extension* $K/\mathbb{Q}$ *is Kneser.*
(4) *The extension* $K/\mathbb{Q}$ *is Cogalois.*

PROOF. (1) $\implies$ (4) and (2) $\implies$ (4): The extension $K/\mathbb{Q}$ is clearly separable, radical by hypothesis, and pure since $K \subseteq \mathbb{R}$, hence it is Cogalois by the Greither-Harrison Criterion.

(4) $\implies$ (3) $\implies$ (2) are obvious.

(4) $\implies$ (1): Let $\{x_1, \ldots, x_r\}$ be a set of representatives of the finite group $\mathrm{Cog}(K/\mathbb{Q}) = T(K/\mathbb{Q})/\mathbb{Q}^*$. Since $x_i \equiv -x_i \pmod{\mathbb{Q}^*}$, we may assume that $x_i > 0$ for all i, $1 \leqslant i \leqslant r$. Then $K = \mathbb{Q}(x_1, \ldots, x_r)$, and for every i, $1 \leqslant i \leqslant r$, there exists $n_i \in \mathbb{N}^*$ such that $x_i^{n_i} = a_i \in \mathbb{Q}$. Clearly, $a_i > 0$ for all i. Then

$$K = \mathbb{Q}(\sqrt[n_1]{a_1}, \ldots, \sqrt[n_r]{a_r}).$$

Of course, we may also assume that all $a_i \in \mathbb{N}^*$. □

COROLLARY 5.3.16. *The following assertions are equivalent for an algebraic number* $\alpha \in \mathbb{R}_+^*$.

(1) α *can be written as a finite sum of real numbers of type* $\pm \sqrt[n_i]{a_i}$, $1 \leqslant i \leqslant r$, *where* $r, n_1, \ldots, n_r, a_1, \ldots, a_r \in \mathbb{N}^*$.
(2) *The extension* $\mathbb{Q}(\alpha)/\mathbb{Q}$ *is radical.*
(3) *The extension* $\mathbb{Q}(\alpha)/\mathbb{Q}$ *is Kneser.*
(4) *The extension* $\mathbb{Q}(\alpha)/\mathbb{Q}$ *is Cogalois.*

PROOF. (1) $\implies$ (4): If α is a number as in (1), then $\mathbb{Q}(\alpha)$ is a subfield of the field $\mathbb{Q}(\sqrt[n_1]{a_1}, \ldots, \sqrt[n_r]{a_r}) \subseteq \mathbb{R}$. By Proposition 3.2.2 (2), the extension $\mathbb{Q}(\sqrt[n_1]{a_1}, \ldots, \sqrt[n_r]{a_r})/\mathbb{Q}$ is Cogalois, hence, so also is its subextension $\mathbb{Q}(\alpha)/\mathbb{Q}$.

(4) $\iff$ (3) $\iff$ (2) follow from Proposition 5.3.15.

(4) $\implies$ (1): Apply Proposition 5.3.15 to the real algebraic number field $K = \mathbb{Q}(\alpha)$ to deduce that

$$\mathbb{Q}(\alpha) = \mathbb{Q}(\sqrt[m_1]{b_1}, \ldots, \sqrt[m_s]{b_s}),$$

for some $s, m_1, \ldots, m_s, b_1, \ldots, b_s \in \mathbb{N}^*$. Then α is a sum with coefficients in $\mathbb{Q}^*$ of products of powers of $\sqrt[m_i]{b_i}$, hence it has the desired form. □

COROLLARY 5.3.17. *The following assertions hold.*

(1) *Each of the numbers* $\sqrt{1 + \sqrt{2}}$, $\sqrt{2 + \sqrt{2 + \sqrt{2 + \ldots + \sqrt{2}}}}$ *cannot be written as a finite sum of real numbers of type* $\pm \sqrt[n_i]{a_i}$, $1 \leqslant i \leqslant r$, *where* $r, n_1, \ldots, n_r, a_1, \ldots, a_r \in \mathbb{N}^*$.

(2) *Let* $d \geqslant 2$ *be a square-free integer, and let* $r \in \mathbb{Z}^*$ *be such that* $r > -\sqrt{d}$ *and* $\sqrt{r^2 - d} \notin \mathbb{Q}(\sqrt{d})$. *Then* $\sqrt{r + \sqrt{d}}$ *cannot be written as a finite sum of real numbers of type* $\pm \sqrt[n_i]{a_i}$, $1 \leqslant i \leqslant r$, *with* $r, n_1, \ldots, n_r, a_1, \ldots, a_r \in \mathbb{N}^*$.

PROOF. Apply Corollary 5.3.16, Proposition 3.2.6 (e), Remark 5.3.2, and Proposition 5.3.14 (3).   □

5.4. Exercises to Chapter 5

1. Prove that the following statements hold for a Galois G-Cogalois extension E/F, with $H_1/F^*, \ldots, H_r/F^*$ the Sylow subgroups of G/F^*
    (a) $F(H_i)/F$ is a Galois extension for every i, $1 \leqslant i \leqslant r$.
    (b) Gal (E/F) is a nilpotent group.

2. Let E/F be a G-Cogalois extension, and let E_1/F, E_2/F be subextensions of E/F. Establish the possible implications between the statements below.
    (a) $E_1 \cap E_2 = F$.
    (b) The fields E_1 and E_2 are linearly disjoint over F.
    (c) $\mathrm{Kne}(E_1 E_2/F) \cong \mathrm{Kne}(E_1/F) \times \mathrm{Kne}(E_2/F)$.

3. Calculate $\left[\mathbb{Q}\left(\sqrt{\frac{13}{4} + \sqrt{3}} \right) : \mathbb{Q} \right]$.

4. Calculate $\left[\mathbb{Q}\left(\sqrt{\frac{r^5 + 4}{4r^2} + \sqrt{r}} \right) : \mathbb{Q} \right]$, where $r \in \mathbb{Q}_+^*$.

5. Let $n \in \mathbb{Z}$, and let ξ be a not specified complex root ξ of the polynomial $X^4 - nX - 1$. Prove that the field $\mathbb{Q}(\xi)$ is a quartic field having only two subfields if and only if $n \in \mathbb{Z} \setminus \{ -4, 0, 4 \}$.

6. Let F be any field, let Ω be an algebraically closed overfield of F, and let $f \in F[X]$ be a quartic polynomial having distinct roots x_1, x_2, x_3, $x_4 \in \Omega$. Denote

$$y_1 = x_1 x_2 + x_3 x_4, \ y_2 = x_1 x_3 + x_2 x_4, \ y_3 = x_1 x_4 + x_2 x_3,$$

and call the polynomial $r = (X - y_1)(X - y_2)(X - y_3)$ the *cubic resolvent* of f. Prove the following statements.

(a) If $f = X^4 + aX^3 + bX^2 + cX + d$, then

$$r = X^3 - bX^2 + (ac - 4d)X - (a^2d - 4bd + c^2).$$

(b) If $f = X^4 + mX + n$, then $r = X^3 - 4nX - m^2$.

(c) If $f \in K[X]$ is an irreducible polynomial, and ξ is any of the four roots $x_1,\ x_2,\ x_3,\ x_4 \in \Omega$ of f, then the extension $F(\xi)/F$ has no proper subextension if and only if the cubic resolvent of f is irreducible in $F[X]$.

7. (*Kappe and Warren* [**75**]). Let F be any field, let $f = X^4 + aX^3 + bX^2 + cX + d \in F[X]$ be a separable irreducible quartic polynomial, and let r be the cubic resolvent of f. Let E be a splitting field of f, let K be a splitting field of r, and let D be the discriminant of f. Denote by $\mathfrak{A}_4$ the alternating group of degree 4 and by $\mathfrak{D}_4$ the dihedral group of order 8. Prove that the following statements hold.

(a) $\mathrm{Gal}\,(E/F) \cong \mathfrak{S}_4 \iff r$ is irreducible in $F[X]$ and $D \notin F^2$.

(b) $\mathrm{Gal}\,(E/F) \cong \mathfrak{A}_4 \iff r$ is irreducible in $F[X]$ and $D \in F^2$.

(c) $\mathrm{Gal}\,(E/F) \cong \mathbb{Z}_2 \times \mathbb{Z}_2 \iff r$ splits into linear factors over F.

(d) $\mathrm{Gal}\,(E/F) \cong \mathbb{Z}_4 \iff r$ has exactly one root $y \in F$, and the polynomial $(X^2 - yX + d)(X^2 + aX + (b - y))$ splits over K.

(e) $\mathrm{Gal}\,(E/F) \cong \mathfrak{D}_4 \iff r$ has exactly one root $y \in F$ and the polynomial $(X^2 - yX + d)(X^2 + aX + (b - y))$ does not split over K.

8. Deduce the result of Exercise 5 from Exercises 6 and 7.

9. Let $d < 0$ be a square-free integer, and let $n \in \mathbb{Z}$. Show that $\mathbb{Q}_{n,d}/\mathbb{Q}$ is a Galois extension if and only if $n^2 - d \in \mathbb{N}^{*2}$.

10. Show that the following statements hold for a given square-free positive integer $d \geqslant 2$.

(a) If d is even, then there is no $n \in \mathbb{Z}$ such that $\mathbb{Q}_{n,d}/\mathbb{Q}$ is a Galois extension with Galois group isomorphic to $\mathbb{Z}_2 \times \mathbb{Z}_2$.

(b) If d is odd, then there exist only finitely many $n \in \mathbb{Z}$ such that $\mathbb{Q}_{n,d}/\mathbb{Q}$ is a Galois extension with Galois group isomorphic to $\mathbb{Z}_2 \times \mathbb{Z}_2$.

(c) There exists at least an $n \in \mathbb{Z}$ such that $\mathbb{Q}_{n,d}/\mathbb{Q}$ is a quartic cyclic extension if and only if the Pell-Fermat equation $x^2 - dy^2 = -1$ has a solution $(x, y) \in \mathbb{Z}^* \times \mathbb{Z}^*$. In this case $n = dy$, and in fact there exist infinitely many such n.

11. Let $r \in \mathbb{Q}$ and $d \in \mathbb{Q} \setminus \mathbb{Q}^2$. Prove the following statements.
    (a) If $c := \sqrt{r^2 - d} \in \mathbb{Q}_+^*$ and $2(r - c) \in \mathbb{Q}^2$, then

    $$\mathrm{Cog}(\mathbb{Q}_{r,d}/\mathbb{Q}) = \{\widehat{1}, \widehat{\sqrt{2(r + c)}}\} \cong \mathbb{Z}_2.$$

    (b) If $c := \sqrt{r^2 - d} \in \mathbb{Q}_+^*$ and $2(r + c) \in \mathbb{Q}^2$, then

    $$\mathrm{Cog}(\mathbb{Q}_{r,d}/\mathbb{Q}) = \begin{cases} \langle \widehat{1 + i} \rangle \cong \mathbb{Z}_4 & \text{if } 2(c - r) \in \mathbb{Q}^2 \\ \langle \widehat{i\sqrt{3} \cdot (1 + i\sqrt{3})} \rangle \cong \mathbb{Z}_6 & \text{if } 6(c - r) \in \mathbb{Q}^2 \\ \{\widehat{1}, \widehat{\sqrt{2(r - c)}}\} \cong \mathbb{Z}_2 & \text{otherwise.} \end{cases}$$

    (c) If $c := \sqrt{r^2 - d} \in \mathbb{Q}_+^*$, $2(r \pm c)$, $2(-r \pm c)$, $6(-r \pm c)$, $-d$, $-3d$ $\notin \mathbb{Q}^2$, then

    $$\mathrm{Cog}(\mathbb{Q}_{r,d}/\mathbb{Q}) = \{\widehat{1}, \widehat{\sqrt{2(r + c)}}, \widehat{\sqrt{2(r - c)}}, \widehat{\sqrt{d}}\} \cong \mathbb{Z}_2 \times \mathbb{Z}_2.$$

12. Let $r \in \mathbb{Q}$ and $d \in \mathbb{Q} \setminus \mathbb{Q}^2$. Prove the following statements.
    (a) If $\sqrt{r^2 - d} \in \mathbb{Q}(\sqrt{d}) \setminus \mathbb{Q}$, $[\mathbb{Q}_{r,d} : \mathbb{Q}] = 4$, and the extension $\mathbb{Q}_{r,d}/\mathbb{Q}$ is Cogalois, then

    $$\mathrm{Cog}(\mathbb{Q}_{r,d}/\mathbb{Q}) = \langle \widehat{1 + i} \rangle \cong \mathbb{Z}_4.$$

    (b) If $[\mathbb{Q}_{r,d} : \mathbb{Q}] = 4$, and the extension $\mathbb{Q}_{r,d}/\mathbb{Q}$ is not radical, then

    $$\mathrm{Cog}(\mathbb{Q}_{r,d}/\mathbb{Q}) = \{\widehat{1}, \widehat{\sqrt{d}}\} \cong \mathbb{Z}_2.$$

13. Let $r \in \mathbb{Q}$, $d \in \mathbb{Q} \setminus \mathbb{Q}^2$ be such that $\sqrt{r^2 - d} \in \mathbb{Q}$, $[\mathbb{Q}_{r,d} : \mathbb{Q}] = 4$, and such that the extension $\mathbb{Q}_{r,d}/\mathbb{Q}$ is not Cogalois. Calculate $\mathrm{Cog}(\mathbb{Q}_{r,d}/\mathbb{Q})$.

14. Calculate $\mathrm{Cog}(\mathbb{Q}(\sqrt{d_1}, \sqrt{d_2})/\mathbb{Q})$, where $d_1, d_2 \in \mathbb{Q} \setminus \mathbb{Q}^2$.

15. For which $d_1, d_2, d_3 \in \mathbb{Q} \setminus \mathbb{Q}^2$ is $\mathbb{Q}(\sqrt{d_1}, \sqrt{d_2} \sqrt{d_3})/\mathbb{Q}$ a Cogalois extension?

16. Calculate $\mathrm{Cog}(\mathbb{Q}(\sqrt{d_1}, \sqrt{d_2} \sqrt{d_3})/\mathbb{Q})$, where $d_1, d_2, d_3 \in \mathbb{Q} \setminus \mathbb{Q}^2$.

17. Investigate whether or not the conditions in Theorem 5.3.11 (3)-(4) are also necessary for $\mathbb{Q}_{r,d}/\mathbb{Q}$ being a non Cogalois extension.

18. Investigate whether or not the condition "$r + \sqrt{d} > 0$" can be removed in Proposition 5.3.14.

19. If $r \in \mathbb{N}^*$, then show that

$$2\cos(\pi/2^{r+1}) = \zeta_{2^{r+2}} + \zeta_{2^{r+2}}^{-1} = \underbrace{\sqrt{2 + \sqrt{2 + \sqrt{2 + \ldots + \sqrt{2}}}}}_{r \text{ radicals}}.$$

20. Let $r \in \mathbb{N}^*$, $r \geqslant 3$, and set $q = 2^r$. Prove that the following statements hold.

(a) $\mathbb{Q}(\zeta_q) = \mathbb{Q}(i, \eta)$, where $\eta = \zeta_q + \zeta_q^{-1} = 2\cos(\pi/2^{r-1})$.

(b) $\mathrm{Gal}(\mathbb{Q}(\zeta_q)/\mathbb{Q}) = G_1 \oplus G_2$, where G_1 is the cyclic subgroup of order 2^{r-2} of $\mathrm{Gal}(\mathbb{Q}(\zeta_q)/\mathbb{Q})$ generated by the automorphism $\sigma_1 \in \mathrm{Gal}(\mathbb{Q}(\zeta_q)/\mathbb{Q})$ defined by $\sigma_1(\zeta_q) = \zeta_q^5$, and G_2 is the cyclic subgroup of order 2 of $\mathrm{Gal}(\mathbb{Q}(\zeta_q)/\mathbb{Q})$ generated by the automorphism $\sigma_2 \in \mathrm{Gal}(\mathbb{Q}(\zeta_q)/\mathbb{Q})$ defined by $\sigma_2(\zeta_q) = \zeta_q^{-1}$.

(c) $\mathrm{Gal}(\mathbb{Q}(\cos(\pi/2^{r-1}))/\mathbb{Q}) = G_1$.

21. Let $n \in \mathbb{N}$. Prove that $\cos(\pi/2^n)$ can be written as a sum of real numbers of type $\pm \sqrt[n_i]{a_i}$, $1 \leqslant i \leqslant r$, where $r, n_1, \ldots, n_r, a_1, \ldots, a_r \in \mathbb{N}^*$, if and only if $n \in \{0, 1, 2\}$.

22. Prove that the following statements are equivalent for an $n \in \mathbb{N}$.

(a) $\mathbb{Q}(\cos(\pi/2^n))/\mathbb{Q}$ is a Cogalois extension.

(b) $\mathbb{Q}(\cos(\pi/2^n))/\mathbb{Q}$ is a Kneser extension.

(c) $\mathbb{Q}(\cos(\pi/2^n))/\mathbb{Q}$ is a radical extension.

(d) $n \in \{0, 1, 2\}$.

23. For which $n \in \mathbb{N}$ is $\mathbb{Q}(\sin(\pi/2^n))/\mathbb{Q}$ a Cogalois extension?

24. Show that $\mathrm{Cog}\,(\mathbb{Q}(\cos(\pi/2^n))/\mathbb{Q}) = \{\widehat{1}, \widehat{\sqrt{2}}\}$ for any $n \in \mathbb{N}$, $n \geqslant 2$.

5.5. Bibliographical comments to Chapter 5

Section 5.1. The results of this section are taken from Albu [8], Albu and Nicolae [19], and Albu and Ţena [25].

Section 5.2. Theorem 5.2.2, due to Albu and Nicolae [19], generalizes Corollary 5.2.3 due to Barrera-Mora, Rzedowski-Calderón, and Villa-Salvador [30].

Section 5.3. The results of this section are taken from Albu [**8**], [**7**], and Albu and Panaitopol [**24**].

CHAPTER 6

RADICAL EXTENSIONS AND CROSSED HOMOMORPHISMS

In this chapter we investigate finite Galois extensions which are radical, Kneser, or G-Cogalois, in terms of crossed homomorphisms. The results of this chapter are based on the description, provided in Section 6.1, of the Cogalois group $\mathrm{Cog}(E/F)$ of any finite Galois extension E/F by means of crossed homomorphisms of the Galois group $\mathrm{Gal}(E/F)$ with coefficients in the group $\mu(E)$ of all roots of unity in E. This description, which is actually a reformulation of the Hilbert's Theorem 90 in terms of Cogalois groups, states that there exists a canonical group isomorphism

$$\mathrm{Cog}(E/F) \cong Z^1(\mathrm{Gal}(E/F), \mu(E)).$$

A consequence of this result, of the uniqueness of the Kneser group of a G-Cogalois extension, as well as of the n-Purity Criterion, is the description of the Kneser group of any finite Galois G-Cogalois extension E/F with the aid of crossed homomorphisms of $\mathrm{Gal}(E/F)$ with coefficients in the group $\mu_n(E)$ of all n-th roots of unity in E, where n is the exponent of the Kneser group G/F^* of E/F.

Another nice application of the Cogalois-like reformulation of Hilbert's Theorem 90 is the finiteness of the Cogalois group of any finite extension of algebraic number fields.

In Section 6.2 we characterize via crossed homomorphisms finite Galois extensions which are radical, Kneser, or G-Cogalois, and provide sufficient conditions under which the property of a Galois extension E/F being radical, Kneser, or G-Cogalois is preserved when one changes the base field F.

6.1. Galois extensions and crossed homomorphisms

In this section we describe first the Cogalois group of a finite Galois extension E/F as the group of crossed homomorphisms of the Galois group $\mathrm{Gal}\,(E/F)$ with coefficients in the group $\mu(E)$ of all roots of unity contained in E. A consequence of this fact is the description of the Kneser group $\mathrm{Kne}\,(E/F)$ of any Galois G-Cogalois extension by means of crossed homomorphisms. As an application of this result, one deduces very easily that the Cogalois group of any finite extension of algebraic number fields is a finite group.

Recall first only those basic facts on *Galois Cohomology* which will be used in the sequel. Let E/F be an arbitrary extension with Galois group Γ, and let $M \leqslant E^*$ be such that $\sigma(M) \subseteq M$ for every $\sigma \in \Gamma$.

A *crossed homomorphism* (or an *1-cocycle*) of Γ with coefficients in M is a map $f : \Gamma \to M$ satisfying the condition

$$f(\sigma\tau) = f(\sigma) \cdot \sigma(f(\tau)),$$

for every $\sigma, \tau \in \Gamma$. The set of all crossed homomorphisms of Γ with coefficients in M is an Abelian group, which will be denoted by $Z^1(\Gamma, M)$.

For every $\alpha \in M$ we shall denote by f_α the *1-coboundary* $f_\alpha : \Gamma \to M$, defined as

$$f_\alpha(\sigma) = \sigma(\alpha) \cdot \alpha^{-1}, \; \sigma \in \Gamma.$$

The set $B^1(\Gamma, M) = \{\, f_\alpha \mid \alpha \in M \,\}$ is a subgroup of $Z^1(\Gamma, M)$. The quotient group $Z^1(\Gamma, M)/B^1(\Gamma, M)$ is called the first *cohomology group* of Γ with coefficients in M, and is denoted by $H^1(\Gamma, M)$. Note that for any group G, for any G-module A, and for any $n \in \mathbb{N}$ one can define the more general concept of *n-th cohomology group* $H^n(G, A)$ of G with coefficients in A (see e.g., Cassels and Fröhlich [**46**, Chapter IV] or Karpilovsky [**76**, p. 369]).

The famous *Hilbert's Theorem 90* asserts that $H^1(\mathrm{Gal}(E/F), E^*) = \mathbf{1}$. for any finite Galois extension E/F (see e.g., Cassels and Fröhlich [**46**, Proposition 2.2, Chapter V] or Karpilovsky [**76**, Theorem 9.2, Chapter 6]).

Recall that for any extension E/F we use throughout this monograph the following notation.

$$
\begin{aligned}
\mu(E) &= \{\, x \in E^* \mid x^n = 1 \text{ for some } n \in \mathbb{N}^* \,\}, \\
T(E/F) &= \{\, x \in E^* \mid x^n \in F^* \text{ for some } n \in \mathbb{N}^* \,\}, \\
\mathrm{Cog}(E/F) &= T(E/F)/F^*, \\
\widehat{x} &= \text{the coset } xF^* \in E^*/F^* \text{ of any } x \in E.
\end{aligned}
$$

For every positive integer $n \geqslant 1$ we shall define the following subgroups of $T(E/F)$ and $\mathrm{Cog}(E/F)$, respectively.

$$T_n(E/F) = \{\, x \in E^* \mid x^n \in F^* \,\},$$
$$\mathrm{Cog}_n(E/F) = T_n(E/F)/F^*.$$

For an arbitrary extension E/F we consider the following map

$$f : \mathrm{Gal}\,(E/F) \times \mathrm{Cog}\,(E/F) \longrightarrow \mu(E),$$
$$f(\sigma, \widehat{\alpha}) = f_\alpha(\sigma) = \sigma(\alpha) \cdot \alpha^{-1}, \ \sigma \in \mathrm{Gal}(E/F), \ \alpha \in T(E/F).$$

Note that f is well-defined. Clearly, for every integer $n \geqslant 1$, the restriction of f to $\mathrm{Cog}_n(E/F)$ induces a map

$$f_n : \mathrm{Gal}\,(E/F) \times \mathrm{Cog}_n(E/F) \longrightarrow \mu_n(E).$$

For every fixed $\sigma \in \mathrm{Gal}\,(E/F)$, the partial map $f(\sigma, -)$ is clearly multiplicative on $\mathrm{Cog}\,(E/F)$, and for every fixed $\widehat{\alpha} \in \mathrm{Cog}(E/F)$, the partial map $f(-, \widehat{\alpha})$ is precisely the 1-coboundary $f_\alpha \in Z^1(\mathrm{Gal}\,(E/F), \mu(E))$, so f and f_n induce the group morphisms

$$\psi : \mathrm{Cog}\,(E/F) \longrightarrow Z^1(\mathrm{Gal}\,(E/F), \mu(E)), \ \ \psi(\widehat{\alpha}) = f_\alpha,$$

and

$$\psi_n : \mathrm{Cog}_n(E/F) \longrightarrow Z^1(\mathrm{Gal}\,(E/F), \mu_n(E)), \ \ \psi_n(\widehat{\alpha}) = f_\alpha,$$

respectively.

LEMMA 6.1.1. *For any finite Galois extension E/F and for any $n \in$ $\mathbb{N}^*$, the morphism*

$$\psi_n : \mathrm{Cog}_n(E/F) \longrightarrow Z^1(\mathrm{Gal}\,(E/F), \mu_n(E))$$

defined above is a group isomorphism.

PROOF. Denote by Γ the group $\mathrm{Gal}\,(E/F)$, and let $\alpha \in T_n(E/F)$. We have

$$\widehat{\alpha} \in \mathrm{Ker}(\psi_n) \Longleftrightarrow f_\alpha(\sigma) = 1, \ \forall\, \sigma \in \Gamma \Longleftrightarrow \sigma(\alpha) = \alpha, \ \forall\, \sigma \in \Gamma$$
$$\Longleftrightarrow \alpha \in \mathrm{Fix}\,(\Gamma) = F,$$

and consequently, ψ_n is a monomorphism.

We are going to prove that ψ_n is surjective. Let $h \in Z^1(\Gamma, \mu_n(E))$. Since clearly $h \in Z^1(\Gamma, E^*)$, by Hilbert's Theorem 90 there exists $\alpha \in E^*$ such that $h = f_\alpha$. Hence $\sigma(\alpha) \cdot \alpha^{-1} \in \mu_n(E)$, so $(\sigma(\alpha) \cdot \alpha^{-1})^n = 1$ for every $\sigma \in \Gamma$. We deduce that $\sigma(\alpha^n) = \alpha^n$ for every $\sigma \in \Gamma$, and consequently

$\alpha^n \in F^*$ since E/F is a Galois extension. Thus, $\alpha \in T_n(E/F)$. Hence $h = \psi_n(\widehat{\alpha})$, with $\widehat{\alpha} \in \mathrm{Cog}_n(E/F)$. This shows that ψ_n is surjective. $\square$

THEOREM 6.1.2. *For any finite Galois extension E/F, the assignment $\widehat{\alpha} \mapsto f_\alpha$ establishes a group isomorphism*

$$\mathrm{Cog}(E/F) \cong Z^1(\mathrm{Gal}(E/F), \mu(E)).$$

PROOF. Clearly, for every $n \geqslant 1$, the restriction of the morphism

$$\psi : \mathrm{Cog}(E/F) \longrightarrow Z^1(\mathrm{Gal}(E/F), \mu(E))$$

to $\mathrm{Cog}_n(E/F)$ is ψ_n. On the other hand, since $\Gamma = \mathrm{Gal}(E/F)$ is finite, $Z^1(\Gamma, \mu(E))$ is the union of all $Z^1(\Gamma, \mu_n(E))$, $n \geqslant 1$, and since every ψ_n is surjective by Lemma 6.1.1, we deduce that ψ is also surjective. $\square$

REMARK 6.1.3. We will see in Section 15.1 that if E/F is an infinite Galois extension, then the group $\mathrm{Cog}(E/F)$ is isomorphic to the group of all *continuous* crossed homomorphisms of the compact topological group $\mathrm{Gal}(E/F)$ (endowed with the Krull topology) with coefficients in the discrete group $\mu(E)$. $\square$

COROLLARY 6.1.4. *Let E/F be a finite Galois extension with $\mu(E)$ finite. Then $\mathrm{Cog}(E/F)$ is a finite group. In particular, for any extension K/L of algebraic number fields, which is not necessarily Galois, the group $\mathrm{Cog}(K/L)$ is finite.*

PROOF. Since $\mathrm{Gal}(E/F)$ and $\mu(E)$ are finite groups, it is obvious that the group $Z^1(\mathrm{Gal}(E/F), \mu(E))$ is finite, hence $\mathrm{Cog}(E/F)$ is also finite by Theorem 6.1.2.

Since $K/\mathbb{Q}$ and $L/\mathbb{Q}$ are both finite extensions, it follows that the extension K/L is a finite separable extension. Consider the normal closure $\widetilde{K}/L$ of the extension K/L, which is a finite Galois extension (see 1.2.7). Then, $\mathrm{Cog}(\widetilde{K}/L)$ is a finite group since $\mu(N)$ is a finite group for any algebraic number field N. Now, observe that $\mathrm{Cog}(K/L)$ is a subgroup of the finite group $\mathrm{Cog}(\widetilde{K}/L)$, hence it is also finite. $\square$

COROLLARY 6.1.5. *If E/F is a finite Galois extension with Galois group Γ, then the map*

$$\varphi : \{\, H \mid F^* \leqslant H \leqslant T(E/F) \,\} \longrightarrow \{\, U \mid U \leqslant Z^1(\Gamma, \mu(E)) \,\},$$
$$\varphi(H) = \{\, f_\alpha \in Z^1(\Gamma, \mu(E)) \mid \alpha \in H \,\},$$

is a lattice isomorphism, which induces a canonical lattice isomorphism

$$\underline{\mathrm{Subgroups}}(\mathrm{Cog}\,(E/F)) \cong \underline{\mathrm{Subgroups}}\,(Z^1(\Gamma,\mu(E))).$$

For every cyclic subgroup C of $Z^1(\Gamma,\mu(E))$ there exists $\alpha \in T(E/F)$ such that $\varphi(F^*\langle\alpha\rangle) = \langle\,f_\alpha\,\rangle = C$. Moreover, $H/F^* \cong \varphi(H)$ for every H with $F^* \leqslant H \leqslant T(E/F)$. $\qquad\square$

LEMMA 6.1.6. *Let E/F be a G-Cogalois extension, let $x \in \Omega^*$ be such that $x^m \in F$ for some $m \in \mathbb{N}^*$, and let $n = \exp(G/F^*)$. Suppose that one of the following two conditions is satisfied.*

(1) $\mathcal{P}_m \subseteq \mathcal{P}_n$ *(in particular, this holds if $m\,|\,n$).*
(2) $\mu_m(E) \subseteq F$ *(in particular, this holds if $\zeta_m \in F$).*

Then, we have $F(x) \subseteq E \Longleftrightarrow x \in G$.

PROOF. Suppose that $F(x) \subseteq E$. Set $K = F(x)$ and $k = \mathrm{ord}(\widehat{x})$. Then clearly $k = \exp(F^*\langle x\rangle/F^*)$ and $k\,|\,m$. Let $p \in \mathcal{P}_k$. Then $p\,|\,m$, hence $p \in \mathcal{P}_m$, and so, $p\,|\,n$ if condition (1) is satisfied. We deduce that $\mu_p(K) \subseteq \mu_p(E) \subseteq F$, since E/F is n-pure by Theorem 4.3.2. If condition (2) is satisfied, then we have $\mu_p(K) \subseteq \mu_p(E) \subseteq \mu_m(E) \subseteq F$. Hence, in both cases, K/F is $F^*\langle x\rangle$-Cogalois, again by Theorem 4.3.2. But, the extension K/F is also $G \cap K^*$-Cogalois by Proposition 4.3.5, so $F^*\langle x\rangle = G \cap K^*$ by Corollary 4.4.2. Thus $x \in G$, as desired. The other implication is obvious. $\qquad\square$

THEOREM 6.1.7. *Let E/F be a Galois G-Cogalois extension with $n = \exp(G/F^*)$. Then, there exists a canonical group isomorphism*

$$\mathrm{Kne}(E/F) \cong Z^1(\mathrm{Gal}(E/F),\mu_n(E)).$$

PROOF. Denote by Γ the Galois group of E/F. Let $\alpha \in T_n(E/F)$. By Lemma 6.1.6, we deduce that $\alpha \in G$, so $G = T_n(E/F)$. Thus,

$$\mathrm{Cog}_n(E/F) = T_n(E/F)/F^* = G/F^* = \mathrm{Kne}(E/F),$$

and so, the desired isomorphism follows at once from Lemma 6.1.1. $\qquad\square$

For any Galois G-Cogalois extension E/F with $n = \exp(G/F^*)$, the map

$$f : \mathrm{Gal}(E/F) \times \mathrm{Cog}(E/F) \longrightarrow \mu(E),$$

considered at the beginning of this section yields by restriction the map

$$g : \mathrm{Gal}(E/F) \times \mathrm{Kne}(E/F) \longrightarrow \mu_n(E),$$
$$g(\sigma,\widehat{\alpha}) = f_\alpha(\sigma) = \sigma(\alpha)\cdot\alpha^{-1}.$$

For any $\Delta \leqslant \mathrm{Gal}\,(E/F)$ and any $W \leqslant \mathrm{Kne}\,(E/F)$ let denote

$$\Delta^\top = \{\, c \in \mathrm{Kne}\,(E/F) \mid g(\sigma, c) = 1,\ \forall\, \sigma \in \Delta \,\},$$

$$W^\top = \{\, \sigma \in \mathrm{Gal}\,(E/F) \mid g(\sigma, c) = 1,\ \forall\, c \in W \,\}.$$

PROPOSITION 6.1.8. *For any Galois G-Cogalois extension E/F, the assignments $(-)^\top$ define mutually inverse anti-isomorphisms between the lattices* Subgroups $(\mathrm{Gal}\,(E/F))$ *and* Subgroups $(\mathrm{Kne}(E/F))$.

PROOF. For simplicity, denote by Γ the group $\mathrm{Gal}\,(E/F)$, by $\mathcal{C}$ the lattice of all subgroups of Γ, by $\mathcal{H}$ the lattice of all subgroups of $G/F^* = \mathrm{Kne}\,(E/F)$, and by $\mathcal{E}$ the lattice of all intermediate fields of the extension E/F.

Since E/F is a G-Cogalois extension, the maps

$$\mathcal{E} \longrightarrow \mathcal{H},\ K \mapsto (K \cap G)/F^*, \ \text{ and } \ \mathcal{H} \longrightarrow \mathcal{E},\ H/F^* \mapsto F(H),$$

are isomorphisms of lattices, inverse to one another.

On the other hand, since E/F is a Galois extension, the maps

$$\mathcal{C} \longrightarrow \mathcal{E},\ \Delta \mapsto \mathrm{Fix}\,(\Delta), \ \text{ and } \ \mathcal{E} \longrightarrow \mathcal{C},\ K \mapsto \mathrm{Gal}\,(E/K),$$

are anti-isomorphisms of lattices, inverse to one another.

If $\Delta \leqslant \Gamma$ and $W = H/F^* \leqslant G/F^*$, where $F^* \leqslant H \leqslant G$, then it is easily proved that

$$\Delta^\top = \{\, \widehat{\alpha} \in G/F^* \mid \sigma(\alpha) = \alpha,\ \forall\, \sigma \in \Delta \,\} = (\mathrm{Fix}\,(\Delta) \cap G)/F^*$$

and

$$W^\top = \{\, \sigma \in \Gamma \mid \sigma(\beta) = \beta,\ \forall\, \beta \in H \,\} = \mathrm{Gal}\,(E/F(H)).$$

It follows that each of the composed maps

$$\mathcal{H} \longrightarrow \mathcal{E} \longrightarrow \mathcal{C} \ \text{ and } \ \mathcal{C} \longrightarrow \mathcal{E} \longrightarrow \mathcal{H}$$

of the canonical bijections considered above gives rise to the maps $(-)^\top$ between $\mathcal{C}$ and $\mathcal{H}$, which finishes the proof. $\square$

6.2. Radical extensions via crossed homomorphisms

In this section we investigate via crossed homomorphisms when a finite Galois extension is radical, Kneser, or G-Cogalois. We also give sufficient conditions under which the property of a Galois extension E/F being radical, Kneser, or G-Cogalois is preserved if one changes the base field F.

Let E/F be a finite Galois extension with Galois group Γ. By Theorem 6.1.2, there exists a canonical isomorphism $\operatorname{Cog}(E/F) \cong Z^1(\Gamma, \mu(E))$, hence the canonical map

$$f : \operatorname{Gal}(E/F) \times \operatorname{Cog}(E/F) \longrightarrow \mu(E),$$

$$f(\sigma, \widehat{\alpha}) = \sigma(\alpha) \cdot \alpha^{-1},$$

considered in Section 6.1 yields, by replacing $\operatorname{Cog}(E/F)$ with its isomorphic copy $Z^1(\Gamma, \mu(E))$, precisely the *evaluation map*

$$\langle -, - \rangle : \Gamma \times Z^1(\Gamma, \mu(E)) \longrightarrow \mu(E), \quad \langle \sigma, h \rangle = h(\sigma).$$

For any $\Delta \leqslant \Gamma$, $U \leqslant Z^1(\Gamma, \mu(E))$, and $\chi \in Z^1(\Gamma, \mu(E))$ denote

$$\begin{aligned}
\Delta^\perp &= \{\, h \in Z^1(\Gamma, \mu(E)) \mid \langle \sigma, h \rangle = 1,\ \forall \sigma \in \Delta \,\}, \\
U^\perp &= \{\, \sigma \in \Gamma \mid \langle \sigma, h \rangle = 1,\ \forall h \in U \,\}, \\
\chi^\perp &= \{\, \sigma \in \Gamma \mid \langle \sigma, \chi \rangle = 1 \,\}.
\end{aligned}$$

One verifies easily that $\Delta^\perp \leqslant Z^1(\Gamma, \mu(E))$, $U^\perp \leqslant \Gamma$, and $\chi^\perp = \langle \chi \rangle^\perp$. Note that in the previous section we have also used the notation $(-)^\perp$, but with a different meaning.

The next result characterizes radical subextensions of a given Galois extension E/F by means of subgroups of $Z^1(\operatorname{Gal}(E/F), \mu(E))$.

THEOREM 6.2.1. *Let E/F be a finite Galois extension with Galois group Γ, and let K be an intermediate field of E/F. Then K/F is a radical extension (resp. a simple radical extension) if and only if there exists $U \leqslant Z^1(\Gamma, \mu(E))$ (resp. $\chi \in Z^1(\Gamma, \mu(E))$) such that $\operatorname{Gal}(E/K) = U^\perp$ (resp. $\operatorname{Gal}(E/K) = \chi^\perp$).*

PROOF. If K/F is a radical extension, then there exists a group G, not necessarily unique, such that $F^* \leqslant G \leqslant T(E/F)$ and $K = F(G)$. If we set

$$U := \{f_\alpha \mid \alpha \in G\} \leqslant Z^1(\Gamma, \mu(E)),$$

then we have

$$U^\perp = \{\, \sigma \in \Gamma \mid f_\alpha(\sigma) = 1,\ \forall \alpha \in G \,\} = \{\, \sigma \in \Gamma \mid \sigma(\alpha) = \alpha,\ \forall \alpha \in G \,\}$$

$$= \{\, \sigma \in \Gamma \mid \sigma(x) = x,\ \forall\, x \in F(G)\,\} = \mathrm{Gal}\,(E/F(G)) = \mathrm{Gal}\,(E/K).$$

Conversely, suppose that there exists an $U \leqslant Z^1(\Gamma, \mu(E))$ such that $\mathrm{Gal}\,(E/K) = U^\perp$. Let $G = \{\, \alpha \in E^* \mid f_\alpha \in U \,\} \leqslant T(E/F)$ be the group of radicals associated with U by Corollary 6.1.5. We have just seen that $U^\perp = \mathrm{Gal}\,(E/F(G))$, hence $\mathrm{Gal}\,(E/K){=}\mathrm{Gal}\,(E/F(G))$. By the Fundamental Theorem of Galois Theory we obtain

$$K = \mathrm{Fix}\,(\mathrm{Gal}\,(E/K)) = \mathrm{Fix}\,(\mathrm{Gal}\,(E/F(G))) = F(G),$$

which shows that K/F is a G-radical extension. The case of simple radical extensions now easily follows from Corollary 6.1.5. $\qquad\square$

REMARK 6.2.2. Denote by $\underline{\mathrm{Radical}}\,(E/F)$ the set of all subextensions K/F of E/F which are radical. With the notation and hypotheses of Theorem 6.2.1, the map

$$\underline{\mathrm{Subgroups}}\,(Z^1(\Gamma, \mu(E))) \longrightarrow \underline{\mathrm{Radical}}\,(E/F),$$

$$U \mapsto F(\{\, \alpha \in E^* \mid f_\alpha \in U \,\})/F,\ U \leqslant Z^1(\Gamma, \mu(E)),$$

is surjective in view of Corollary 6.1.5. In general, this map is not injective (see Exercise 4). This may happen because distinct subgroups G and G' of $T(E/F)$ containing F^*, may define the same field $F(G) = F(G')$. $\qquad\square$

The result below provides characterizations, in terms of crossed homomorphisms, of Kneser and G-Cogalois subextensions of finite Galois extensions.

COROLLARY 6.2.3. *Let E/F be a finite Galois extension with Galois group Γ, let K/F be a finite G-radical subextension of E/F with G/F^* a finite group, and denote $U = \{\, f_\alpha \mid \alpha \in G \,\} \leqslant Z^1(\Gamma, \mu(E))$. Then*

(1) *The extension K/F is G-Kneser if and only if $(\Gamma : U^\perp) = |U|$.*

(2) *The extension K/F is G-Cogalois if and only if it is G-Kneser and the map*

$$\{\, V \mid V \leqslant U \,\} \longrightarrow \{\, \Delta \mid U^\perp \leqslant \Delta \leqslant \Gamma \,\},\ V \mapsto V^\perp,$$

is bijective, or equivalently, an anti-isomorphism of lattices.

PROOF. (1) The extension K/F is G-Kneser if and only if $[K : F] = |G/F^*|$. If we set $\Delta = \mathrm{Gal}(E/K)$, then $[K : F] = (\Gamma : \Delta)$ by Galois Theory, and $G/F^* \cong U$, so $|G/F^*| = |U|$, by Corollary 6.1.5. On the other hand, $\Delta = U^\perp$ according to Theorem 6.2.1. Summing up, we deduce that

$$[K : F] = |G/F^*| \iff (\Gamma : \Delta) = |U| \iff (\Gamma : U^\perp) = |U|.$$

(2) By Theorem 4.3.2, the extension K/F is G-Cogalois if and only if it is G-Kneser, and the map

$$\{\, H \mid F^* \leqslant H \leqslant G \,\} \longrightarrow \text{Intermediate}\,(K/F),\ H \mapsto F(H),$$

is a lattice isomorphism. Since $U \cong G/F^*$, the lattice $\{\, H \mid F^* \leqslant H \leqslant G \,\}$ is canonically isomorphic to the lattice $\{\, V \mid V \leqslant U \,\}$. On the other hand, the canonical map

$$\text{Intermediate}\,(K/F) \longrightarrow \{\, \Delta \mid U^{\perp} \leqslant \Delta \leqslant \Gamma \,\},\ L \mapsto \text{Gal}(E/L),$$

is an anti-isomorphism of lattices by Galois Theory.

Summing up, we deduce that the G-Kneser extension K/F is G-Cogalois if and only if the canonical map

$$\delta : \{\, V \mid V \leqslant U \,\} \longrightarrow \{\, \Delta \mid U^{\perp} \leqslant \Delta \leqslant \Gamma \,\}$$

obtained by composing the three canonical maps considered above is bijective, or equivalently, an anti-isomorphism of lattices. Observe that the last part of the proof of Theorem 6.2.1 shows that δ is precisely the map $V \mapsto V^{\perp}$. $\square$

Next, we give sufficient conditions under which the property of a Galois extension E/F being radical, Kneser, or G-Cogalois is preserved when one changes the base field F. Recall that Ω denotes an algebraically closed field containing the given field F as a subfield. Any overfield L of F which will be considered in the sequel is supposed to be a subfield of Ω.

Let E/F be a finite Galois extension with Galois group Γ, let L/F be any extension with $L \cap E = F$, and consider their compositum EL. According to the Galois Theory (see 1.2.9), the restriction map

$$\text{Gal}\,(EL/L) \xrightarrow{\sim} \text{Gal}\,(E/F),\ \sigma \mapsto \sigma_{|E},$$

is an isomorphism of groups. Consequently, by the Fundamental Theorem of Galois Theory we deduce that the maps

$$\varepsilon : \text{Subextensions}\,(E/F) \longrightarrow \text{Subextensions}\,(EL/L),\ K/F \mapsto LK/L,$$

and

$$\lambda : \text{Subextensions}\,(EL/L) \longrightarrow \text{Subextensions}\,(E/F),\ K_1/L \mapsto (K_1 \cap E)/F,$$

are isomorphisms of lattices, inverse to one another.

We have denoted in Remark 6.2.2 by $\underline{\text{Radical}}(E/F)$ the set of all subextensions K/F of E/F which are radical. For every $K/F \in \underline{\text{Radical}}\,(E/F)$ there exists a group G, not necessarily unique, with $F^* \leqslant G \leqslant T(E/F)$

and $K = F(G)$. If we set $G_1 = GL^*$, then clearly $LK = L(G_1)$ and $L^* \leqslant G_1 \leqslant T(EL/L)$. It follows that $\varepsilon(K/F) \in \underline{\text{Radical}}\,(EL/L)$, and consequently, the restriction of ε to radical extensions gives rise to the injective map

$$\rho : \underline{\text{Radical}}\,(E/F) \longrightarrow \underline{\text{Radical}}\,(EL/L),$$

$$F(G)/F \mapsto (F(G)L)/L = L(GL^*)/L,\ F^* \leqslant G \leqslant T(E/F),$$

which is not necessarily bijective.

THEOREM 6.2.4. *Let E/F be a finite Galois extension with Galois group Γ, and let L/F be an arbitrary extension such that $E \cap L = F$. If $\mu(EL) = \mu(E)$, then the following assertions hold.*

(1) $GL^* \cap E^* = G$ *for every G with $F^* \leqslant G \leqslant T(E/F)$.*
(2) $G_1 = (G_1 \cap E^*)L^*$ *for every G_1 with $L^* \leqslant G_1 \leqslant T(EL/L)$.*
(3) *The map*

$$\rho : \underline{\text{Radical}}\,(E/F) \longrightarrow \underline{\text{Radical}}\,(EL/L),$$

$$F(G)/F \mapsto L(GL^*)/L,\ F^* \leqslant G \leqslant T(E/F),$$

is bijective, and the map

$$\underline{\text{Radical}}\,(EL/L) \longrightarrow \underline{\text{Radical}}\,(E/F),$$

$$L(G_1)/L \mapsto F(G_1 \cap E^*)/F,\ L^* \leqslant G_1 \leqslant T(EL/L),$$

is its inverse.

PROOF. (1) follows at once from the equality $E \cap L = F$.

(2) Denote $\Gamma_1 = \text{Gal}\,(EL/L)$. We have seen that the map

$$\Gamma_1 \longrightarrow \Gamma,\ \sigma_1 \mapsto \sigma_1|_E,$$

is an isomorphism of groups. Since $\mu(EL) = \mu(E)$, this isomorphism induces the group isomorphism

$$v : Z^1(\Gamma, \mu(E)) \longrightarrow Z^1(\Gamma_1, \mu(EL)),$$

$$v(h)(\sigma_1) = h(\sigma_1|_E),\ h \in Z^1(\Gamma, \mu(E)),\ \sigma_1 \in \Gamma_1.$$

Let G_1 with $L^* \leqslant G_1 \leqslant T(EL/L)$. The inclusion $(G_1 \cap E^*)L^* \subseteq G_1$ is obvious. Now let $\alpha_1 \in G_1$. Then $f_{\alpha_1} \in Z^1(\Gamma_1, \mu(EL))$, hence there exists $\alpha \in T(E/F)$ such that $f_{\alpha_1} = v(f_\alpha)$ since v is an isomorphism and $Z^1(\Gamma, \mu(E)) = B^1(\Gamma, T(E/F))$ by Theorem 6.1.2. Thus, $f_{\alpha_1}(\sigma_1) = f_\alpha(\sigma_1|_E)$ for all $\sigma_1 \in \Gamma_1$. It follows that $\sigma_1(\alpha_1) \cdot \alpha_1^{-1} = \sigma_1(\alpha) \cdot \alpha^{-1}$ for all $\sigma_1 \in \Gamma_1$, i.e., $\sigma_1(\alpha_1 \cdot \alpha^{-1}) = \alpha_1 \cdot \alpha^{-1}$ for all $\sigma_1 \in \Gamma_1$, hence $\alpha_1 \cdot \alpha^{-1} \in$

$\mathrm{Fix}(\mathrm{Gal}\,(EL/L)) = L$. We deduce that $\alpha_1 = \alpha \cdot y$ for some $y \in L^* \subseteq G_1$. Thus $\alpha = \alpha_1 \cdot y^{-1} \in G_1 \cap E^*$, which proves the inclusion $G_1 \subseteq (G_1 \cap E^*)L^*$.

(3) Though apparently the definition of ρ as given in the statement of the theorem seems to depend on the group G, which is not necessarily uniquely associated with a given radical subextension $F(G)/F$ of E/F, we have seen above that ρ is a well-defined injective map.

To prove that ρ is a surjective map, let $K_1/L \in \underline{\mathrm{Radical}}\,(EL/L)$. Then $K_1 = L(G_1)$ for some G_1 with $L^* \leqslant G_1 \leqslant T(EL/L)$, hence, if we set $G = G_1 \cap E^*$, then $F(G) \in \underline{\mathrm{Radical}}\,(E/F)$, and

$$\rho(F(G)/F) = L(F(G))/L = L(F(G_1 \cap E^*))/L = L((G_1 \cap E^*)L^*)/L,$$

$$L((G_1 \cap E^*)L^*) = L(G_1) = L(GL^*) = K_1,$$

which shows that ρ is surjective, hence bijective, and its inverse map ρ^{-1} is that described in the statement of the theorem. Observe that

$$\rho^{-1} : \underline{\mathrm{Radical}}\,(EL/L) \longrightarrow \underline{\mathrm{Radical}}\,(E/F),$$

can be also defined as

$$\rho^{-1}(K_1/L) = (K_1 \cap E)/F.$$

$\square$

REMARKS 6.2.5. (1) The isomorphism v defined in the proof of Theorem 6.2.4 induces the isomorphism of lattices

$$(*) \qquad \{\, U \mid U \leqslant Z^1(\Gamma, \mu(E)) \,\} \longrightarrow \{\, U_1 \mid U_1 \leqslant Z^1(\Gamma_1, \mu(EL)) \,\},$$

$$U \mapsto U_1 = v(U).$$

By Corollary 6.1.5, there exist lattice isomorphisms

$$\{\, U \mid U \leqslant Z^1(\Gamma, \mu(E)) \,\} \longrightarrow \{\, G \mid F^* \leqslant G \leqslant T(E/F) \,\},$$
$$U \mapsto G = \{\, \alpha \in E^* \mid f_\alpha \in U \,\},$$

and

$$\{\, U_1 \mid U_1 \leqslant Z^1(\Gamma_1, \mu(EL)) \,\} \longrightarrow \{\, G_1 \mid L^* \leqslant G_1 \leqslant T(EL/L) \,\},$$
$$U_1 \mapsto G_1 = \{\, \alpha_1 \in (EL)^* \mid f_{\alpha_1} \in U_1 \,\}.$$

Now, using $(*)$, we obtain an isomorphism of lattices

$$\nu : \{\, G \mid F^* \leqslant G \leqslant T(E/F) \,\} \longrightarrow \{\, G_1 \mid L^* \leqslant G_1 \leqslant T(EL/L) \,\},$$

$$G \mapsto G_1 = \{\, \alpha_1 \in (EL)^* \mid f_{\alpha_1} \in v(\{\, f_\alpha \mid \alpha \in G \,\}) \,\}.$$

We claim that $\nu(G) = GL^*$ for every G with $F^* \leqslant G \leqslant T(E/F)$. Indeed, if $G_1 = \nu(G)$, then it is easily seen by the above considerations that for an $\alpha_1 \in (EL)^*$ one has

$$
\begin{aligned}
\alpha_1 \in G_1 \quad &\Longleftrightarrow \quad \exists\, \alpha \in G,\ \forall\, \sigma_1 \in \Gamma_1,\ f_{\alpha_1}(\sigma_1) = f_\alpha(\sigma_1|_E) \\
&\Longleftrightarrow \quad \exists\, \alpha \in G,\ \forall\, \sigma_1 \in \Gamma_1,\ \sigma_1(\alpha_1)/\alpha_1 = \sigma_1(\alpha)/\alpha \\
&\Longleftrightarrow \quad \exists\, \alpha \in G,\ \forall\, \sigma_1 \in \Gamma_1,\ \sigma_1(\alpha_1/\alpha) = \alpha_1/\alpha \\
&\Longleftrightarrow \quad \exists\, \alpha \in G,\ \alpha_1/\alpha \in \mathrm{Fix}(\mathrm{Gal}\,(EL/L)) = L.
\end{aligned}
$$

Hence $G_1 = \nu(G) = GL^*$. From Theorem 6.2.4 (2) we deduce that the inverse ν^{-1} of ν can be described explicitly by

$$
\nu^{-1}(G_1) = G_1 \cap E^*,\ L^* \leqslant G_1 \leqslant T(EL/L).
$$

(2) Another argument for the surjectivity of ρ is the following one. If $K_1/L \in \underline{\mathrm{Radical}}\,(EL/L)$, then there exists $U_1 \leqslant Z^1(\mathrm{Gal}\,(EL/L), \mu(EL))$ such that $\mathrm{Gal}\,(EL/K_1) = U_1^\perp$ by Theorem 6.2.1. Let

$$
v\ :\ Z^1(\Gamma, \mu(E)) \longrightarrow Z^1(\Gamma_1, \mu(EL))
$$

be the isomorphism considered in the proof of Theorem 6.2.4 (2), and denote $U = v^{-1}(U_1)$. It is easily verified that $\mathrm{Gal}\,(E/(K_1 \cap E)) = U^\perp$. Again by Theorem 6.2.1, we deduce that $(K_1 \cap E)/F \in \underline{\mathrm{Radical}}\,(E/F)$, and by Galois Theory we have $\rho((K_1 \cap E)/F) = K_1/L$. $\square$

COROLLARY 6.2.6. *Let E/F be a finite Galois extension, and let L/F be an arbitrary extension such that $L \cap E = F$ and $\mu(EL) = \mu(E)$. Let G be a group such that $F^* \leqslant G \leqslant T(E/F)$, and denote $G_1 = GL^*$. Then*

(1) *$G/F^* \cong G_1/L^*$ and $[\,F(G) : F\,] = [\,L(G_1) : L\,]$.*
(2) *The extension $F(G)/F$ is G-Kneser if and only if the extension $L(G_1)/L$ is G_1-Kneser.*
(3) *The extension $F(G)/F$ is G-Cogalois if and only if the extension $L(G_1)/L$ is G_1-Cogalois.*

PROOF. (1) We have

$$
G_1/L^* = (GL^*)/L^* \cong G/(G \cap L^*) = G/F^*
$$

since $G \cap L^* = F$.

Let $U := \{f_\alpha \,|\, \alpha \in G\}$ be the subgroup of $Z^1(\Gamma, \mu(E))$ which corresponds via Theorem 6.2.1 to the radical subextension $F(G)/F$ of the Galois extension E/F, and preserve the notation from the proof of Theorem 6.2.4. If we denote $U_1 = v(U)$, then by Theorem 6.2.1, we have

$$[F(G):F] = (\Gamma:U^{\perp}) = (\Gamma_1:v(U)^{\perp}) = (\Gamma_1:U_1^{\perp}) = [L(G_1):L].$$

(2) By Corollary 6.2.3 (1), the extension $F(G)/F$ is G-Kneser if and only if $(\Gamma:U^{\perp}) = |U|$, and similarly, the extension $L(G_1)/L$ is G_1-Kneser if and only if $(\Gamma_1:U_1^{\perp}) = |U_1|$. But $U_1 = v(U)$, so $|U_1| = |U|$. To conclude, apply (1).

(3) By Corollary 6.2.3 (2), $F(G)/F$ is G-Cogalois if and only if it is G-Kneser and the map

$$\{\, V \mid V \leqslant U \,\} \longrightarrow \{\, \Delta \mid U^{\perp} \leqslant \Delta \leqslant \Gamma \,\}, \quad V \mapsto V^{\perp},$$

is bijective. Using (2) and the canonical isomorphism $G_1 \cong G$, these conditions are equivalent to the conditions that $L(G_1)/L$ is a G_1-Kneser extension and the map

$$\{\, V_1 \mid V_1 = v(V) \leqslant v(U) = U_1 \,\} \longrightarrow \{\, \Delta_1 \mid U_1^{\perp} \leqslant \Delta_1 \leqslant \Gamma_1 \,\}, \quad V_1 \mapsto V_1^{\perp},$$

is bijective. Again by Corollary 6.2.3 (2), these last conditions are equivalent to the fact that the extension $L(G_1)/L$ is G_1-Cogalois. $\square$

COROLLARY 6.2.7. *Let E/F be a finite Galois extension, and let L/F be an arbitrary extension such that $L \cap E = F$. Let K be an intermediate field of the extension E/F such that KL/L is a G_1-radical extension, with G_1/L^* a finite group of exponent n.*

If $\zeta_n \in E$, then K/F is G-radical and $G/F^* \cong G_1/L^*$, where $G = G_1 \cap E^*$.

PROOF. We have already noticed that the restriction map to E gives rise to an isomorphism

$$\Gamma_1 = \mathrm{Gal}\,(EL/L) \xrightarrow{\sim} \mathrm{Gal}\,(E/F) = \Gamma.$$

Let $U_1 = \{\, f_{\alpha_1} \mid \alpha_1 \in G_1 \,\}$ be the subgroup of $Z^1(\Gamma_1, \mu(EL))$ which corresponds by Theorem 6.2.1 to the radical subextension $L(G_1)/L$ of the Galois extension EL/L. Since $\alpha_1^n \in L^*$ for every $\alpha_1 \in G_1$, it follows that $f_{\alpha_1}(\sigma_1) \in \mu_n(\Omega)$ for every $\sigma_1 \in \Gamma_1$, and consequently, $U_1 \leqslant Z^1(\Gamma_1, \langle\zeta_n\rangle)$. By Theorem 6.2.1, we have $U_1^{\perp} = \mathrm{Gal}(EL/L(G_1))$.

As in the proof of Theorem 6.2.4, the above considered isomorphism

$$\Gamma_1 \longrightarrow \Gamma, \; \sigma \mapsto \sigma_{|E},$$

yields the group isomorphism

$$v\,:\, Z^1(\Gamma, \langle\zeta_n\rangle) \longrightarrow Z^1(\Gamma_1, \langle\zeta_n\rangle),$$
$$v(f)(\sigma) = f(\sigma_{|E}), \; f \in Z^1(\Gamma, \langle\zeta_n\rangle), \; \sigma \in \Gamma_1.$$

Denote $U = v^{-1}(U_1)$. Since $\zeta_n \in E$, it follows that $U \leqslant Z^1(\Gamma, \mu(E))$. If we denote $G = \{\, \alpha \in E^* \mid f_\alpha \in U \,\}$, then, as in Remark 6.2.5 (1), we deduce that $G = G_1 \cap E^*$, $G_1 = GL^*$, and so,

$$L(G_1) \cap E = L(GL^*) \cap E = F(G)L \cap E = F(G) - KL \cap E = K.$$

Finally, by the proof of Corollary 6.2.6 (1) we have $G/F^* \cong G_1/L^*$.   □

COROLLARY 6.2.8. *With the notation and hypotheses from Corollary 6.2.7, the extension K/F is G-Kneser (resp. G-Cogalois) if and only if KL/L is G_1-Kneser (resp. G_1-Cogalois).*

PROOF. Adapt the proof of Corollary 6.2.6. The details are left to the reader.   □

6.3. Exercises to Chapter 6

1. (*Dummit* [55]). Let E/F be a finite Galois extension, let $\Gamma = \mathrm{Gal}(E/F)$, and let $\Delta \lhd \Gamma$. Show that the canonical sequence of Abelian groups

$$0 \longrightarrow Z^1(\Gamma/\Delta, \mu(E)^\Delta) \longrightarrow Z^1(\Gamma, \mu(E)) \longrightarrow Z^1(\Delta, \mu(E))$$

   is exact, where $\mu(E)^\Delta = \{\, \zeta \in \mu(E) \mid \sigma(\zeta) = \zeta,\ \forall\, \sigma \in \Delta \,\}$.
   Show that the right-hand map is not generally surjective, even if Γ is a cyclic group.

2. Let E/F be a finite Galois extension, and let K be any intermediate field of E/F. Show that the diagram below

$$
\begin{array}{ccc}
\mathrm{Cog}(E/F) & \longrightarrow & \mathrm{Cog}(E/K) \\
\Big\downarrow{\scriptstyle \wr} & & {\scriptstyle \wr}\Big\downarrow \\
Z^1(\mathrm{Gal}(E/F), \mu(E)) & \longrightarrow & Z^1(\mathrm{Gal}(E/K), \mu(E))
\end{array}
$$

   is commutative, where all the arrows are canonical morphisms.

3. Let E/F be a finite Galois Cogalois extension with Galois group Γ. Prove that the canonical morphism

$$Z^1(\Gamma, \mu(E)) \longrightarrow Z^1(\Delta, \mu(E))$$

   is surjective for every $\Delta \leqslant \Gamma$. (*Hint:* Use Exercise 2.)

4. Let $F = \mathbb{Q}$, $E = \mathbb{Q}(\zeta_3)$, and $\Gamma = \mathrm{Gal}(E/F)$. Show that
$$Z^1(\Gamma, \mu(E)) \cong \mathrm{Cog}(E/F) \cong \mathbb{Z}_6$$
and $|\underline{\mathrm{Radical}}(E/F)| = 2$.

5. Show that the result of Lemma 6.1.6 may fail if neither condition (1) nor condition (2) is satisfied. (*Hint*: Take the $\mathbb{Q}^*\langle\sqrt{-3}\rangle$-Cogalois extension $\mathbb{Q}(\zeta_3)/\mathbb{Q}$ and the element ζ_3.)

6. Let E/F be a G-Cogalois extension with $n = \exp(G/F^*)$, and let $x \in T(E/F)$ with $m = \mathrm{ord}(\widehat{x})$. Prove that $x \in G$ if and only if $\mathcal{P}_m \subseteq \mathcal{P}_n$.

7. Show that Lemma 6.1.6 holds even if $\mathcal{P}_m = \varnothing$, i.e., if $m = 1$ or $m = 2$.

8. Let K be an algebraic number field, let $r, n_1, \ldots, n_r \in \mathbb{N}^*$ with $n_1, \ldots, n_r \geqslant 2$, and set $n = n_1 \cdot \ldots \cdot n_r$. Prove that there exist $a_1, \ldots, a_r \in \mathbb{N}^*$ such that $a_i^{n_i} \in K^{*n}$, $K^{*n}\langle a_i \rangle / K^{*n} \cong \mathbb{Z}/n_i\mathbb{Z}$ for every $i \in \{1, \ldots, r\}$, and
$$K^{*n}\langle a_1, \ldots, a_r \rangle / K^{*n} = \bigoplus_{i=1}^{r} (K^{*n}\langle a_i \rangle / K^{*n}),$$
that is, for any $k_1, \ldots, k_r \in \mathbb{N}$ one has
$$a_1^{k_1} \cdot \ldots \cdot a_r^{k_r} \in K^{*n} \iff n_i \,|\, k_i \text{ for all } i \in \{1, \ldots, r\}.$$
(*Hint*: Let δ_K be the discriminant of K, and let $p_1, \ldots, p_r$ be distinct prime numbers which do not divide δ_K. Then $p_1, \ldots, p_r$ are unramified primes in K, and take $a_i := p_i^{n/n_i}$, $i = 1, \ldots, r$.)

9. Let K/F be a separable extension of degree $n \geqslant 2$ with $\gcd(n, e(F)) = 1$, and set $L = F(\zeta_n)$ and $E = K(\zeta_n)$. Suppose that $K \cap L = F$, E/L is an Abelian extension, and E/F is a Galois extension. Prove the following statements.

    (a) E/L is an Abelian extension with $\exp(\mathrm{Gal}(E/L)\,|\,n$, so E/L is a classical n-Kummer extension.

    (b) There exist $r, n_1, \ldots, n_r \in \mathbb{N}^*$ with $n_1, \ldots, n_r \geqslant 2$ and $\alpha_1, \ldots, \alpha_r \in L^*$ such that, if we set $n = n_1 \cdot \ldots \cdot n_r$, then $[L(\sqrt[n]{\alpha_i}) : L] = n_i = \mathrm{ord}(\widehat{\sqrt[n]{\alpha_i}})$ for every i, $1 \leqslant i \leqslant r$, $E = L(\sqrt[n]{\alpha_1}, \ldots, \sqrt[n]{\alpha_r})$, $E^{*n} \cap L^* = L^{*n}\langle \alpha_1, \ldots, \alpha_r \rangle$, $\alpha_1, \ldots, \alpha_r$ are independent modulo L^{*n}, that is,
$$\alpha_1^{k_1} \cdot \ldots \cdot \alpha_r^{k_r} \in L^{*n} \iff n_i \,|\, k_i \text{ for all } i = 1, \ldots, r,$$

and
$$\mathrm{Gal}(E/L) = \langle \tau_1 \rangle \oplus \cdots \oplus \langle \tau_r \rangle,$$
where
$$\tau_i(\sqrt[n]{\alpha_i}) = \zeta_{n_i} \sqrt[n]{\alpha_i} \quad \text{and} \quad \tau_i(\sqrt[n]{\alpha_j}) = \sqrt[n]{\alpha_j}$$
for every $i \ne j$ in $\{1, \dots, r\}$.

(c) For every $\sigma \in \mathrm{Gal}(L/F)$ and every i, $1 \leqslant i \leqslant r$, one has $\sigma(\alpha_i) \in L^{*n} \langle \alpha_1, \dots, \alpha_r \rangle$, hence there exist uniquely determined elements $\gamma_i^\sigma \in L$ and $0 \leqslant b_{i1}^\sigma < n_1, \dots, 0 \leqslant b_{ir}^\sigma < n_r$ such that
$$\sigma(\alpha_i) = (\gamma_i^\sigma)^n \cdot \alpha_1^{b_{i1}^\sigma} \cdot \ldots \cdot \alpha_r^{b_{ir}^\sigma}.$$

(*Hint*: Use Hasse [**69**, Satz 152, p. 223].)

10. Let K/F be a separable extension of degree $n \geqslant 2$ with $\gcd(n, e(F)) = 1$, and suppose that $K \cap F(\zeta_n) = F$. Prove the following statements.

    (a) If $K(\zeta_n)/F(\zeta_n)$ is an Abelian extension, then $K(\zeta_n)/F$ is a Galois extension

    (b) There exist $s \in \mathbb{N}^*$ and $a_1, \dots, a_s \in F^*$ such that $K = F(\sqrt[n]{a_1}, \dots, \sqrt[n]{a_s})$ if and only if $K(\zeta_n)/F(\zeta_n)$ is an Abelian extension and, with the notation of Exercise 9, the following condition is satisfied:

    (†) $b_{ij}^\sigma = \delta_{ij}$ for all $\sigma \in \mathrm{Gal}(F(\zeta_n)/F)$ and $i, j \in \{1, \dots, r\}$.

    (*Hint*: Use Exercises 8 and 9, and Theorem 6.2.1.)

11. (*Barrera-Mora and Vélez* [**32**]). Let E/F be a separable extension of degree $n \geqslant 2$ with $\gcd(n, e(F)) = 1$, and let $\widetilde{E}/F$ be the normal closure of E/F. Suppose that there exists a finite extension L/F such that $\widetilde{E}(\zeta_n) \cap L = F$ and $LE = L(\sqrt[n]{\alpha})$ for some $\alpha \in L^*$. Prove that $E = F(\sqrt[n]{\alpha})$. (*Hint*: Apply Theorem 6.2.1.)

12. (*Barrera-Mora and Vélez* [**32**]). Let K/F be a separable extension of degree $n \geqslant 2$ with $\gcd(n, e(F)) = 1$, and set $L = F(\zeta_n)$ and $E = K(\zeta_n)$. Suppose that $K \cap L = F$, E/L is a cyclic extension, and E/F is a Galois extension. Prove the following statements.

    (a) There exists $\alpha \in L^*$ such that $E = L(\sqrt[n]{\alpha})$ and $\mathrm{Gal}(E/L) = \langle \tau \rangle$, where $\tau(\sqrt[n]{\alpha}) = \zeta_n \sqrt[n]{\alpha}$.

    (b) For every $\sigma \in \mathrm{Gal}(L/F)$ there exist $\gamma_\sigma \in L$ and $b_\sigma \in \mathbb{N}$ such that $\gcd(b_\sigma, n) = 1$ and $\sigma(\alpha) = \gamma_\sigma^n \cdot \alpha^{b_\sigma}$.

(*Hint*: Apply Exercise 9.)

13. (*Barrera-Mora and Vélez* [**32**]). Let E_1/F and E_2/F be finite separable subextensions of an extension E/F, with $E_1 \cap E_2 = F$. Suppose that the extensions E_2/F and E_1E_2/E_2 are both Galois. Prove that E_1E_2/F is a Galois extension.

14. (*Barrera-Mora and Vélez* [**32**]). Let K/F be a separable extension of degree $n \geqslant 2$ with $\gcd(n, e(F)) = 1$, and suppose that $K \cap F(\zeta_n) = F$ and $K(\zeta_n)/F(\zeta_n)$ is a cyclic extension. Prove the following statements.
    (a) $K(\zeta_n)/F$ is a Galois extension. Moreover, $K(\zeta_n)/F$ is an Abelian extension if and only if K/F is a cyclic extension.
    (b) K/F is a simple radical extension if and only if $b_\sigma \equiv 1 \,(\mathrm{mod}\ n)$ for all $\sigma \in \mathrm{Gal}(F(\zeta_n)/F)$, where the integers b_σ were defined in Exercise 12.
    (*Hint*: Apply Exercises 12 and 13, and Theorem 6.2.1)

15. (*Barrera-Mora and Vélez* [**32**]). Let E/F be a separable extension of degree a prime number p different from $\mathrm{Char}(F)$, and suppose that the following conditions are satisfied.
    (a) $F(\zeta_p)/F$ is a quadratic extension,
    (b) The normal closure of E/F is $E(\zeta_p)/F$,
    (c) $\mathrm{Gal}(E(\zeta_p)/F)$ is isomorphic to the dihedral group $\mathfrak{D}_p$ of order $2p$.
    Prove that E/F is a simple radical extension. (*Hint*: Use Exercise 14.)

16. (*Barrera-Mora and Vélez* [**32**]). Prove that for any field F of characteristic $\neq 3$, and for any separable cubic extension E/F such that $E(\zeta_3)/F$ is the normal closure of E/F, there exists $a \in F^*$ such that $E = F(\sqrt[3]{a})$. (*Hint*: Use Exercise 15.)

17. (*Barrera-Mora and Vélez* [**32**]). This example shows that condition (c) in Exercise 15 cannot be omitted. Let $u = (1+\sqrt{5})/2$, $\alpha = \sqrt[5]{u}$, $F = \mathbb{Q}(\zeta_5)$, and $E = F(\alpha)$. Prove the following statements.
    (a) $u \in F$, $\mathrm{Min}(\alpha, F) = X^5 - u$, and $\mathrm{Min}(\alpha, \mathbb{Q}) = X^{10} - X^5 - 1$.
    (b) $E/\mathbb{Q}$ is a Galois extension with $\mathrm{Gal}(E/F) = \langle \sigma, \tau \rangle$, where $\sigma(\zeta_5) = \zeta_5^2$, $\tau(\zeta_5) = \zeta_5$, $\sigma^4 = \tau^5 = 1_E$, and $\sigma \circ \tau = \tau^3 \circ \sigma$.
    (c) If K is the fixed field of the 2-Sylow group of $\mathrm{Gal}(E/\mathbb{Q})$, then $[K : \mathbb{Q}] = 5$ and $E/\mathbb{Q}$ is the normal closure of $K/\mathbb{Q}$.
    (d) $K/\mathbb{Q}$ is not a simple radical extension.

18. (*Barrera-Mora* [**29**]). An extension E/F is said to be a *repeated radical extension* if there exists a tower of fields

$$F = F_0 \subseteq F_1 \subseteq F_2 \subseteq \ldots \subseteq F_n = E$$

so that F_{i+1}/F_i is a simple radical extension for all $i = 0, \ldots, n-1$.

Show that a subextension of a repeated radical extension is not necessarily a repeated radical extension. (*Hint*: Any cubic cyclic extension $K/\mathbb{Q}$ is not a repeated radical extension, but it is a subextension of the repeated radical extension $K(\zeta_3)/\mathbb{Q}$.)

19. (*Barrera-Mora* [**29**]). Show that there exists a repeated radical extension E/F having an intermediate field K such that the extension K/F is not a repeated radical extension. Furthermore, the extension E/F can be chosen to be Galois and $E \setminus F$ contains no roots of unity.

20. (*Barrera-Mora* [**29**]). Let E/F be a repeated radical Galois extension such that $E \setminus F$ contains no p-th root of unity for any prime number p. Prove that F contains a primitive p-th root of unity for every prime divisor p of $[E : F]$.

21. (*Barrera-Mora* [**29**]). Let E/F be a repeated radical extension so that $E \setminus F$ contains no p-th root of unity for any prime number p. Prove that there exists a tower of fields

$$F = F_0 \subseteq F_1 \subseteq F_2 \subseteq \ldots \subseteq F_n = E$$

such that $[F_{i+1} : F_i]$ is a prime number p_i and $F_{i+1} = F_i(\sqrt[p_i]{a_i})$ for some $a_i \in F_i$, $i = 0, \ldots, n-1$.

22. (*Barrera-Mora* [**29**]). A *radical tower* is a repeated radical extension E/F such that there exists a tower of fields

$$F = F_0 \subseteq F_1 \subseteq F_2 \subseteq \ldots \subseteq F_n = E$$

with $F_{i+1} = F_i(\sqrt[n_i]{a_i})$, $a_i \in F_i$, and $[F_{i+1} : F_i] = n_i$ for all $i = 0, \ldots, n-1$.

Show that in a radical tower, the n_i's can always be taken to be prime.

23. (*Barrera-Mora* [**29**]). Let E/F be a separable extension of degree $n \geqslant 2$, and let $n = p_1^{k_1} \cdot \ldots \cdot p_r^{k_r}$, with $p_1, \ldots, p_r$ mutually distinct prime numbers and $k_1, \ldots, k_r \in \mathbb{N}^*$. Assume that the following conditions are satisfied.

(a) $\zeta_{p_i} \notin E \setminus F$ for all $i = 1, \ldots, r$.

(b) There exist intermediate fields $E_1, \ldots, E_r$ of E/F such that E_i/F is a radical tower and $[E_i : F] = p_i^{k_i}$ for all $i = 1, \ldots, r$.

Prove that every subextension of E/F is a radical tower.

24. (*Barrera-Mora* [**29**]). Prove that under the assumptions of Exercise 23, the fields $E_1, \ldots, E_r$ are unique.

6.4. Bibliographical comments to Chapter 6

Section 6.1. Theorem 6.1.2 was established in the particular case of finite Galois extensions of algebraic number fields by Dummit [**55**], and for arbitrary Galois extensions which are not necessarily finite by Barrera-Mora, Rzedowski-Calderón, and Villa-Salvador [**30**] (see Theorem 15.1.2). Our approach, based on Lemma 6.1.1 is a bit different, and can be also easily adjusted for infinite Galois extensions.

Corollary 6.1.4 is essentially due to Greither and Harrison [**63**], who proved it for finite extensions of algebraic number fields by using a different approach. The idea to use Theorem 6.1.2 for the proof of the generalization of Greither and Harrison's result to arbitrary finite Galois extensions E/F with $\mu(E)$ finite is due to Barrera-Mora, Rzedowski-Calderón, and Villa-Salvador [**30**].

Theorem 6.1.7 is due to Albu and Nicolae [**22**], and is based on an older version of Lemma 6.1.6, proved by Albu, Nicolae, and Ţena [**23**]. Lemma 6.1.6, as stated in this section is due to Albu [**7**]. Corollary 6.1.5 and Proposition 6.1.8 were established by Albu and Nicolae [**22**]. Particular cases of Proposition 6.1.8, for E/F a Galois Cogalois extension or a so-called "neat presentation" were proved by Greither and Harrison [**63**] using a quite sophisticated technique including the Lyndon-Hochschild spectral sequence.

Section 6.2. All the results of this section are due to Albu and Nicolae [**22**]. Theorems 6.2.1 and 6.2.4 generalize to arbitrary finite radical extensions similar results of Barrera-Mora and Vélez [**32**] established for simple radical extensions.

CHAPTER 7

EXAMPLES OF G-COGALOIS EXTENSIONS

The fundamental property of Cogalois extensions says that any such extension E/F is an extension with $T(E/F)/F^*$-Cogalois correspondence (Theorem 3.2.3 (1)); in other words, an extension with $T(E/F)$-Cogalois extension. We show in this chapter that this is an immediate consequence of the n-Purity Criterion we proved in Section 4.3.

We deduce the essential part of the finite Kummer Theory from the fact that any finite *classical Kummer extension* is G-Cogalois, which in turn, is an immediate consequence of the n-Purity Criterion. Moreover, this criterion allows us to provide other large classes of G-Cogalois extensions which generalize or are closely related to classical finite Kummer extensions: *finite generalized Kummer extensions, finite Kummer extensions with few roots of unity*, and *finite quasi-Kummer extensions*.

7.1. Classical Kummer extensions

In this section we present first various characterizations of classical Kummer extensions which are not necessarily finite. Then, we show that finite classical Kummer extensions are G-Cogalois. As an immediate consequence of this fact we deduce the whole finite classical Kummer Theory.

Recall that Ω is a fixed algebraically closed field containing the fixed base field F as a subfield; any considered overfield of F is supposed to be a subfield of Ω.

For any nonempty subset A of F^* and any $n \in \mathbb{N}^*$ we will denote by $\sqrt[n]{A}$ the subset of $T(\Omega/F)$ defined by

$$\sqrt[n]{A} := \{\, x \in \Omega \,|\, x^n \in A \,\}.$$

In particular, if A is a singleton $\{a\}$, then $\sqrt[n]{\{a\}}$ is precisely the set of all roots (in Ω) of the polynomial $X^n - a \in F[X]$. We shall use throughout this monograph the notation $\sqrt[n]{a}$ to designate a root, which in general is not specified, of this polynomial. Thus, $\sqrt[n]{a} \in \sqrt[n]{\{a\}}$. More precisely, for any choice of the root $\sqrt[n]{a}$, we have

$$\sqrt[n]{\{a\}} = \{\, \zeta_n^k \sqrt[n]{a} \mid 0 \leqslant k \leqslant n - 1 \,\}.$$

In particular, if $\zeta_n \in F$, then $F(\sqrt[n]{\{a\}}) = F(\sqrt[n]{a})$.

However, in certain cases, for instance when F is a subfield of the field $\mathbb{R}$ of all real numbers and $a > 0$, then $\sqrt[n]{a}$ will always mean the unique positive root in $\mathbb{R}$ of the polynomial $X^n - a$.

DEFINITION 7.1.1. *A classical n-Kummer extension, where $n \in \mathbb{N}^*$, is an Abelian extension E/F such that $\gcd(n, e(F)) = 1$, $\mu_n(\Omega) \subseteq F$ and $\mathrm{Gal}(E/F)$ is a group of exponent a divisor of n.*

A classical Kummer extension, or just a Kummer extension is any extension which is a classical n-Kummer extension for a certain integer $n \geqslant 1$. If E/F is a classical Kummer extension, we also say that E is a classical Kummer extension of F. $\square$

The next result gives a more precise form of classical Kummer extensions which are not necessarily finite.

THEOREM 7.1.2. *The following assertions are equivalent for an extension E/F and a natural number $n \geqslant 1$.*

(1) *E/F is a classical n-Kummer extension.*

(2) *$\gcd(n, e(F)) = 1$, $\mu_n(\Omega) \subseteq F$, and $E = F(\sqrt[n]{A})$ for some $\varnothing \neq A \subseteq F^*$.*

(3) *$\gcd(n, e(F)) = 1$, $\mu_n(\Omega) \subseteq F$, and $E = F(B)$ for some $\varnothing \neq B \subseteq E^*$ with $B^n \subseteq F$.*

PROOF. (1) $\Longrightarrow$ (2): The extension E/F is the compositum of all its finite subextensions E'/F. Observe that the finite Abelian group $\mathrm{Gal}(E'/F)$ is an internal direct sum of a finite family $(H_i)_{1 \leqslant i \leqslant m}$ of cyclic subgroups, where each H_i is the Galois group of a cyclic subextension K_i/F of E/F. We deduce that each such finite extension E'/F is a compositum of cyclic extensions (see 1.2.9), hence E/F is the compositum of all its cyclic subextensions.

Let K/F be an arbitrary finite cyclic subextension of E/F. Since $\mathrm{Gal}(E/F)$ is by hypothesis an Abelian group of exponent a divisor of n, it follows that $\sigma^n = 1_K$ for all $\sigma \in \mathrm{Gal}(K/F)$, hence $|\mathrm{Gal}(K/F)|$ is a divisor of n, and so $[K : F] \mid n$. But, it is known that if $\zeta_n \in F$, then for any finite cyclic extension K/F of degree a divisor of n, there exists an $x \in K^*$ such that $x^n \in F$ and $K = F(x)$ (see 1.2.11), in other words, K/F is a simple radical extension $F(\sqrt[n]{a})/F$ generated by an n-th radical $\sqrt[n]{a}$ of some $a \in F^*$.

Set $B = \{ x \mid x \in E^*, x^n \in F \}$ and denote by B' the set of all elements in B that generate those cyclic subextensions K/F of E/F whose compositum is the given extension E/F. Since $B' \subseteq B$, we have $E = F(B') \subseteq F(B) \subseteq E$, hence $E = F(B)$. If we set $A = B^n$, then clearly $E = F(\sqrt[n]{A})$.

$(2) \implies (1)$: Obviously, condition (2) implies that E is the compositum of the family of fields $(F(\sqrt[n]{a}))_{a \in A}$. As is well-known, each extension $F(\sqrt[n]{a})/F$ is a cyclic extension of degree a divisor of n, hence E/F is an Abelian extension.

Let $\sigma \in \mathrm{Gal}(E/F)$. Then, for every $a \in A$, the restriction of σ^n to $F(\sqrt[n]{a})$ is the identity map, which implies that $\sigma^n = 1_E$. This shows that the group $\mathrm{Gal}(E/F)$ has exponent a divisor of n.

$(2) \iff (3)$ is obvious. $\qquad\square$

The form of finite classical Kummer extensions is an immediate consequence of Theorem 7.1.2.

COROLLARY 7.1.3. *The following assertions are equivalent for a finite extension E/F and a natural number $n \geqslant 1$.*

(1) *E/F is a classical n-Kummer extension.*
(2) *$\gcd(n, e(F)) = 1$, $\mu_n(\Omega) \subseteq F$, and $E = F(\sqrt[n]{a_1}, \ldots, \sqrt[n]{a_r})$ for some $r \in \mathbb{N}^*$ and $\{a_1, \ldots, a_r\} \subseteq F^*$.* $\qquad\square$

We are now going to show that the whole finite Kummer Theory can be very easily derived from Cogalois Theory. A similar result also holds for infinite Kummer extensions and will be presented in Section 13.1.

LEMMA 7.1.4. *Let E/F be a separable G-radical extension with G/F^* finite, and let $n \in \mathbb{N}^*$ be such that $G^n \subseteq F^*$. If the extension E/F is n-pure, then E/F is G-Cogalois.*

PROOF. The finite group G/F^* has a finite exponent, say m, and $m \mid n$ by Lemma 1.4.6. If $p \in \mathcal{P}_m$, then $p \mid n$, and $\mu_p(E) \subseteq F$ by hypothesis. Consequently, the extension E/F is also m-pure, hence it is G-Cogalois by Theorem 4.3.2. $\qquad\square$

LEMMA 7.1.5. *The following statements hold for any extension E/F such that $\mu_n(E) \subseteq F$ for a given $n \in \mathbb{N}^*$, and any group G such that $F^* \leqslant G \leqslant E^*$.*

(1) *The map $G/F^* \longrightarrow G^n/F^{*n}$, $xF^* \mapsto x^n F^{*n}$, is a group isomorphism.*

(2) *The maps $H \mapsto H^n$ and $M \mapsto \sqrt[n]{M} \cap G$ establish lattice isomorphisms, inverse to one another, between the lattices*

$$\{\, H \mid F^* \leqslant H \leqslant G \,\} \quad and \quad \{\, M \mid F^{*n} \leqslant M \leqslant G^n \,\}.$$

(3) *If $\mu_n(\Omega) \subseteq F$, then $\sqrt[n]{X} \subseteq G$ for every $\varnothing \neq X \subseteq G^n$.*

PROOF. (1) The surjective group morphism $G \longrightarrow G^n$, $x \mapsto x^n$, induces clearly a surjective group morphism

$$\varphi : G/F^* \longrightarrow G^n/F^{*n}, \quad xF^* \mapsto x^n F^{*n}.$$

For simplicity, denote $\widehat{x} = xF^*$ and $\widehat{\widehat{x}} = xF^{*n}$ for any $x \in E^*$.

We are going to show that φ is also injective. Let $x \in G$ be such that $\varphi(\widehat{x}) = \widehat{1}$. Then, $\widehat{\widehat{x^n}} = \widehat{\widehat{1}} \Rightarrow x^n \in F^{*n} \Rightarrow x^n = y^n$ for some $y \in F^* \Rightarrow (xy^{-1})^n = 1 \Rightarrow xy^{-1} \in \mu_n(E) \Rightarrow xy^{-1} \in F^* \Rightarrow x \in F^* \Rightarrow \widehat{x} = \widehat{1}$.

(2) Recall that for any nonempty subset A of F, we denoted $\sqrt[n]{A} = \{\, x \mid x \in \Omega,\, x^n \in A \,\}$, where Ω is an algebraically closed field containing E as a subfield. Using again the fact that $\mu_n(E) \subseteq F$, one shows easily that

$$\sqrt[n]{H^n} \cap G = H \quad and \quad (\sqrt[n]{M} \cap G)^n = M$$

for every $F^* \leqslant H \leqslant G$ and every $F^{*n} \leqslant M \leqslant G^n$, which proves (2).

(3) Clearly, $X \subseteq G^n$ implies that $\sqrt[n]{X} \subseteq \sqrt[n]{G^n}$. But $\sqrt[n]{G^n} = G$. Indeed, the inclusion $G \subseteq \sqrt[n]{G^n}$ is clear. For the opposite inclusion, let $x \in \sqrt[n]{G^n}$. Then $x^n = g^n$ for some $g \in G$, hence $(xg^{-1})^n = 1$, i.e., $xg^{-1} \in \mu_n(\Omega) \subseteq F$. Consequently, $x \in G$ since $F^* \subseteq G$. $\qquad\square$

THEOREM 7.1.6. *Let E/F be a finite classical n-Kummer extension, with $E = F(\sqrt[n]{a_1}, \dots, \sqrt[n]{a_r})$, $n, r \in \mathbb{N}^*$, and $\{a_1, \dots, a_r\} \subseteq F^*$. Then, the following statements hold.*

(1) *E/F is an $F^*\langle \sqrt[n]{a_1}, \dots, \sqrt[n]{a_r}\rangle$-Cogalois extension.*

(2) *The maps $H \mapsto F(\sqrt[n]{H})$ and $K \mapsto K^n \cap (F^{*n}\langle a_1, \dots, a_r\rangle)$ establish isomorphisms of lattices, inverse to one another, between the lattice of all subgroups H of $F^{*n}\langle a_1, \dots, a_r\rangle$ containing F^{*n} and the lattice of all intermediate fields K of E/F. Moreover, any subextension K/F of E/F is a classical n-Kummer extension.*

(3) *If* H *is any subgroup of* $F^{*n}\langle a_1,\dots,a_r\rangle$ *containing* F^{*n}, *then any set of representatives of the group* $\sqrt[n]{H}/F^*$ *is a vector space basis of* $F(\sqrt[n]{H})$ *over* F, *and* $[F(\sqrt[n]{H}):F] = |H/F^{*n}|$. *In particular, one has*

$$[E:F] = |F^*\langle \sqrt[n]{a_1},\dots,\sqrt[n]{a_r}\rangle/F^*| = |F^{*n}\langle a_1,\dots a_r\rangle/F^{*n}|.$$

(4) *There exists a canonical group isomorphism*

$$F^*\langle \sqrt[n]{a_1},\dots,\sqrt[n]{a_r}\rangle/F^* \cong \operatorname{Hom}(\operatorname{Gal}(E/F),\mu_n(F)).$$

PROOF. (1) Set $A = \{a_1,\dots,a_r\}$, $B = \sqrt[n]{A}$ and $G = F^*\langle B\rangle$. Since $\mu_n(\Omega) \subseteq F$ it follows that the extension E/F is n-pure. Clearly $G^n \subseteq F$, hence the extension E/F is G-Cogalois by Lemma 7.1.4. But $G = F^*\langle B\rangle = F^*\langle \sqrt[n]{a_1},\dots,\sqrt[n]{a_r}\rangle$, and the result follows.

(2) By (1) and Theorem 4.3.2, the maps

$$H \mapsto F(H) \text{ and } K \mapsto K \cap (F^*\langle \sqrt[n]{A}\rangle)$$

are lattice isomorphisms, inverse to one another, between the lattice of all subgroups H of $F^*\langle \sqrt[n]{A}\rangle$ containing F^* and the lattice of all subextensions K/F of E/F. If K/F is any subextension of E/F, then it is an Abelian extension and its Galois group is a quotient group of $\operatorname{Gal}(E/F)$, hence a group having exponent a divisor of n. By definition, K/F is a classical n-Kummer extension. Now, apply Lemma 7.1.5 for $G = F^*\langle \sqrt[n]{A}\rangle$ and observe that $G^n = F^{*n}\langle A\rangle$.

(3) The extension E/F is $F^*\langle \sqrt[n]{A}\rangle$-Cogalois by (1), hence the extension $F(\sqrt[n]{H})/F$ is $\sqrt[n]{H}$-Kneser by Corollary 2.1.10. Use again Lemma 7.1.5 and Corollary 2.1.10 to deduce the desired statements.

(4) We have $\operatorname{Kne}(E/F) = G/F^* = F^*\langle \sqrt[n]{a_1},\dots,\sqrt[n]{a_r}\rangle/F^*$ by (1). On the other hand, by Theorem 6.1.7, there exists a canonical group isomorphism

$$\operatorname{Kne}(E/F) \cong Z^1(\operatorname{Gal}(E/F),\mu_m(E)),$$

where $m = \exp(G/F^*)$. Since E/F is a classical n-Kummer extension we have $\mu_n(\Omega) = \mu_n(E) = \mu_n(F) \subseteq F$, and since $m \mid n$ we also have $\mu_m(E) \subseteq \mu_n(E) \subseteq F$. But $\operatorname{Gal}(E/F)) \cong \operatorname{Kne}(E/F)$ by (1) and Theorem 5.2.2, hence

$$Z^1(\operatorname{Gal}(E/F),\mu_m(E)) = \operatorname{Hom}(\operatorname{Gal}(E/F),\mu_m(E))$$
$$= \operatorname{Hom}(\operatorname{Gal}(E/F),\mu_n(F))$$

and the desired result follows. $\square$

7.2. Generalized Kummer extensions

In this section we present another class of G-Cogalois extensions, which is larger than the class of Kummer extensions, namely the class of *generalized Kummer extensions*. This new class also includes the class of *Kummer extension with few roots of unity* which will be discussed in the next section.

We show that a theory of finite generalized Kummer extensions, which is very similar to that of finite classical Kummer extensions, can be developed using the properties of G-Cogalois extensions; since, in general, they are not Galois extensions, no other approach (e.g., via Galois Theory, as in the case of classical Kummer extensions) is applicable. The case of infinite generalized Kummer extensions will be examined in Section 13.2.

DEFINITION 7.2.1. *We say that a finite extension E/F is a* generalized *n-Kummer extension,   where $n \in \mathbb{N}^*$, if $\gcd(n, e(F)) = 1$, $\mu_n(E) \subseteq F$, and there exist $r \in \mathbb{N}^*$, $a_1, \ldots, a_r \in F^*$ such that $E = F(\sqrt[n]{a_1}, \ldots, \sqrt[n]{a_r})$.*

A generalized Kummer extension *is an extension which is a generalized n-Kummer extension for some integer $n \geqslant 1$.*   □

Clearly, any classical n-Kummer extension is a generalized n-Kummer extension. By Corollary 5.1.4, any generalized Kummer extension is a separable extension, but not necessarily a Galois extension.

LEMMA 7.2.2. *Let E/F be a finite generalized n-Kummer extension, with $E = F(\sqrt[n]{a_1}, \ldots, \sqrt[n]{a_r})$, $r \in \mathbb{N}^*$, and $a_1, \ldots, a_r \in F^*$. Then E/F is an $F^*\langle \sqrt[n]{a_1}, \ldots, \sqrt[n]{a_r} \rangle$-Cogalois extension.*

PROOF. Observe that the extension E/F is G-radical, where $G = F^*\langle \sqrt[n]{a_1}, \ldots, \sqrt[n]{a_r} \rangle$, and $G^n \subseteq F$; it is also n-pure, since for every $p \in \mathcal{P}_n$ one has $\mu_p(E) \subseteq \mu_n(E) \subseteq F$. It follows that E/F is a G-Cogalois extension by Lemma 7.1.4, as desired.   □

THEOREM 7.2.3. *Let F be a field, and let $r, n \in \mathbb{N}^*$, $a_1, \ldots, a_r \in F^*$. Denote $G = F^*\langle \sqrt[n]{a_1}, \ldots, \sqrt[n]{a_r} \rangle$ and $E = F(\sqrt[n]{a_1}, \ldots, \sqrt[n]{a_r})$. If $\gcd(n, e(F)) = 1$ and $\mu_n(E) \subseteq F$, then the following statements hold.*

(1) *The generalized n-Kummer extension E/F is G-Cogalois.*

(2) *The maps $H \mapsto F(\sqrt[n]{H} \cap G)$ and $K \mapsto K^n \cap (F^{*n}\langle a_1, \ldots, a_r \rangle)$ establish isomorphisms of lattices, inverse to one another, between the lattice of all subgroups H of $F^{*n}\langle a_1, \ldots, a_r \rangle$ containing F^{*n} and the lattice of all intermediate fields K of E/F. Moreover, any subextension K/F of E/F is a generalized n-Kummer extension.*

(3) *If H is any subgroup of $F^{*n}\langle a_1, \dots, a_r \rangle$ containing F^{*n}, then any set of representatives of the group $(\sqrt[n]{H} \cap G)/F^*$ is a vector space basis of $F(\sqrt[n]{H} \cap G)$ over F, and $[F(\sqrt[n]{H} \cap G) : F] = |H/F^{*n}|$. In particular, one has*

$$[E : F] = |F^*\langle \sqrt[n]{a_1}, \dots, \sqrt[n]{a_r} \rangle/F^*| = |F^{*n}\langle a_1, \dots a_r \rangle/F^{*n}|.$$

PROOF. Proceed as in the proof of Theorem 7.1.6, by applying Lemma 7.1.5 and Lemma 7.2.2.

To see that any subextension K/F of E/F is still a generalized n-Kummer extension, observe that, by (1), $K = F(U)$ for some subgroup U of G containing F^*. Clearly $U^n \leqslant G^n \subseteq F^*$, and so, any element of the finite group U/F^* is a finite product $\widehat{\sqrt[n]{a_1}}^{k_1} \cdot \ldots \cdot \widehat{\sqrt[n]{a_r}}^{k_r}$, where $k_1, \dots, k_r \in \mathbb{N}$. Since clearly $\mu_n(K) \subseteq \mu_n(E) \subseteq F$, one concludes that K/F is a generalized n-Kummer extension. $\square$

The next result establishes when a generalized Kummer extension is a classical Kummer extension.

PROPOSITION 7.2.4. *A finite generalized Kummer extension is a classical Kummer extension if and only if it is a Galois extension.*

PROOF. Let E/F be a finite generalized n-Kummer extension, with $E = F(\sqrt[n]{a_1}, \dots, \sqrt[n]{a_r})$, $r \in \mathbb{N}^*$, $a_1, \dots, a_r \in F^*$, and $\mu_n(E) \subseteq F$. Denote $G = F^*\langle \sqrt[n]{a_1}, \dots, \sqrt[n]{a_r} \rangle$ and let $m = \exp(G/F^*)$.

If E/F is a Galois extension, then $\gcd(m, e(F)) = 1$ and $\zeta_m \in E$ by Corollary 5.1.7. But $m \mid n$, hence

$$\zeta_m \in E \cap \mu_m(\Omega) = \mu_m(E) \subseteq \mu_n(E) \subseteq F.$$

Thus, E/F is a classical m-Kummer extension.

Conversely, any classical Kummer extension is by definition a Galois extension. $\square$

REMARK 7.2.5. Let E/F be an extension such that there exist $r \in \mathbb{N}^*$, $n_1, \dots, n_r \in \mathbb{N}^*$ and $a_1, \dots, a_r \in F^*$ with $E = F(\sqrt[n_1]{a_1}, \dots, \sqrt[n_r]{a_r})$. Let $G = F^*\langle \sqrt[n_1]{a_1}, \dots, \sqrt[n_r]{a_r} \rangle$ and $n = \operatorname{lcm}(n_1, \dots, n_r)$.

We claim that if $\gcd(e(F), n) = 1$ and $\mu_n(E) \subseteq F$, then E/F is a generalized n-Kummer extension and a G-Cogalois extension. Indeed, for each i, $1 \leqslant i \leqslant r$, there exists $q_i \in \mathbb{N}^*$ such that $n = n_i q_i$. Hence we can write $\sqrt[n_i]{a_i} = \sqrt[n]{a_i^{q_i}}$, and then $E = F(\sqrt[n]{c_1}, \dots, \sqrt[n]{c_r})$, with $c_i = a_i^{q_i}$ for all i, $1 \leqslant i \leqslant r$. By Lemma 7.2.2, E/F is an $F^*\langle \sqrt[n]{c_1}, \dots, \sqrt[n]{c_r} \rangle$-Cogalois extension, so, it is also a G-Cogalois extension. $\square$

EXAMPLES 7.2.6. (1) A generalized Kummer extension is not necessarily a classical Kummer extension (see Exercise 2).

(2) Observe that any generalized n-Kummer extension E/F is n-pure since for any $p \in \mathcal{P}_n$ one has $\mu_p(E) \subseteq \mu_n(E) \subseteq F$. However, the converse is not true according to Exercise 3. $\qquad\square$

7.3. Kummer extensions with few roots of unity

In this section we deal with finite Kummer extensions with few roots of unity, which are very particular cases of generalized Kummer extensions discussed in the previous section.

DEFINITION 7.3.1. *A finite extension E/F is said to be an n-*Kummer *extension with few roots of unity, where $n \in \mathbb{N}^*$, if $\gcd(n, e(F)) = 1$, $\mu_n(E) \subseteq \{-1, 1\}$, and there exist $r \in \mathbb{N}^*$, and $a_1, \ldots, a_r \in F^*$ such that $E = F(\sqrt[n]{a_1}, \ldots, \sqrt[n]{a_r})$.*

A Kummer extension with few roots of unity *is an extension which is an n-Kummer extensions with few roots of unity for some positive integer $n \geqslant 1$.* $\qquad\square$

The next result establishing the basic properties of Kummer extensions with few roots of unity is an obvious consequence of Theorem 7.2.3. We state it for the reader's convenience.

THEOREM 7.3.2. *Let F be a field, and let $r, n \in \mathbb{N}^*$, $a_1, \ldots, a_r \in F^*$. Denote $G = F^*\langle \sqrt[n]{a_1}, \ldots, \sqrt[n]{a_r} \rangle$ and $E = F(\sqrt[n]{a_1}, \ldots, \sqrt[n]{a_r})$. If $\gcd(n, e(F)) = 1$ and $\mu_n(E) \subseteq \{-1, 1\}$, then the following statements hold.*

(1) *The n-Kummer extension with few roots of unity E/F is G-Cogalois.*

(2) *The maps $H \mapsto F(\sqrt[n]{H} \cap G)$ and $K \mapsto K^n \cap (F^{*n}\langle a_1, \ldots, a_r \rangle)$ establish isomorphisms of lattices, inverse to one another, between the lattice of all subgroups H of $F^{*n}\langle a_1, \ldots, a_r \rangle$ containing F^{*n} and the lattice of all intermediate fields K of E/F. Moreover, any subextension K/F of E/F is an n-Kummer extension with few roots of unity.*

(3) *If H is any subgroup of $F^{*n}\langle a_1, \ldots, a_r \rangle$ containing F^{*n}, then any set of representatives of the group $(\sqrt[n]{H} \cap G)/F^*$ is a vector space*

basis of $F(\sqrt[n]{H} \cap G)$ *over* F, *and* $[F(\sqrt[n]{H} \cap G) : F] = |H/F^{*n}|$.
In particular, one has

$$[E : F] = |F^*\langle \sqrt[n]{a_1}, \ldots, \sqrt[n]{a_r} \rangle / F^*| = |F^{*n}\langle a_1, \ldots a_r \rangle / F^{*n}|.$$

□

Recall that whenever $a \in \mathbb{R}^*_+$ and $n \in \mathbb{N}^*$, then $\sqrt[n]{a}$ will always mean the unique positive root in $\mathbb{R}$ of the polynomial $X^n - a$.

COROLLARY 7.3.3. *Let* F *be a subfield of* $\mathbb{R}$, *let* $r, n \in \mathbb{N}^*$, *and let* $a_1, \ldots, a_r \in F^*$ *be positive real numbers. Denote* $G = F^*\langle \sqrt[n]{a_1}, \ldots, \sqrt[n]{a_r} \rangle$ *and* $E = F(\sqrt[n]{a_1}, \ldots, \sqrt[n]{a_r})$. *Then, the following statements hold.*

(1) *The extension* E/F *is* G-*Cogalois.*

(2) *The maps* $H \mapsto F(\sqrt[n]{H} \cap G)$ *and* $K \mapsto K^n \cap (F^{*n}\langle a_1, \ldots, a_r \rangle)$ *establish isomorphisms of lattices, inverse to one another, between the lattice of all subgroups* H *of* $F^{*n}\langle a_1, \ldots, a_r \rangle$ *containing* F^{*n} *and the lattice of all intermediate fields* K *of* E/F.

(3) *If* H *is any subgroup of* $F^{*n}\langle a_1, \ldots, a_r \rangle$ *containing* F^{*n}, *then any set of representatives of the group* $(\sqrt[n]{H} \cap G)/F^*$ *is a vector space basis of* $F(\sqrt[n]{H} \cap G)$ *over* F, *and* $[F(\sqrt[n]{H} \cap G) : F] = |H/F^{*n}|$. *In particular, one has*

$$[E : F] = |F^*\langle \sqrt[n]{a_1}, \ldots, \sqrt[n]{a_r} \rangle / F^*| = |F^{*n}\langle a_1, \ldots a_r \rangle / F^{*n}|.$$

PROOF. The result follows immediately from Theorem 7.3.2 since any subfield K of $\mathbb{R}$ is obviously a field with few roots of unity, i.e., $\mu(K) \subseteq \{-1, 1\}$. □

REMARK 7.3.4. The extension $\mathbb{Q}(\sqrt{-3}, \sqrt[3]{2})/\mathbb{Q}(\sqrt{-3})$ is an example of a classical Kummer extension which is not a Kummer extension with few roots of unity (see Exercise 6). □

7.4. Quasi-Kummer extensions

The aim of this section is to present another class of G-Cogalois extensions which is close to the class of classical Kummer extensions.

DEFINITION 7.4.1. *We say that a finite extension* E/F *is a* quasi-Kummer extension *if there exist* $r \in \mathbb{N}^*$, $n_1, \ldots, n_r \in \mathbb{N}^*$, *and* $a_1, \ldots, a_r \in F^*$ *such that* $E = F(\sqrt[n_1]{a_1}, \ldots, \sqrt[n_r]{a_r})$, $\gcd(n, e(F)) = 1$, *and* $\mu_p(\Omega) \subseteq F$ *for any* $p \in \mathcal{P}_n$, *where* $n = \mathrm{lcm}\,(n_1, \ldots, n_r)$. □

Observe that the condition $\mu_p(\Omega) \subseteq F$ is equivalent to the condition $\zeta_p \in F$, and that any finite classical Kummer extension is a quasi-Kummer extension. However, the class of quasi-Kummer extensions is strictly larger than the class of classical Kummer extensions (see Exercise 9).

THEOREM 7.4.2. *A quasi-Kummer extension* $F(\sqrt[n_1]{a_1}, \ldots, \sqrt[n_r]{a_r})/F$ *as in the Definition 7.4.1 is* $F^*\langle \sqrt[n_1]{a_1}, \ldots, \sqrt[n_r]{a_r} \rangle$-*Cogalois.*

PROOF. Let $G = F^*\langle \sqrt[n_1]{a_1}, \ldots, \sqrt[n_r]{a_r} \rangle$, $n = \mathrm{lcm}(n_1, \ldots, n_r)$, and $m = \exp(G/F^*)$. Then $m \mid n$, hence for any $p \in \mathcal{P}_m$ we have $p \mid n$, and so, $\mu_p(E) \subseteq \mu_p(\Omega) \subseteq F$. This shows that E/F is m-pure. On the other hand, since $\gcd(m, e(F)) = 1$, the $F^*\langle \sqrt[n_1]{a_1}, \ldots, \sqrt[n_r]{a_r} \rangle$-radical extension E/F is separable by Corollary 5.1.4. Now apply Theorem 4.3.2, to conclude that the extension E/F is G-Cogalois.   □

THEOREM 7.4.3. *Any finite Galois* G-*Cogalois extension is a quasi-Kummer extension.*

PROOF. Let E/F be a finite Galois G-Cogalois extension, and consider a set $\{b_1, \ldots, b_r\}$ of representatives of the finite group G/F^*. If $n = \exp(G/F^*)$ then $b_i^n = a_i \in F$ for all i, $1 \leqslant i \leqslant r$, hence

$$E = F(b_1, \ldots, b_r) = F(\sqrt[n]{a_1}, \ldots, \sqrt[n]{a_r}).$$

Since the G-radical extension E/F is Galois, it follows that $\gcd(n, e(F)) = 1$ and $\zeta_n \in E$ by Corollary 5.1.7, so $\mu_n(\Omega) \subseteq \mu_n(E)$. By Theorem 4.3.2, the G-Cogalois extension E/F is n-pure, hence $\mu_p(E) \subseteq F$ for all $p \in \mathcal{P}_n$. Thus, $\mu_p(\Omega) \subseteq \mu_p(E) \subseteq F$ for all $p \in \mathcal{P}_n$. This shows that the extension E/F is quasi-Kummer.   □

COROLLARY 7.4.4. *The following assertions are equivalent for an algebraic number field* E.

(1) $E/\mathbb{Q}$ *is a Galois* G-*Cogalois extension for some group* G.
(2) $E/\mathbb{Q}$ *is an Abelian* G-*Cogalois extension for some group* G.
(3) *There exist finitely many nonzero rational integers* $a_1, \ldots, a_r$ *such that* $E = \mathbb{Q}(\sqrt{a_1}, \ldots, \sqrt{a_r})$.
(4) $E/\mathbb{Q}$ *is a classical 2-Kummer extension.*

PROOF. (1) $\Longrightarrow$ (3): By Theorem 7.4.3, $E/\mathbb{Q}$ is a quasi-Kummer extension, hence there exist $r, n_1, \ldots, n_r \in \mathbb{N}^*$, and $a_1, \ldots, a_r \in \mathbb{Q}^*$ such that $E = \mathbb{Q}(\sqrt[n_1]{a_1}, \ldots, \sqrt[n_r]{a_r})$ and $\mu_p(\mathbb{C}) \subseteq \mathbb{Q}$ for every $p \in \mathcal{P}_n$, where $n = \mathrm{lcm}(n_1, \ldots, n_r)$. But $\zeta_m \in \mathbb{Q}$ if and only if $m = 1$ or $m = 2$. Consequently, we must have $n \leqslant 2$, hence $n_i \leqslant 2$ for all $i = 1, \ldots r$. Of course, we may suppose that $a_i \in \mathbb{Z}^*$ for all $i = 1, \ldots, r$, and we may ignore those a_i for which $n_i = 1$.

$(3) \Longrightarrow (4)$ follows from Corollary 7.1.3.

$(4) \Longrightarrow (2)$ and $(2) \Longrightarrow (1)$ are obvious. $\qquad\square$

COROLLARY 7.4.5. *The cyclotomic extension* $\mathbb{Q}(\zeta_n)/\mathbb{Q}$ *is* G-*Cogalois for some group* G *if and only if* $n \in \{1, 2, 3, 4, 6, 8, 12, 24\}$.

PROOF. Assume that the extension $\mathbb{Q}(\zeta_n)/\mathbb{Q}$ is G-Cogalois for some group G. As is well-known, this extension is an Abelian extension with

$$\mathrm{Gal}(\mathbb{Q}(\zeta_n)/\mathbb{Q}) \cong U(\mathbb{Z}_n),$$

where, as usually, $U(\mathbb{Z}_n)$ denotes the group of units of the ring $\mathbb{Z}_n$ of integers modulo n.

By Theorem 5.2.2, we have

$$G/\mathbb{Q}^* \cong \mathrm{Gal}(\mathbb{Q}(\zeta_n)/\mathbb{Q}),$$

hence $G/\mathbb{Q}^* \cong U(\mathbb{Z}_n)$. On the other hand, by Corollary 7.4.4, $\mathbb{Q}(\zeta_n)/\mathbb{Q}$ has to be a classical 2-Kummer extension, which is H-Galois with respect to a group H with $\exp(H/\mathbb{Q}^*) \leqslant 2$. It follows that $G = H$ by Corollary 4.4.2. Consequently, we must have $\exp(G/\mathbb{Q}^*) \leqslant 2$, i.e., every element of $U(\mathbb{Z}_n)$ must have order $\leqslant 2$.

Of course, we may assume that $n \geqslant 2$. Let $n = 2^k \cdot p_1^{k_1} \cdot \ldots \cdot p_s^{k_s}$ with k, s, $k_1, \ldots, k_s \in \mathbb{N}$ and $p_1 < \cdots < p_s$ positive odd prime numbers. According to the Chinese Reminder Theorem, there exists a ring isomorphism

$$\mathbb{Z}_n \cong \mathbb{Z}_{2^k} \times \mathbb{Z}_{p_1^{k_1}} \times \cdots \times \mathbb{Z}_{p_s^{k_s}},$$

which yields a group isomorphism

$$U(\mathbb{Z}_n) \cong U(\mathbb{Z}_{2^k}) \times U(\mathbb{Z}_{p_1^{k_1}}) \times \cdots \times U(\mathbb{Z}_{p_1^{k_s}}).$$

It is known that

$$U(\mathbb{Z}_2) \cong \{0\},\ U(\mathbb{Z}_4) \cong \mathbb{Z}_2,\ U(\mathbb{Z}_{2^k}) \cong \mathbb{Z}_2 \times \mathbb{Z}_{2^{k-2}},\ k \geqslant 3,$$

and

$$U(\mathbb{Z}_{p^s}) \cong \mathbb{Z}_{p^{s-1}(p-1)},\ p \text{ odd prime},\ s \in \mathbb{N}^*.$$

All these facts imply that there exists a group isomorphism

$$G/\mathbb{Q}^* \cong U(\mathbb{Z}_{2^k}) \times \mathbb{Z}_{p_1^{k_1-1}(p_1-1)} \times \cdots \times \mathbb{Z}_{p_s^{k_s-1}(p_s-1)},$$

where $U(\mathbb{Z}_{2^k})$ is one of the groups $\{0\}$, $\mathbb{Z}_2$, or $\mathbb{Z}_2 \times \mathbb{Z}_{2^{k-2}}$ when $k \geqslant 3$.

But the group $U(\mathbb{Z}_{2^k})$ should not contain any element of order > 2, hence necessarily $k - 2 \leqslant 1$, i.e., $k \in \{0, 1, 2, 3\}$.

Assume that $s \geqslant 1$ and let i, $1 \leqslant i \leqslant s$. Since the group $\mathbb{Z}_{p_i^{k_i-1}(p_i-1)}$ should not contain any element of order > 2, it follows that $k_i \leqslant 1$ and $p_i - 1 = 2$, hence $p_i = 3$. But $p_1 < \cdots < p_s$, hence necessarily $s = 1$ and $p_1 = 3$. Consequently,

$$n \in \{ 2^k \cdot 3^{k_1} \mid k = 0, 1, 2, 3; \ k_1 = 0, 1 \} = \{1, 2, 3, 4, 6, 8, 12, 24\}.$$

Conversely, for any of the listed values of n we are going to express $\mathbb{Q}(\zeta_n)/\mathbb{Q}$ as a quasi-Kummer extension and indicate the group G for which the extension is G-Cogalois.

For $n = 1$ and $n = 2$ we have $\mathbb{Q}(\zeta_1) = \mathbb{Q}(\zeta_2) = \mathbb{Q}$, and then $G = \mathbb{Q}^*$.

For $n = 3$ we have $\mathbb{Q}(\zeta_3) = \mathbb{Q}(\sqrt{-3})$ and $G = \mathbb{Q}^*\langle \sqrt{-3} \rangle$.

For $n = 4$ we have $\mathbb{Q}(\zeta_4) = \mathbb{Q}(i)$ and $G = \mathbb{Q}^*\langle i \rangle$.

For $n = 6$ we have $\mathbb{Q}(\zeta_6) = \mathbb{Q}(\sqrt{-3})$ and $G = \mathbb{Q}^*\langle \sqrt{-3} \rangle$.

For $n = 8$ we have $\mathbb{Q}(\zeta_8) = \mathbb{Q}(\sqrt{2} + i\sqrt{2}) = \mathbb{Q}(\sqrt{2}, i)$, and then

$$G = \mathbb{Q}^*\langle \sqrt{2}, i \rangle.$$

For $n = 12$ we have $\mathbb{Q}(\zeta_{12}) = \mathbb{Q}(\sqrt{3} + i) = \mathbb{Q}(\sqrt{3}, i)$, and then

$$G = \mathbb{Q}^*\langle \sqrt{3}, i \rangle.$$

For $n = 24$ we have

$$\mathbb{Q}(\zeta_{24}) = \mathbb{Q}\left(\cos\frac{\pi}{12} + i\sin\frac{\pi}{12} \right) = \mathbb{Q}((\sqrt{6} + \sqrt{2}) + i(\sqrt{6} - \sqrt{2}))$$
$$= \mathbb{Q}(\sqrt{2}, \sqrt{6}, i),$$

and then $G = \mathbb{Q}^*\langle \sqrt{2}, \sqrt{6}, i \rangle$. The proof is complete. $\square$

7.5. Cogalois extensions

We gave in Section 3.2 a straightforward proof of one of the basic properties of Cogalois extensions, namely that saying that any such extension E/F is an extension with $T(E/F)/F^*$-Cogalois correspondence. We show in this section that, with the aid of the n-Purity Criterion (Theorem 4.3.2), this property can be deduced immediately by placing the Cogalois extensions in the general framework of G-Cogalois extensions. We also investigate when a given finite Abelian group can be realized as the Cogalois group of a Cogalois or a Galois Cogalois extension.

THEOREM 7.5.1. *Any finite Cogalois extension* E/F *is* $T(E/F)$-*Cogalois.*

PROOF. By the Greither-Harrison Criterion (Theorem 3.1.7), the extension E/F is separable and pure, so, a fortiori, n-pure, where $n = \exp(T(E/F)/F^*)$. By Theorem 4.3.2 we deduce that E/F is $T(E/F)$-Cogalois. $\square$

We end this section by discussing an analogue of the *Inverse Galois Theory Problem* for Cogalois extensions.

PROPOSITION 7.5.2. *Given any finite Abelian group* A, *there exists an algebraic number field* E *such that* $E/\mathbb{Q}$ *is a Cogalois extension and* $\mathrm{Cog}(E/\mathbb{Q}) \cong A$.

PROOF. According to the structure theorem of finitely generated Abelian groups, we have

$$A \cong \prod_{k=1}^{r} \mathbb{Z}_{n_k}$$

for some $r, n_1, \ldots, n_r \in \mathbb{N}^*$.

Choose r different positive prime integers $p_1, \ldots, p_r$, take

$$E = \mathbb{Q}(\sqrt[n_1]{p_1}, \ldots, \sqrt[n_r]{p_r}),$$

and let

$$G = \mathbb{Q}^*\langle \sqrt[n_1]{p_1}, \ldots, \sqrt[n_r]{p_r}\rangle \text{ and } n = \mathrm{lcm}(n_1, \ldots, n_r).$$

Then, $E/\mathbb{Q}$ is a classical n-*Kummer* extension with few roots of unity, so, by Remark 7.2.5, it is G-Cogalois. The extension $E/\mathbb{Q}$ is also Cogalois, and $\mathrm{Cog}(E/\mathbb{Q}) = G/\mathbb{Q}^*$ by Corollary 4.4.3. Since $p_1, \ldots, p_r$ are different prime numbers, we deduce that $\langle \widehat{\sqrt[n_1]{p_1}}\rangle, \ldots, \langle \widehat{\sqrt[n_r]{p_r}}\rangle$ is an independent family of subgroups of the group $\mathbb{Q}^*\langle \sqrt[n_1]{p_1}, \ldots, \sqrt[n_r]{p_r}\rangle/\mathbb{Q}^*$ (use e.g., Lemma 9.2.4), hence

$$\mathbb{Q}^*\langle \sqrt[n_1]{p_1}, \ldots, \sqrt[n_r]{p_r}\rangle/\mathbb{Q}^* \cong \langle \widehat{\sqrt[n_1]{p_1}}\rangle \times \ldots \times \langle \widehat{\sqrt[n_r]{p_r}}\rangle.$$

But $\langle \widehat{\sqrt[n_i]{p_i}}\rangle$ is a cyclic group of order n_i for every $i = 1, \ldots, r$, so we have

$$\mathrm{Cog}(E/\mathbb{Q}) \cong \mathbb{Z}_{n_1} \times \ldots \times \mathbb{Z}_{n_r} \cong A,$$

and we are done. $\square$

PROPOSITION 7.5.3. *Let* Γ *be an arbitrary finite group. Then, there exists a Galois Cogalois extension* $E/\mathbb{Q}$ *with* $\mathrm{Gal}(E/\mathbb{Q}) \cong \Gamma$ *if and only if* $\Gamma \cong (\mathbb{Z}_2)^r$ *for some* $r \in \mathbb{N}$.

PROOF. If $E/\mathbb{Q}$ is a Galois Cogalois extension, then it is a Galois $T(E/\mathbb{Q})$-Cogalois extension, and so, by Corollary 7.4.4, we deduce that $E/\mathbb{Q}$ is a classical 2-Kummer extension. Consequently, $\mathrm{Gal}(E/\mathbb{Q}) \cong (\mathbb{Z}_2)^r$ for some $r \in \mathbb{N}$.

Conversely, suppose that $\Gamma \cong (\mathbb{Z}_2)^r$ for some $r \in \mathbb{N}$. Of course we may assume that $r \geqslant 1$. Choose r different positive prime numbers $p_1, \dots, p_r$, and let $E = \mathbb{Q}(\sqrt{p_1}, \dots, \sqrt{p_r})$. Then $E/\mathbb{Q}$ is a Galois Cogalois extension such that $\mathrm{Gal}(E/\mathbb{Q}) \cong (\mathbb{Z}_2)^r$.   □

REMARKS 7.5.4. (1) For any real number field E, the extension $E/\mathbb{Q}$ is radical if and only if it is G-Cogalois for some group G. Indeed, any such extension has the form $\mathbb{Q}(\sqrt[n_1]{a_1}, \dots, \sqrt[n_r]{a_r})/\mathbb{Q}$ for some $a_1, \dots, a_r \in \mathbb{Q}_+^*$, which is $\mathbb{Q}^*\langle \sqrt[n_1]{a_1}, \dots, \sqrt[n_r]{a_r}\rangle$-Cogalois. He already noticed in Examples 3.2.1 (1) that the extension $\mathbb{Q}(\sqrt[n_1]{a_1}, \dots, \sqrt[n_r]{a_r})/\mathbb{Q}$ is Cogalois and $\mathrm{Cog}(E/F) = \mathbb{Q}^*\langle \sqrt[n_1]{a_1}, \dots, \sqrt[n_r]{a_r}\rangle/\mathbb{Q}$.

(2) Any quadratic imaginary extension $\mathbb{Q}(\sqrt{d})/\mathbb{Q}$ with $d < 0$ a square-free integer is $\mathbb{Q}^*\langle\sqrt{d}\rangle$-Cogalois, being a classical 2-Kummer extension.

(3) The determination of all imaginary algebraic number fields E which are G-Cogalois for some group G seems to be not so easy.

(4) The relations among the classes of extensions discussed up to this point are given in Exercises 2 - 13.   □

7.6. Exercises to Chapter 7

1. Give an example of an Abelian extension E/F with $\mathrm{Gal}(E/F)$ a bounded group of exponent $n \geqslant 2$, such that $\gcd(n, e(F)) \neq 1$ and $\mu_n(\Omega) \not\subseteq F$.

2. Let $F = \mathbb{Q}(i)$ and $E = F(\sqrt[8]{3})$. Prove that E/F is a generalized 8-Kummer extension, but not a classical Kummer extension. (*Hint:* Show that $\mu_8(E) = \{1, -1, i, -i\}$ and $\sqrt{2} \notin E$.)

3. Show that for any odd prime p, the extension $\mathbb{Q}(\zeta_{p^3})/\mathbb{Q}(\zeta_p)$ is p^2-pure, but it is not a generalized p^2-Kummer extension. (*Hint:* Use Exercise 1, Chapter 3.)

4. Give an example of a Cogalois Abelian extension which is not a classical Kummer extension. (*Hint:* Use Exercise 3.)

5. Show that any finite classical 2-Kummer extension E/F is a 2-Kummer extension with few roots of unity, but it may happen that $\mu_n(E) \not\subseteq F$ for some $n \in \mathbb{N}^*$, $n \neq 2$.

6. Let $F = \mathbb{Q}(\zeta_3)$ and $E = F(\sqrt[3]{2})$. Show that E/F is a classical Kummer extension which is not a Kummer extension with few roots of unity.

7. Give an example of a Cogalois extension which is not quasi-Kummer, and an example of a quasi-Kummer extension which is not Cogalois.

8. Give an example of a finite classical Kummer extension which is not Cogalois.

9. Give an example of a finite quasi-Kummer extension which is not a classical Kummer extension. (*Hint*: Use Theorem 7.4.3 and Exercise 3.)

10. Let $n \in \mathbb{N}^*$. We say that a finite separable extension E/F is an *n-G-Kneser extension* if E/F is a G-radical extension with G/F^* a finite group of exponent a divisor of n, such that $\mu_n(G) \subseteq F^*$, and $\zeta_4 \in F$ whenever $1 + \zeta_4 \in G$.

    Prove that any finite separable n-G-Kneser extension is G-Kneser, but in general, a finite separable G-Kneser extension with G/F^* a finite group of exponent a divisor of n is not necessarily n-G-Kneser.

11. Show that the p^2-pure extension considered in Exercise 3 is not p^2-$\mathbb{Q}^* \langle \zeta_{p^3} \rangle$-Kneser.

12. Show that any finite generalized n-Kummer extension E/F is an n-G-Kneser extension for some group G, but in general, not conversely. (*Hint*: Take $F = \mathbb{Q}(i)$ and $E = F(\sqrt[6]{3})$.)

13. Let $n \in \mathbb{N}^*$. Show that an n-G-Kneser extension is not necessarily n-pure, in particular, is not necessarily G-Cogalois.

14. Provide infinitely many examples of finite separable radical extensions E/F which are not G-Cogalois for any group G. (*Hint*: Use Corollary 7.4.5.)

15. (*Tena* [**105**]). Let $n \in \mathbb{N}^*$. Prove that there exist the following expressions by linear quadratic radicals:
$$\cos \frac{2\pi}{n} = \lambda_1 \sqrt{a_1} + \cdots + \lambda_r \sqrt{a_r}$$

$$\sin \frac{2\pi}{n} = \mu_1 \sqrt{a_1} + \cdots + \mu_r \sqrt{a_r}$$

for some $a_i \in \mathbb{N}^*$, λ_i, $\mu_i \in \mathbb{Q}$, $i = 1, \ldots, r$, if and only if $n \in \{1, 2, 3, 4, 6, 8, 12, 24\}$. (*Hint*: Use Corollary 7.4.5.)

16. (*Barrera-Mora, Rzedowski-Calderón, and Villa-Salvador* [**30**]). Let E/F be a finite Galois Cogalois extension such that $\mu(E) = \mu(F)$. Show that E/F is a classical Kummer extension; in particular, E/F is an Abelian extension and $\text{Cog}(E/F) \cong \text{Gal}(E/F)$. (*Hint*: Apply Proposition 7.2.4.)

17. Provide an example of a finite classical Kummer extension E/F which is Cogalois and $\mu(E) \neq \mu(F)$.

18. (*Barrera-Mora, Rzedowski-Calderón, and Villa-Salvador* [**30**]). Let F be an algebraic number field, and let E/F be a Galois Cogalois extension with $\Gamma = \text{Gal}(E/F)$. Prove that there exists $\Delta \lhd \Gamma$ such that Δ is Abelian and $\Gamma/\Delta \cong \mu(E)/\mu(F)$. Furthermore, show that the extension $E/\text{Fix}(\Delta)$ is a classical Kummer extension. In particular, Γ is a metabelian group.

19. (*Barrera-Mora, Rzedowski-Calderón, and Villa-Salvador* [**30**]). Let E/F be an extension satisfying the assumptions of Exercise 18. Show that Γ is not necessarily the semidirect product of Δ and $\mu(E)/\mu(F)$.

20. (*Barrera-Mora, Rzedowski-Calderón, and Villa-Salvador* [**30**]). Let A be any finite Abelian group. Prove that for any given algebraic number field F there exists a finite extension E/F such that E/F is a Cogalois extension and $\text{Cog}(E/F) \cong A$.

21. (*Barrera-Mora, Rzedowski-Calderón, and Villa-Salvador* [**30**]). Let E/F be a Galois Cogalois extension with Galois group Γ. Prove the following statements.
    (a) If $p \in \mathbb{P}$ is such that $\zeta_p \notin F$, then $p \nmid |\Gamma|$.
    (b) If $\zeta_4 \notin F$, then the 2-Sylow subgroup of Γ is isomorphic to $(\mathbb{Z}_2)^r$ for some $r \in \mathbb{N}$.

22. (*Barrera-Mora, Rzedowski-Calderón, and Villa-Salvador* [**30**]). Let F be any field, and let $p \in \mathbb{P}$ be such that $\zeta_p \notin F$. Prove that there is no Galois Cogalois extension E/F with $p \mid [E : F]$.

23. (*Barrera-Mora, Rzedowski-Calderón, and Villa-Salvador* [**30**]). Let $r, n_1, \ldots, n_r \in \mathbb{N}^*$ with $n_1 \mid n_2 \mid \ldots \mid n_r$, and let F be an algebraic

number field such that $\zeta_{n_r} \in F$. Prove that there exist infinitely many Galois Cogalois extensions E/F such that

$$\mathrm{Gal}(E/F) \cong \mathrm{Cog}(E/F) \cong \prod_{k=1}^{r} \mathbb{Z}_{n_k}.$$

24. (*Barrera-Mora, Rzedowski-Calderón, and Villa-Salvador* [**30**]). Let F be an algebraic number field, and let A be a finite Abelian group with $A \cong \prod_{k=1}^{r} \mathbb{Z}_{n_k}$ and $n_1 \mid n_2 \mid \ldots \mid n_r$. Prove that the following two assertions are equivalent.
   (a) There exists a Galois Cogalois extension E/F such that
   $$\mathrm{Gal}(E/F) \cong \mathrm{Cog}(E/F) \cong A.$$
   (b) $\zeta_{n_{r-1}} \in F$ and $F(\zeta_{n_r})/F$ is a pure extension.

25. Consider the following statements for an algebraic number field E.
   (a) $E/\mathbb{Q}$ is a Galois Cogalois extension.
   (b) $E/\mathbb{Q}$ is an Abelian Cogalois extension.
   (c) There exist $r \in \mathbb{N}$ and square-free rational integers $a_1, \ldots, a_r$ $\in \mathbb{Z} \setminus \{0, -1, -3\}$ such that $a_i/a_j \notin \{1, -1, 3, -3\}$ for all $i \neq j$ in $\{1, \ldots, r\}$ and $E = \mathbb{Q}(\sqrt{a_1}, \ldots, \sqrt{a_r})$.
   Prove that $(a) \Longleftrightarrow (b) \Longrightarrow (c)$. Is the implication $(c) \Longrightarrow (b)$ true?

7.7. Bibliographical comments to Chapter 7

Section 7.1. The concept of *Kummer extension* is a classical one in Field Theory. To distinguish such extensions from various generalizations, we refer to them as *classical n-Kummer extensions* (note that Artin [**26**], Karpilovsky [**76**], Lang [**80**], etc. refer to them also as *Kummer extensions of exponent n*, and Bourbaki [**40**] as *Kummer extensions of exponent dividing n*). All these sources provide a good account of the Kummer Theory. The idea to place finite classical Kummer extensions in the framework of G-Cogalois extensions goes back to Albu and Nicolae [**19**], and was also exploited by Albu [**10**] and Albu and Ţena [**25**] for infinite Kummer extensions (see Chapter 13).

Section 7.2. Generalized Kummer extensions were introduced and investigated by Albu and Nicolae [**19**] for finite extensions, and by Albu and Ţena [**25**] for infinite extensions.

Section 7.3. The concept of Kummer extension with few roots of unity was introduced by Albu [3] for finite extensions, and by Albu and Ţena [25] for infinite extensions. A systematic study of finite Kummer extensions with few roots of unity was undertaken by Albu [3]. Note that a series of results in Albu [3] can be deduced more directly, more simply, and more generally using the Kneser Criterion and the concept and properties of a G-Cogalois extension.

Section 7.4. The notion of *neat presentation* was introduced and investigated by Greither and Harrison [63]. A neat presentation is a Galois extensions E/F of the following type: $E = F(x_1, \ldots, x_r)$, with $\mathrm{Char}(F) = 0$ and $x_1^n, \ldots, x_r^n \in F$ for some $n \in \mathbb{N}^*$, such that $\mu_p(\Omega) \subseteq F$ for every $p \in \mathcal{P}_n$. Using heavy cohomological machinery, which includes the Lyndon-Hochschild spectral sequence, Greither and Harrison proved that any such extension is an extension with $F^*\langle x_1, \ldots, x_r \rangle / F^*$-Cogalois correspondence, i.e., in our terminology, is an $F^*\langle x_1, \ldots, x_r \rangle$-Cogalois extension.

Albu and Nicolae [19] introduced the more general concept of *generalized neat presentation* by dropping from Greither and Harrison's definition the condition "E/F is Galois" and by weakening the condition "$\mathrm{Char}(F) = 0$" to "$\gcd(e(F), n) = 1$", and proved in a very simple manner (see also Theorem 7.4.2) that any such extension is still $F^*\langle x_1, \ldots, x_r \rangle$-Cogalois. In this monograph, the generalized neat presentations were renamed *quasi-Kummer extensions*.

Theorem 7.4.3 is due to Albu, Nicolae, and Ţena [23]. A particular case of Corollary 7.4.4 for Galois Cogalois extensions was established by Barrera-Mora, Rzedowski-Calderón, and Villa-Salvador [30] using the description of the Cogalois group of a Galois extension by means of crossed homomorphisms. The idea to apply Theorem 7.4.3 to prove Corollary 7.4.4 is due to Ţena [105], to whom also belongs Corollary 7.4.5.

Section 7.5. The fact that any finite Cogalois extension E/F is an extension with $\mathrm{Cog}(E/F)$-Cogalois correspondence was first proved by Greither and Harrison [63]. The very short proof of this property provided in this section, based on the n-Purity Criterion, is due to Albu and Nicolae [19].

Propositions 7.5.2 and 7.5.3, related to the dual of the *Inverse Galois Theory Problem*, are due to Barrera-Mora, Rzedowski-Calderón, and Villa-Salvador [30].

CHAPTER 8

G-COGALOIS EXTENSIONS AND PRIMITIVE ELEMENTS

By a well-known result, any finite separable extension E/F has a primitive element, i.e., an element $u \in E$ such that $E = F(u)$. However, in general, there is no practical procedure to find effectively such a u for a given extension E/F.

The aim of this chapter is to show that for a fairly large class of finite separable extensions, namely, the G-Cogalois extensions, it is possible to provide a simple and easily manageable method of finding primitive elements.

Section 8.1 contains the main result of this chapter (Theorem 8.1.2) which essentially establishes an easily verifiable necessary and sufficient condition for a finite sum of elements of G to be a primitive element of a G-Cogalois extension E/F. In particular, it follows that for any set of representatives S of the (finite) Kneser group G/F^* of any G-Cogalois extension E/F, the sum $\sum_{s \in S} s$ is a primitive element of E/F.

Applications of results of Sections 8.1 and 7.2 to elementary Field Arithmetic are given in Section 8.2.

8.1. Primitive elements for *G*-Cogalois extensions

Any G-Cogalois extension is by definition separable and finite, hence, by a classical result, it has in any case a primitive element. The main result of this section gives an easy way to provide such an element: it can be obtained by summing up any elements $x_1, \ldots, x_n$ of G such that $G = F^*\langle x_1, \ldots, x_n \rangle$ and $\widehat{x_i} \neq \widehat{x_j}$ for every $i \neq j$ in $\{1, \ldots, n\}$.

PROPOSITION 8.1.1. *Let E/F be a G-Kneser extension, let $n \geqslant 2$ be a natural number, and let $x_1, \ldots, x_n \in G$ be such that $\widehat{x_i} \neq \widehat{x_j}$ for every $i, j \in \{1, \ldots, n\}$, $i \neq j$. Then $x_1 + \ldots + x_n \notin F$.*

PROOF. Suppose that $x_1 + \ldots + x_n = a \in F$. By Corollary 2.1.10, $\{x_1, \ldots, x_n\}$ is linearly independent over F, hence we cannot have $a = 0$. It follows, again by Corollary 2.1.10, that there exists an i, $1 \leqslant i \leqslant n$, such that $\widehat{x_i} = \widehat{a} = \widehat{1}$, i.e., $x_i \in F$. Then $x_1 + \ldots + x_{i\ 1} + x_{i+1} + \ldots + x_n \in F$, and, as above we deduce that $x_j \in F^*$ for some $j \in \{1, \ldots, n\} \setminus \{i\}$, i.e., $\widehat{x_i} = \widehat{x_j} = \widehat{1}$, which is a contradiction. $\qquad\square$

THEOREM 8.1.2. *Let E/F be a G-Cogalois extension, let $n \in \mathbb{N}^*$, and let $(x_i)_{1 \leqslant i \leqslant n}$ be a finite family of elements of G such that $\widehat{x_i} \neq \widehat{x_j}$ for every $i, j \in \{1, \ldots, n\}$, $i \neq j$. Then $x_1 + \ldots + x_n$ is a primitive element of E/F if and only if $G = F^*\langle x_1, \ldots, x_n \rangle$.*

PROOF. If $E = F(x_1 + \ldots + x_n)$, then $E = F[x]$, where $x = x_1 + \ldots + x_n$, hence any element of E, in particular any element $g \in G$, has the form $f(x)$ for some polynomial $f \in F[X]$, and consequently $g = h(x_1, \ldots, x_n)$ for some polynomial $h \in F[X_1, \ldots, X_n]$. Thus, $g = h_1 + \ldots + h_m$, with $h_i \in F^*\langle x_1, \ldots, x_n \rangle$, and $\widehat{h_i} \neq \widehat{h_j}$ for every $i, j \in \{1, \ldots, m\}$, $i \neq j$. It follows that there exists i, $1 \leqslant i \leqslant m$, such that $\widehat{g} = \widehat{h_i}$, for otherwise, by Corollary 2.1.10, $\{g, h_1, \ldots, h_m\}$ would be linearly independent over F, which is a contradiction. Hence $g \in F^*\langle x_1, \ldots, x_n \rangle$ for every $g \in G$, i.e., $G = F^*\langle x_1, \ldots, x_n \rangle$.

Now suppose that $G = F^*\langle x_1, \ldots, x_n \rangle$, where $x_1, \ldots, x_n$ are as in the statement of the theorem, and consider the subextension $F(x_1 + \ldots + x_n)/F$ of E/F. By Theorem 4.3.2, there exists H such that $F^* \leqslant H \leqslant G$ and $F(x_1 + \ldots + x_n) = F(H)$. We are going to prove by induction on n that

$$x_1 + \cdots + x_n \in F(H) \implies x_1, \ldots, x_n \in H.$$

If $n = 1$, then $F(G) = F(x_1) = F(H)$, and again by Theorem 4.3.2, we deduce that $G = H$, hence $x_1 \in H$.

Now assume that $n \geqslant 2$ and the result holds for $n - 1$. We are going to prove that it also holds for n. Since $x_1 + \cdots + x_n \in F(H)$, we have

$$x_1 + \ldots + x_n = h_1 + \ldots + h_m$$

for some $h_1, \ldots, h_m \in H$, with $m \geqslant 1$ and $\widehat{h_i} \neq \widehat{h_j}$ for every $i, j \in \{1, \ldots, m\}$, $i \neq j$. By Corollary 2.1.10, we deduce that $\{x_1, \ldots, x_n\}$ and $\{h_1, \ldots, h_m\}$ are both linearly independent sets over F, and that $\{\widehat{x_1}, \ldots, \widehat{x_n}\} \cap \{\widehat{h_1}, \ldots, \widehat{h_m}\} \neq \varnothing$. For simplicity, suppose that $\widehat{x_n} = \widehat{h_m}$. Then $x_n = \lambda h_m$ for some $\lambda \in F^*$, and so, we have $x_n \in H$. Thus

$$x_1 + \cdots + x_{n-1} = h_1 + \cdots + h_{m-1} + (1 - \lambda)h_m,$$

which shows that $x_1 + \cdots + x_{n-1} \in F(H)$. By induction, we deduce that $x_1, \ldots, x_{n-1} \in H$, which finishes our induction. So, $x_1, \ldots, x_n \in H$, hence

$$G = F^*\langle x_1, \ldots, x_n \rangle \leqslant H \leqslant G,$$

i.e., $H = G$. Then

$$F(x_1 + \cdots + x_n) = F(H) = F(G) = E,$$

and we are done. $\hfill\square$

COROLLARY 8.1.3. *Let E/F be a G-Cogalois extension, let $n \in \mathbb{N}^*$, and let $(x_i)_{1 \leqslant i \leqslant n}$ be a finite family of elements of G such that $\widehat{x_i} \neq \widehat{x_j}$ for every $i \neq j$ in $\{1, \ldots, n\}$. If $x_1 + \ldots + x_n$ is a primitive element of E/F, then $\lambda_1 x_1 + \ldots + \lambda_n x_n$ is also a primitive element of E/F for any family $(\lambda_i)_{1 \leqslant i \leqslant n}$ of elements of F^*.*

PROOF. Set $y_i = \lambda_i x_i$ for every $i = 1, \ldots, n$. Then $F^*\langle x_1, \ldots, x_n \rangle = F^*\langle y_1, \ldots, y_n \rangle$ and $\widehat{y_i} \neq \widehat{y_j}$ for every $i, j \in \{1, \ldots, n\}$, $i \neq j$. Now apply Theorem 8.1.2. $\hfill\square$

The next result shows that the condition "$\widehat{x_i} \neq \widehat{x_j}$ for every $i \neq j$ in $\{1, \ldots, n\}$" in the statement of Theorem 8.1.2 is not essential.

COROLLARY 8.1.4. *Let E/F be a G-Cogalois extension, let $n \in \mathbb{N}^*$, and let $(x_i)_{1 \leqslant i \leqslant n}$ be a finite family of elements of G. Consider a set of representatives $\{y_1, \ldots, y_r\} \subseteq \{x_1, \ldots, x_n\}$ of the subset $\{\widehat{x_1}, \ldots, \widehat{x_n}\}$ of G/F^*. For each k, $1 \leqslant k \leqslant r$, let $Y_k = \{x_i \mid \widehat{x_i} = \widehat{y_k}, 1 \leqslant i \leqslant n\}$ and $z_k = \sum_{x \in Y_k} x$. Then $x_1 + \ldots + x_n$ is a primitive element of E/F if and only if $G = F^*\langle \{z_j \mid j \in J\}\rangle$, where $J = \{j \mid 1 \leqslant j \leqslant r, z_j \neq 0\}$.*

In particular, if $\{u_1, \ldots, u_r\}$ is any set of representatives of G/F^*, then $u_1 + \ldots + u_r$ is a primitive element of the extension E/F.

PROOF. Clearly, $\sum_{1 \leqslant i \leqslant n} x_i = \sum_{1 \leqslant k \leqslant r} z_k = \sum_{j \in J} z_j$ and $z_j \in G$ for every $j \in J$, since $z_j = \lambda_j y_j$ for some $\lambda_j \in F^*$. Moreover, $\widehat{z_j} \neq \widehat{z_k}$ for all $j \neq k$ in J. Now apply Theorem 8.1.2. $\hfill\square$

LEMMA 8.1.5. *Let E/F be an extension, let $n \in \mathbb{N}^*$, and let $x_1, \ldots, x_n \in E^*$. If $F(x_1, \ldots, x_n)/F$ is an $F^*\langle x_1, \ldots, x_n \rangle$-Kneser extension, then the following assertions are equivalent.*

(1) $[F(x_1, \ldots, x_n) : F] = \prod_{i=1}^n [F(x_i) : F]$.

(2) *The canonical morphism of F-algebras*

$$F(x_1) \otimes_F \ldots \otimes_F F(x_n) \longrightarrow F(x_1, \ldots, x_n),$$

$$z_1 \otimes \ldots \otimes z_n \mapsto z_1 \cdot \ldots \cdot z_n,$$

is an isomorphism.

(3) *The family* $\langle \widehat{x_1} \rangle, \ldots, \langle \widehat{x_n} \rangle$ *of subgroups of* $F^*\langle x_1, \ldots, x_n \rangle / F^*$ *is independent, i.e., if* $i_1, \ldots, i_n \in \mathbb{N}$, *then*

$$x_1^{i_1} \cdot \ldots \cdot x_n^{i_n} \in F^* \Longrightarrow d_1 | i_1, \ldots, d_n | i_n,$$

where $d_j = \mathrm{ord}(\widehat{x_j})$ *for every* j, $1 \leqslant j \leqslant n$.

(4) *If* $i_1, \ldots, i_n \in \mathbb{N}$, *then* $x_1^{i_1} \cdot \ldots \cdot x_n^{i_n} \in F^* \Longrightarrow x_k^{i_k} \in F^*$ *for every* k, $1 \leqslant j \leqslant n$.

(5) *There exists a group isomorphism*

$$F^*\langle x_1, \ldots, x_r \rangle / F^* \cong \prod_{1 \leqslant i \leqslant r} (F^*\langle x_i \rangle / F^*).$$

(6) $| F^*\langle x_1, \ldots, x_r \rangle / F^* | = \prod_{1 \leqslant i \leqslant r} | F^*\langle x_i \rangle / F^* |.$

PROOF. (1) $\Longleftrightarrow$ (2): The canonical morphism of finite dimensional F-algebras

$$F(x_1) \otimes_F \ldots \otimes_F F(x_n) \longrightarrow F(x_1, \ldots, x_n)$$

is surjective, and it is injective if and only if

$$[\, F(x_1) \otimes_F \ldots \otimes_F F(x_n) : F \,] = [\, F(x_1, \ldots, x_n) : F \,],$$

i.e., if and only if

$$[\, F(x_1, \ldots, x_n) : F \,] = \prod_{i=1}^{n} [\, F(x_i) : F \,].$$

(1) $\Longrightarrow$ (5): Since the extension $F(x_1, \ldots, x_n)/F$ is $F^*\langle x_1, \ldots, x_n \rangle$-Kneser, it follows that the extension $F(x_j)/F$ is $K^*\langle x_j \rangle$-Kneser for every j, $1 \leqslant j \leqslant n$ by Corollary 2.1.10, hence

$$[\, F(x_j) : F \,] = | F^*\langle x_j \rangle / F^* | = \mathrm{ord}(\widehat{x_j}) = d_j.$$

Using again the fact that $F(x_1, \ldots, x_n)/F$ is $F^*\langle x_1, \ldots, x_n \rangle$ Kneser, we have

$$d_1 \cdot \ldots \cdot d_n = [\, F(x_1, \ldots, x_n) : F \,] = | F^*\langle x_1, \ldots, x_n \rangle / F^* |.$$

On the other hand, the group $F^*\langle x_1, \ldots, x_n \rangle / F^*$ is isomorphic to a quotient group of the direct product of groups $\langle \widehat{x_1} \rangle \times \ldots \times \langle \widehat{x_n} \rangle$. Since both groups have the same order $d_1 \cdot \ldots \cdot d_n$, they are isomorphic. This implies that the family $\langle \widehat{x_1} \rangle, \ldots, \langle \widehat{x_n} \rangle$ of subgroups of $F^*\langle x_1, \ldots, x_n \rangle / F^*$ is independent, and also proves the implication (5) $\Longrightarrow$ (3).

(3) $\implies$ (1): The condition in (3) means precisely that $\langle \widehat{x_1}, \ldots, \widehat{x_n} \rangle$ is the internal direct sum of its subgroups $\langle \widehat{x_1} \rangle, \ldots, \langle \widehat{x_n} \rangle$ (see 1.1.2), hence

$$\langle \widehat{x_1}, \ldots, \widehat{x_n} \rangle \cong \langle \widehat{x_1} \rangle \times \ldots \times \langle \widehat{x_n} \rangle.$$

But, as observed in the implication (1) $\implies$ (5), $d_j = [F(x_j) : F]$ and $[F(x_1, \ldots, x_n) : F] = |F^*\langle x_1, \ldots, x_n \rangle / F^*|$. Consequently

$$[F(x_1, \ldots, x_n) : F] = |F^*\langle x_1, \ldots, x_n \rangle / F^*| = |\langle \widehat{x_1}, \ldots, \widehat{x_n} \rangle|$$

$$= |\langle \widehat{x_1} \rangle \times \ldots \times \langle \widehat{x_n} \rangle| = |\langle \widehat{x_1} \rangle| \cdot \ldots \cdot |\langle \widehat{x_n} \rangle| = d_1 \cdot \ldots \cdot d_n = \prod_{i=1}^{n} [F(x_i) : F].$$

(3) $\iff$ (4) and (5) $\implies$ (6) are obvious, and (6) $\implies$ (1) follows from the last part of the proof of (1) $\implies$ (5). $\qquad \square$

PROPOSITION 8.1.6. *Let $n \in \mathbb{N}^*$, and let $x_1, \ldots, x_n \in \Omega^*$ be such that $F(x_1, \ldots, x_n)/F$ is $F^*\langle x_1, \ldots, x_n \rangle$-Cogalois. If*

$$[F(x_1, \ldots, x_n) : F] = \prod_{i=1}^{n} [F(x_i) : F],$$

then

$$F(x_1, \ldots, x_n) = F(x_1 + \ldots + x_n).$$

PROOF. Without loss of generality we may assume that $x_1, \ldots, x_n \notin F$. Set $E = F(x_1, \ldots, x_n)$, $G = F^*\langle x_1, \ldots, x_n \rangle$, and $d_i = \mathrm{ord}(\widehat{x_i})$, $i = 1, \ldots, n$.

By Theorem 8.1.2 we have only to show that the cosets $\langle \widehat{x_1} \rangle, \ldots, \langle \widehat{x_n} \rangle \in \Omega^*/F^*$ are mutually distinct. Assume that this is not the case, hence only r cosets among them are mutually distinct, with $1 \leqslant r < n$. Without loss of generality, we may suppose that these are $\widehat{x_1}, \ldots, \widehat{x_r}$.

The canonical group morphism

$$\varphi : \langle \widehat{x_1} \rangle \times \cdots \times \langle \widehat{x_r} \rangle \longrightarrow \langle \widehat{x_1}, \ldots, \widehat{x_r} \rangle,$$

$$\varphi(\widehat{x_1}^{j_1}, \ldots, \widehat{x_r}^{j_r}) = \widehat{x_1}^{j_1} \cdot \ldots \cdot \widehat{x_r}^{j_r},$$

is obviously surjective, hence the orders of these groups satisfy the inequality

$$|\langle \widehat{x_1}, \ldots, \widehat{x_r} \rangle| \leqslant |\langle \widehat{x_1} \rangle| \cdot \ldots \cdot |\langle \widehat{x_r} \rangle| = d_1 \cdot \ldots \cdot d_r.$$

But $\langle \widehat{x_1}, \ldots, \widehat{x_r} \rangle = \langle \widehat{x_1}, \ldots, \widehat{x_n} \rangle = G/F^*$, and so, the inequality above can be written as

$$|G/F^*| \leqslant d_1 \cdot \ldots \cdot d_r.$$

On the other hand, by the proof of the implication (1) $\implies$ (5) in Lemma 8.1.5 we have

$$|G/F^*| = d_1 \cdot \ldots \cdot d_n.$$

So, $d_{r+1} = \cdots = d_n = 1$, i.e., $x_{r+1}, \ldots, x_n \in F^*$, which contradicts the choice of the elements $x_1, \ldots, x_n$. Thus, the cosets $\langle \widehat{x_1} \rangle, \ldots, \langle \widehat{x_n} \rangle$ are mutually distinct, and we are done. □

8.2. Applications to elementary Field Arithmetic (II)

In Section 5.3 we presented a series of applications of Cogalois Theory to elementary Field Arithmetic. The aim of this section is to provide further such applications, essentially based on Theorem 8.1.2, Proposition 8.1.6, and Theorem 7.2.3.

Recall that Ω is a fixed algebraically closed field containing the fixed base field F as a subfield; any algebraic extension of F is supposed to be a subfield of Ω.

Also, recall that for any $u \in F^*$ and $n \in \mathbb{N}^*$, $\sqrt[n]{a}$ will designate a root (which in general is not specified) in Ω of the polynomial $X^n - a \in F[X]$.

However, in certain cases, for instance when F is a subfield of the field $\mathbb{R}$ of all real numbers and $a > 0$, then $\sqrt[n]{a}$ will always mean the unique positive root in $\mathbb{R}$ of the polynomial $X^n - a$.

PROPOSITION 8.2.1. *Let* $r, n_1, \ldots, n_r \in \mathbb{N}^*$, *let* $a_1, \ldots, a_r \in F^*$, *and let* $n = \mathrm{lcm}\,(n_1, \ldots, n_r)$. *Assume that the following three conditions are satisfied.*

(1) $\gcd(n, e(F)) = 1$.
(2) $\mu_n(F(\sqrt[n_1]{a_1}, \ldots, \sqrt[n_r]{a_r})) \subseteq F$.
(3) $\widehat{\sqrt[n_i]{a_i}} \neq \widehat{\sqrt[n_j]{a_j}}$ *for all* $i \neq j$ *in* $\{1, \ldots, r\}$.

Then

$$F(\sqrt[n_1]{a_1}, \ldots, \sqrt[n_r]{a_r}) = F(\sqrt[n_1]{a_1} + \cdots + \sqrt[n_r]{a_r}).$$

PROOF. Conditions (1) and (2) show that $F(\sqrt[n_1]{a_1}, \ldots, \sqrt[n_r]{a_r})/F$ is a generalized n-Kummer extension, hence according to Theorem 7.2.3, it is $F^*\langle \sqrt[n_1]{a_1}, \ldots, \sqrt[n_r]{a_r} \rangle$-Cogalois. This fact and condition (3) imply the desired equality in view of Theorem 8.1.2. □

COROLLARY 8.2.2. *Let F be a subfield of $\mathbb{R}$, let $r, n_1, \ldots, n_r \in \mathbb{N}^*$, and let $a_1, \ldots, a_r \in F^*$ be positive numbers. Then*

$$F(\sqrt[n_1]{a_1}, \ldots, \sqrt[n_r]{a_r}) = F(\sqrt[n_1]{a_1} + \cdots + \sqrt[n_r]{a_r}).$$

PROOF. For simplicity, denote $x_i = \sqrt[n_i]{a_i}$, $1 \leqslant i \leqslant r$, and $G = F^*\langle x_1, \ldots, x_r \rangle$. By Corollary 7.3.3, the extension $F(\sqrt[n_1]{a_1}, , \ldots, \sqrt[n_r]{a_r})/F$ is G-Cogalois.

Now proceed as in the proof of Corollary 8.1.4 by considering a set of representatives $\{y_1, \ldots, y_m\} \subseteq \{x_1, \ldots, x_r\}$ of the subset $\{\widehat{x_1}, \ldots, \widehat{x_r}\}$ of G/F^*. For each k, $1 \leqslant k \leqslant m$, let

$$Y_k = \{x_i \mid \widehat{x_i} = \widehat{y_k}, \ 1 \leqslant i \leqslant r\} \ \text{ and } \ z_k = \sum_{x \in Y_k} x.$$

Since all elements of Y_k are strictly positive, we deduce that $z_k > 0$ for every k, $1 \leqslant k \leqslant m$. Obviously, $\sum_{1 \leqslant i \leqslant r} x_i = \sum_{1 \leqslant k \leqslant m} z_k$, and for every k, $1 \leqslant k \leqslant m$, there exists a $\lambda_k \in F^*$ such that $z_k = \lambda_k y_k$. Moreover, $\widehat{z_i} \neq \widehat{z_j}$ for all $i \neq j$ in $\{1, \ldots m\}$. Observe that $F(x_1, \ldots, x_r) = F(z_1, \ldots, z_m)$ and $G = F^*\langle z_1, \ldots, z_m \rangle$. Now apply Proposition 8.2.1. $\square$

COROLLARY 8.2.3. *Let $r, n_1, \ldots, n_r \in \mathbb{N}^*$, and let $a_1, \ldots, a_r \in \mathbb{Q}_+^*$. Then*

$$\mathbb{Q}(\sqrt[n_1]{a_1}, \ldots, \sqrt[n_r]{a_r}) = \mathbb{Q}(\sqrt[n_1]{a_1} + \cdots + \sqrt[n_r]{a_r}).$$

$\square$

COROLLARY 8.2.4. *Let F be a subfield of $\mathbb{R}$, let $r, n_1, \ldots, n_r \in \mathbb{N}^*$, and let $a_1, \ldots, a_r \in F^*$ be positive numbers. Then*

$$\sqrt[n_1]{a_1} + \cdots + \sqrt[n_r]{a_r} \in F \iff \sqrt[n_i]{a_i} \in F \ \text{ for all } i, 1 \leqslant i \leqslant r.$$

PROOF. Assume that $\sqrt[n_1]{a_1} + \cdots + \sqrt[n_r]{a_r} \in F$. By Corollary 8.2.2,

$$F(\sqrt[n_1]{a_1}, \ldots, \sqrt[n_r]{a_r}) = F(\sqrt[n_1]{a_1} + \cdots + \sqrt[n_r]{a_r}).$$

But $\sqrt[n_1]{a_1} + \cdots + \sqrt[n_r]{a_r} \in F$ if and only if $F(\sqrt[n_1]{a_1} + \cdots + \sqrt[n_r]{a_r}) = F$, hence $F(\sqrt[n_1]{a_1}, \ldots, \sqrt[n_r]{a_r}) = F$, and then $\sqrt[n_1]{a_1}, \ldots, \sqrt[n_r]{a_r} \in F$. The other implication is obvious. $\square$

COROLLARY 8.2.5. *Let $r, n_1, \ldots, n_r \in \mathbb{N}^*$ and let $a_1, \ldots, a_r \in \mathbb{Q}_+^*$. Then*

$$\sqrt[n_1]{a_1} + \cdots + \sqrt[n_r]{a_r} \in \mathbb{Q} \iff \sqrt[n_i]{a_i} \in \mathbb{Q} \ \text{ for all } i, 1 \leqslant i \leqslant r.$$

$\square$

COROLLARY 8.2.6. *Let* $n, r \in \mathbb{N}^*$, *let* F *be a field with* $\gcd(n, e(F)) = 1$, *and let* $a_1, \ldots, a_r \in F^*$. *Suppose that* $\mu_n(F(\sqrt[n]{a_1}, \ldots, \sqrt[n]{a_r})) \subseteq F$ *and* $[F(\sqrt[n]{a_1}, \ldots, \sqrt[n]{a_r}) : F] = \prod_{i=1}^r [F(\sqrt[n]{a_i}) : F]$. *Then*

$$F(\sqrt[n]{a_1}, \ldots, \sqrt[n]{a_r}) = F(\sqrt[n]{a_1} + \ldots + \sqrt[n]{a_r}).$$

PROOF. The conditions $\mu_n(F(\sqrt[n]{a_1}, \ldots, \sqrt[n]{a_r})) \subseteq F$ and $\gcd(n, e(F)) = 1$ ensure by Theorem 7.2.3 that the extension $F(\sqrt[n]{a_1}, \ldots, \sqrt[n]{a_r})/F$ is $F^*\langle \sqrt[n]{a_1}, \ldots, \sqrt[n]{a_r} \rangle$-Cogalois. Now apply Proposition 8.1.6. $\square$

The example below shows that the condition in Corollary 8.2.6 on the degrees of the involved extensions is essential.

EXAMPLE 8.2.7. We have $\mathbb{Q}(\sqrt{2}, \sqrt[3]{3}, \sqrt[6]{72}) = \mathbb{Q}(\sqrt{2} + \sqrt[3]{3} + \sqrt[6]{72})$ by Corollary 8.2.3, but

$$[\mathbb{Q}(\sqrt{2}, \sqrt[3]{3}, \sqrt[6]{72}) : \mathbb{Q}] = [\mathbb{Q}(\sqrt{2}, \sqrt[3]{3}) : \mathbb{Q}] = 6$$
$$< [\mathbb{Q}(\sqrt{2}) : \mathbb{Q}] \cdot [\mathbb{Q}(\sqrt[3]{3}) : \mathbb{Q}] \cdot [\mathbb{Q}(\sqrt[6]{72}) : \mathbb{Q}].$$

$\square$

Next, we present some applications to elementary Field Arithmetic of the property that generalized Kummer extensions are G-Cogalois (see Theorem 7.2.3 (1)).

PROPOSITION 8.2.8. *Let* $r, n_0, n_1, \ldots, n_r \in \mathbb{N}^*$, *and let* $a_0, a_1, \ldots, a_r \in F^*$. *Let* $n = \mathrm{lcm}(n_0, n_1, \ldots, n_r)$, *and suppose that* $\gcd(n, e(F)) = 1$ *and* $\mu_n(F(\sqrt[n_1]{a_1}, \ldots, \sqrt[n_r]{a_r})) \subseteq F$, *i.e.*, $F(\sqrt[n_1]{a_1}, \ldots, \sqrt[n_r]{a_r})/F$ *is a generalized n-Kummer extension.*

Then $\sqrt[n_0]{a_0} \in F(\sqrt[n_1]{a_1}, \ldots, \sqrt[n_r]{a_r})$ *if and only if there exist* $j_1, \ldots, j_r \in \mathbb{N}$ *and* $c \in F^*$ *such that*

$$\sqrt[n_0]{a_0} = c \cdot \sqrt[n_1]{a_1}^{j_1} \cdot \ldots \cdot \sqrt[n_r]{a_r}^{j_r}.$$

PROOF. Set $E = F(\sqrt[n_1]{a_1}, \ldots, \sqrt[n_r]{a_r})$ and $G = F^*\langle \sqrt[n_1]{a_1}, \ldots, \sqrt[n_r]{a_r} \rangle$. By Theorem 7.2.3, E/F is a G-Cogalois extension. Now, observe that since $n_0 \mid n$, we have $\mu_{n_0}(E) \subseteq \mu_n(E) \subseteq F$. So, we can apply Lemma 6.1.6 to deduce that $\sqrt[n_0]{a_0} \in F(\sqrt[n_1]{a_1}, \ldots, \sqrt[n_r]{a_r})$ if and only if $\sqrt[n_0]{a_0} \in G$, as desired. $\square$

COROLLARY 8.2.9. *Let* $r, n \in \mathbb{N}^*$, *let* $a_0, a_1, \ldots, a_r \in F^*$, *and suppose that* $\gcd(n, e(F)) = 1$, *and* $\mu_n(F(\sqrt[n]{a_0}, \sqrt[n]{a_1}, \ldots, \sqrt[n]{a_r})) \subseteq \{-1, 1\}$ *or* $\mu_n(\Omega) \subseteq F$, *in other words, the extension* $F(\sqrt[n]{a_0}, \sqrt[n]{a_1}, \ldots, \sqrt[n]{a_r})/F$ *is either an n-Kummer extension with few roots of unity or a classical n-Kummer extension.*

Then $\sqrt[n]{a_0} \in F(\sqrt[n]{a_1}, \ldots, \sqrt[n]{a_r})$ if and only if there exist $j_1, \ldots, j_r \in \mathbb{N}$ and $c \in F^*$ such that

$$a_0 = c^n \cdot a_1^{j_1} \cdot \ldots \cdot a_r^{j_r}.$$

PROOF. Set $E = F(\sqrt[n]{a_1}, \ldots, \sqrt[n]{a_r})$ and $G = F^* \langle \sqrt[n]{a_1}, \ldots, \sqrt[n]{a_r} \rangle$. Assume that $\sqrt[n]{a_0} \in F(\sqrt[n]{a_1}, \ldots, \sqrt[n]{a_r})$. By Proposition 8.2.8, we deduce that

$$\sqrt[n]{a_0} = c \cdot \sqrt[n]{a_1}^{j_1} \cdot \ldots \cdot \sqrt[n]{a_r}^{j_r}$$

for some $j_1, \ldots, j_r \in \mathbb{N}$ and $c \in F^*$. Raising this equality to the n-th power, we obtain $a_0 = c^n \cdot a_1^{j_1} \cdot \ldots \cdot a_r^{j_r}$.

Conversely, the equation in the statement of the corollary can be written as

$$(\sqrt[n]{a_0})^n = (c \cdot \sqrt[n]{a_1}^{j_1} \cdot \ldots \cdot \sqrt[n]{a_r}^{j_r})^n.$$

If we denote $\zeta = (c \cdot \sqrt[n]{a_1}^{j_1} \cdot \ldots \cdot \sqrt[n]{a_r}^{j_r})^{-1} \cdot \sqrt[n]{a_0}$, then

$$\zeta \in \mu_n(\Omega) \cap F(\sqrt[n]{a_0}, \sqrt[n]{a_1}, \ldots, \sqrt[n]{a_r})) = \mu_n(F(\sqrt[n]{a_0}, \sqrt[n]{a_1}, \ldots, \sqrt[n]{a})).$$

By hypothesis, it follows that $\zeta \in F$, hence

$$\sqrt[n]{a_0} = (c \cdot \zeta) \cdot \sqrt[n]{a_1}^{j_1} \cdot \ldots \cdot \sqrt[n]{a_r}^{j_r},$$

and $c \cdot \zeta \in F$. This shows that $\sqrt[n]{a_0} \in F(\sqrt[n]{a_1}, \ldots, \sqrt[n]{a_r})$. $\square$

COROLLARY 8.2.10. Let $r, n_0, n_1, \ldots n_r \in \mathbb{N}^*$, and let $a_0, a_1, \ldots a_r \in \mathbb{Q}_+^*$. Then $\sqrt[n_0]{a_0}$ can be written as a finite sum of monomials of form $c \cdot \sqrt[n_1]{a_1}^{j_1} \cdot \ldots \cdot \sqrt[n_r]{a_r}^{j_r}$, with $j_1, \ldots, j_r \in \mathbb{N}$ and $c \in \mathbb{Q}^*$, if and only if $\sqrt[n_0]{a_0}$ is itself such a monomial.

PROOF. The condition about the possibility to express $\sqrt[n_0]{a_0}$ as a sum of monomials of the indicated form means precisely that $\sqrt[n_0]{a_0} \in \mathbb{Q}(\sqrt[n_1]{a_1}, \ldots, \sqrt[n_r]{a_r})$. Since $\mu_m(\mathbb{Q}(\sqrt[n_1]{a_1}, \ldots, \sqrt[n_r]{a_r})) \subseteq \{-1, 1\}$ for every $m \in \mathbb{N}^*$, we can apply Proposition 8.2.8 to obtain the desired result. $\square$

We will now examine the elementary arithmetic of finite G-radical extensions E/F with $\exp(G/F^*)$ a prime number p; in other words, of finite p-bounded radical extensions.

LEMMA 8.2.11. Let F be an arbitrary field, let $p > 0$ be a prime number, other than the characteristic of F, let $r \in \mathbb{N}^*$, let $a_1, \ldots, a_r \in F^*$, and let $\sqrt[p]{a_1}, \ldots, \sqrt[p]{a_r} \in \Omega$ denote any fixed p-th roots. Assume that

$$[F(\sqrt[p]{a_1}, \ldots, \sqrt[p]{a_r}) : F] = p^r.$$

Then, we have either $\zeta_p \in F$ *or* $\mu_p(F(\sqrt[p]{a_1},\ldots,\sqrt[p]{a_r})) = \{1\}$, *in other words,* $F(\sqrt[p]{a_1},\ldots,\sqrt[p]{a_r})/F$ *is either a classical p-Kummer extension or a p-Kummer extension with few roots of unity.*

PROOF. Since $\mu_p(F) = \mu_p(\Omega) \cap F^*$ is a subgroup of the group $\mu_p(\Omega)$ of order p, we have either $\mu_p(F) = \mu_p(\Omega)$ or $\mu_p(F) = \{1\}$. If $\mu_p(F) = \mu_p(\Omega)$ then clearly $\zeta_p \in F$.

Now assume that $\mu_p(F) = \{1\}$. We are going to prove by induction on r that for any field K with $e(K) \neq p$ and $\mu_p(K) = \{1\}$, any $\{b_1,\ldots,b_r\} \subseteq \Omega^*$ such that $b_i^p \in K$ for all i, $1 \leqslant i \leqslant r$, and

$$[K(b_1,\ldots,b_r) : K] = p^r$$

one has

$$\mu_p(K(b_1,\ldots,b_r)) = \{1\}.$$

For $r = 1$ observe that we cannot have $b_1 \in K$ since $[K(b_1) : K] = p \geqslant 2$. Hence $b_1 \notin K$. Assume that $\mu_p(K(b_1)) = \mu_p(\Omega)$. Then $\zeta_p \in K(b_1)$, i.e., $K(\zeta_p) \subseteq K(b_1)$, and consequently,

$$[K(\zeta_p) : K] \mid [K(b_1) : K].$$

But $[K(\zeta_p) : K] \leqslant \varphi(p) = p-1$ and $[K(b_1) : K] = p$, hence necessarily we must have $[K(\zeta_p) : K] = 1$, i.e., $\zeta_p \in K$. Then, $\zeta_p \in \mu_p(\Omega) \cap K = \mu_p(K) = \{1\}$, which is a contradiction. Thus $\mu_p(K(b_1))$ is a proper subgroup of the group $\mu_p(\Omega)$ of prime order p. This implies that $\mu_p(K(b_1)) = \{1\}$, as desired.

Now assume that the assertion is true for a given $r \in \mathbb{N}^*$, and prove it for $r + 1$. So, let $\{b_1,\ldots,b_r,b_{r+1}\} \subseteq \Omega^*$ with $b_i^p \in K$ for every i, $1 \leqslant i \leqslant r+1$, and $[K(b_1,\ldots,b_r,b_{r+1}) : K] = p^{r+1}$. We have to show that $\mu_p(K(b_1,\ldots,b_r,b_{r+1})) = \{1\}$.

Set $E = K(b_1,\ldots,b_r)$. The equality $[K(b_1,\ldots,b_r,b_{r+1}) : K] = p^{r+1}$ clearly implies that $[E(b_{r+1}) : E] = p$ and $[E : K] = p^r$. By the step r of the induction, we have $\mu_p(E) = \{1\}$. Now, by the step $r = 1$ of the induction applied to the field E and the element b_{r+1} we deduce that $\mu_p(E(b_{r+1})) = \{1\}$, i.e., $\mu_p(K(b_1,\ldots,b_r,b_{r+1})) = \{1\}$. $\square$

REMARKS 8.2.12. (1) Lemma 8.2.11 trivially holds for $p = e(F)$ since $\mu_p(\Omega) = \{1\}$.

(2) Exercise 2 shows that condition "$[F(\sqrt[p]{a_1},\ldots,\sqrt[p]{a_r}) : F] = p^r$" in Lemma 8.2.11 is essential.

(3) By Exercise 3, a result similar to that of Lemma 8.2.11 is valid without the condition "$[F(\sqrt[p]{a_1}, \ldots, \sqrt[p]{a_r}) : F] = p^r$", but only for specific p-th roots $\sqrt[p]{a_1}, \ldots, \sqrt[p]{a_r} \in \Omega$. This result involves the condition $C_1(n; a_1, \ldots, a_r)$, which extends the condition $C_1(n; a)$ defined in Section 1.3.

Let $n, r \in \mathbb{N}^*$, and let $a_1, \ldots, a_r \in F^*$. We say that the field F satisfies the condition $C_1(n; a_1, \ldots, a_r)$ if for every i, $1 \leqslant i \leqslant r$, the binomial $X^n - a_i$ possesses a root in Ω, say $\sqrt[n]{a_i}$, such that $\mu_n(F(\sqrt[n]{a_1}, \ldots, \sqrt[n]{a_r})) \subseteq \{-1, 1\}$. If $\gcd(n, e(F)) = 1$, then $F(\sqrt[n]{a_1}, \ldots, \sqrt[n]{a_r})/F$ is a finite n-Kummer extension with few roots of unity. Some properties of fields satisfying condition $C_1(n; a_1, \ldots, a_r)$ are stated in Exercises 5 - 8. $\square$

PROPOSITION 8.2.13. *Let F be an arbitrary field, let $p > 0$ be a prime number, other than the characteristic of F, let $r \in \mathbb{N}^*$, let $a_1, \ldots, a_r \in F^*$, and let $\sqrt[p]{a_1}, \ldots, \sqrt[p]{a_r} \in \Omega$ denote any fixed p-th roots. If*

$$[F(\sqrt[p]{a_1}, \ldots, \sqrt[p]{a_r}) : F] = p^r,$$

then the following assertions hold.

(1) *The extension $F(\sqrt[p]{a_1}, \ldots, \sqrt[p]{a_r})/F$ is $F^*\langle \sqrt[p]{a_1}, \ldots, \sqrt[p]{a_r} \rangle$-Cogalois.*

(2) *The Kneser group $F^*\langle \sqrt[p]{a_1}, \ldots, \sqrt[p]{a_r} \rangle/F^*$ of the extension E/F is isomorphic to the direct product $\mathbb{Z}_p^r$ of r copies of the group $\mathbb{Z}_p$.*

(3) $|\langle \widehat{\sqrt[p]{a_1}}, \ldots, \widehat{\sqrt[p]{a_r}} \rangle| = |\langle \widehat{a_1}, \ldots, \widehat{a_r} \rangle| = p^r$, *where $\widehat{a}$ denotes for any $a \in F^*$ its coset in the group F^*/F^{*p}.*

(4) *If $i_1, \ldots, i_n \in \mathbb{N}$ and $a_1^{i_1} \cdot \ldots \cdot a_r^{i_r} \in F^{*p}$, then $p \,|\, i_1, \ldots, p \,|\, i_r$.*

(5) $F(\sqrt[p]{a_1}, \ldots, \sqrt[p]{a_r}) = F(\sqrt[p]{a_1} + \cdots + \sqrt[p]{a_r})$.

PROOF. (1): By Lemma 8.2.11, the extension $F(\sqrt[p]{a_1}, \ldots, \sqrt[p]{a_r})/F$ is either a classical p-Kummer extension or a p-Kummer extension with few roots of unity, hence a generalized p-Kummer extension, and so, by Theorem 7.2.3 (1), it is $F^*\langle \sqrt[p]{a_1}, \ldots, \sqrt[p]{a_r} \rangle$-Cogalois.

(2) By (1), E/F is in particular an $F^*\langle \sqrt[p]{a_1}, \ldots, \sqrt[p]{a_r} \rangle$-Kneser extension. Now apply Lemma 8.1.5.

(3) follows from Theorem 7.2.3 (3), and (4) follows from Lemma 8.1.5.

(5) follows from (1) and Proposition 8.1.6. $\square$

REMARK 8.2.14. If the condition "$[F(\sqrt[p]{a_1}, \ldots, \sqrt[p]{a_r}) : F] = p^r$" in Proposition 8.2.13 is not fulfilled, then the result may fail. Indeed, the extension $\mathbb{Q}(\sqrt[3]{2}, \zeta_3 \sqrt[3]{2})/\mathbb{Q}$ has degree 6, but the subgroup $\langle \widehat{2}, \widehat{2} \rangle$ of $\mathbb{Q}^*/\mathbb{Q}^{*3}$ has order 3.   □

COROLLARY 8.2.15. *Let p be a prime other than the characteristic of a field F. Let $E = F(u)$ with $u \in \Omega$ any root of an irreducible polynomial $X^p - a \in F[X]$. Then an element $v \in E$ satisfies $v^p \in F$ if and only if $v = bu^n$ for some $b \in F$ and $n \in \mathbb{N}$.*

PROOF. If $v = bu^n$ then $v^p = b^p a^n \in F$. Conversely, assume that $v^p = c \in F$. Clearly, we may assume that $v \neq 0$. Since $[F(u) : F] = p$, E/F is an $F^*\langle u \rangle$-Cogalois extension by Proposition 8.2.13 (1), and $\mu_p(E) \subseteq F$ by Lemma 8.2.11. Hence, by Lemma 6.1.6, we deduce that $v \in F^*\langle u \rangle$, i.e., v has the desired form.   □

COROLLARY 8.2.16. *Let p be a prime other than the characteristic of a field F. Let $u, v \in \Omega$ be any roots of the irreducible polynomials $X^p - a$ and $X^p - b \in F[X]$, respectively. Then $[F(u, v) : F] = p^2$ unless $b = c^p a^n$ for some $c \in F$ and $n \in \mathbb{N}$.*

PROOF. If $b \notin F(u)^p$, then the polynomial $X^p - b$ remains irreducible over $F(u)$ by Lemma 1.3.2, hence $[F(u, v) : F] = p^2$. If $b \in F(u)^p$, then $b = w^p$ for some $w \in F(u)$, so $w = cu^n$ for some $c \in F^*$ and $n \in \mathbb{N}$ by Corollary 8.2.15. Raising this equation to the p-th power, we obtain $b = c^p a^n$.   □

COROLLARY 8.2.17. *Let p be a prime other than the characteristic of a field F. Let $u, v \in \Omega$ be any roots of the irreducible polynomials $X^p - a$ and $X^p - b \in F[X]$. If $[F(u, v) : F] = p^2$ then $F(u, v) = F(u + v)$.*

PROOF. Apply Proposition 8.2.13 (5).   □

LEMMA 8.2.18. *Let F be any field, let $p > 0$ be any prime number, and let K/F be a finite extension such that $[K : F] < p$. Then $K^p \cap F = F^p$.*

PROOF. We will proceed as in the proof of Lemma 1.3.2. The inclusion $F^p \subseteq K^p \cap F$ is obvious. Now, let $a \in K^p \cap F$, i.e., $a = u^p \in F$ for some $u \in K$. Let $f = X^m + b_{m-1}X^{m-1} + \ldots + b_1 X + b_0$ be the minimal polynomial of u over F. Since $F(u) \subseteq K$ and $[K : F] < p$, it follows that $1 \leqslant m < p$, hence $\gcd(m, p) = 1$. Then $1 = ms + pt$ for some $s, t \in \mathbb{Z}$.

Since u is a root of the polynomial $X^p - a$, it follows that $f \mid X^p - a$, hence f is a product in $\Omega[X]$ of m binomials of form $X - \zeta_p^j u$, hence

$$b_0 = \pm \zeta_p^r u^m$$

for some $r \in \mathbb{N}$. Then

$$u = u^{ms+pt} = (u^m)^s (u^p)^t = (\pm b_0 \zeta_p^{-r})^s a^t.$$

Thus $u \zeta_p^{rs} = a^t (\pm b_0)^s \in F$, and so, $a = u^p = (u \zeta_p^{rs})^p \in F^p$, as desired. $\square$

PROPOSITION 8.2.19. *Let F be an arbitrary field, let $p > 0$ be a prime number, other than the characteristic of F, let $r \in \mathbb{N}$, $r \geqslant 2$, let $a_1, \ldots, a_r$ $\in F^*$, and let $\sqrt[p]{a_1}, \ldots, \sqrt[p]{a_r} \in \Omega$ denote any fixed p-th roots. Further, let denote $E = F(\sqrt[p]{a_1}, \ldots, \sqrt[p]{a_{r-1}})$. Then either $[E(\sqrt[p]{a_r}) : E] = p$, or*

$$a_r = c^p \cdot a_1^{j_1} \cdot \ldots \cdot a_{r-1}^{j_{r-1}}$$

for some $j_1, \ldots, j_r \in \mathbb{N}$ and $c \in F^*$.

PROOF. If $a_r \notin E^p$, then the polynomial $X^p - a_r$ is irreducible in $E[X]$ by Lemma 1.3.2, so

$$[E(\sqrt[p]{a_r}) : E] = p.$$

Now, assume that $a_r \in E^p$. Then $a_r = u^p$ for some $u \in E = F(\sqrt[p]{a_1}, \ldots, \sqrt[p]{a_{r-1}})$. If we set $K = F(\zeta_p)$, then $u^p \in K$ and the extension $K(\sqrt[p]{a_1}, \ldots, \sqrt[p]{a_{r-1}})/K$ is a classical p-Kummer extension, so it is G-Cogalois by Theorem 7.1.6 (1), where $G = K^* \langle \sqrt[p]{a_1}, \ldots, \sqrt[p]{a_{r-1}} \rangle$. By Lemma 6.1.6, it follows that $u \in G$, i.e.,

$$u = v \cdot \sqrt[p]{a_1}^{j_1} \cdot \ldots \cdot \sqrt[p]{a_{r-1}}^{j_{r-1}}$$

for some $v \in K^*$ and $j_1, \ldots, j_{r-1} \in \mathbb{N}$.

Raising the last equation to the p-th power, we obtain

$$a_r = v^p \cdot a_1^{j_1} \cdot \ldots \cdot a_{r-1}^{j_{r-1}}.$$

This implies that $v^p = a_r \cdot a_1^{-j_1} \cdot \ldots \cdot a_{r-1}^{-j_{r-1}} \in K^p \cap F$. But $[K : F] = [F(\zeta_p) : F] \leqslant p - 1$, hence $v^p \in F^p$ by Lemma 8.2.18, i.e., $v^p = c^p$ for some $c \in F^*$. Thus

$$a_r = c^p \cdot a_1^{j_1} \cdot \ldots \cdot a_{r-1}^{j_{r-1}},$$

and we are done. $\square$

8.3. Exercises to Chapter 8

1. Does the result in Theorem 8.1.2 hold if one weakens the hypothesis " E/F is a G-Cogalois extension" to " E/F is a G-Kneser extension"?

2. Show that if condition " $[F(\sqrt[p]{a_1}, \ldots, \sqrt[p]{a_r}) : F] = p^r$ " in Lemma 8.2.11 is not satisfied, then, in general, the result does not hold for any choices of p-th roots $\sqrt[p]{a_1}, \ldots, \sqrt[p]{a_r}$. (*Hint*: Take $p = 3$, $r = 2$, $F = \mathbb{Q}$, $a_1 = a_2 = 2$, and consider the 3-th roots of unity $\sqrt[3]{2}$, $\sqrt[3]{2}\,\zeta_3$.)

3. Let F be an arbitrary field and let $p > 0$ be any prime number. Prove that for any $r \in \mathbb{N}^*$ and any $a_1, \ldots, a_r \in F^*$, either there exist specific p-th roots $\sqrt[p]{a_1}, \ldots, \sqrt[p]{a_r} \in \Omega$ such that $\mu_p(F(\sqrt[p]{a_1}, \ldots, \sqrt[p]{a_r})) = \{1\}$, (i.e., the field F satisfies condition $C(p; a_0, \ldots, a_r)$), or $\zeta_p \in F$.

4. Let F be any field, let $p > 0$ be any prime number, let $a \in F^*$, and suppose that $\zeta_p \in F$ and $\sqrt[p]{a} \notin F$. Show that $[F(\sqrt[p]{a}) : F] = p$.

5. Let F be an arbitrary subfield of $\mathbb{R}$. Show that F satisfies the condition $C_1(n; a_1, \ldots, a_r)$, where $r \in \mathbb{N}^*$ is arbitrary, and, either $n \in \mathbb{N}^*$ is an arbitrary odd number and $a_1, \ldots, a_r \in F^*$ are arbitrary, or $n \in \mathbb{N}^*$ is an arbitrary even number and $a_1, \ldots, a_r \in F^*_+$ are arbitrary.

6. Show that a field F satisfying the conditions $C_1(n; a_1)$ and $C_1(n; a_2)$ may not satisfy the condition $C_1(n; a_1, a_2)$.

7. Let $n \in \mathbb{N}^*$ be an odd number, let $r \in \mathbb{N}^*$, $a_1, \ldots, a_r \in F^*$, and suppose that F satisfies the condition $C_1(n; a_1, \ldots, a_r)$. Prove that if E/F is any Abelian extension with $\zeta_n \in L$, then

$$[F(\sqrt[n]{a_1}, \ldots, \sqrt[n]{a_r}) : F] = [L(\sqrt[n]{a_1}, \ldots, \sqrt[n]{a_r}) : L],$$

where $\sqrt[n]{a_i}$ denotes for every $i = 1, \ldots, r$ a specified root in Ω of the polynomial $X^n - a_i$ such that $\mu_n(F(\sqrt[n]{a_1}, \ldots, \sqrt[n]{a_r})) \subseteq \{-1, 1\}$.

8. Let $n \in \mathbb{N}^*$ be an odd number, let $r \in \mathbb{N}^*$, $a_1, \ldots, a_r \in F^*$, and suppose that F satisfies the condition $C_1(n; a_1, \ldots, a_r)$. Prove that if E/F is an Abelian extension with $E \subseteq F(\sqrt[n]{a_1}, \ldots, \sqrt[n]{a_r})$, then $E = F$. As above, $\sqrt[n]{a_i}$ denotes for every $i = 1, \ldots, r$ a specified root in Ω of the polynomial $X^n - a_i$ with $\mu_n(F(\sqrt[n]{a_1}, \ldots, \sqrt[n]{a_r})) \subseteq \{-1, 1\}$.

9. (*Zhou* [**113**]). Let $r, n_1, \ldots, n_r \in \mathbb{N}^*$, let $p_1, \ldots, p_r \in \mathbb{P}$ be distinct primes, and let $x_k \in \mathbb{C}$ be any root of the polynomial $X^{n_k} - p_k$, $k = 1, \ldots, r$. Prove the following statements.
   (a) $[\mathbb{Q}(x_1, \ldots, x_r) : \mathbb{Q}] = n_1 \cdot \ldots \cdot n_r$.
   (b) $\mathbb{Q}(x_1, \ldots, x_r) = \mathbb{Q}(x_1 + \ldots + x_r)$.

10. (*Zhou* [**113**]). Let $r, m, n_1, \ldots, n_r \in \mathbb{N}^*$, and let $p_1, \ldots, p_r \in \mathbb{P}$ be distinct primes such that $p_k \mid m$, $m = p_k^{s_k} m_k$, $\gcd(m_k, p_k) = 1$, and $\gcd(n_k, p_k^{s_k} - p_k^{s_k - 1})$ for every $k = 1, \ldots r$. Let $x_k \in \mathbb{C}$ be any root of the polynomial $X^{n_k} - p_k$, $k = 1, \ldots, r$. Prove the following statements.
   (a) $[\mathbb{Q}(\zeta_m)(x_1, \ldots, x_r) : \mathbb{Q}(\zeta_m)] = n_1 \cdot \ldots \cdot n_r$.
   (b) $\mathbb{Q}(\zeta_m)(x_1, \ldots, x_r) = \mathbb{Q}(\zeta_m)(x_1 + \ldots + x_r)$.

11. (Vélez [**108**]). Let F be an arbitrary field, let $n \in \mathbb{N}^*$ be such that $\gcd(n, e(F)) = 1$, and let $X^n - a$, $X^n - b$ be irreducible polynomials in $F[X]$ with roots $u, v \in \Omega$, respectively, where Ω is an algebraically closed overfield of F. If the extension $F(u)/F$ has the USP, then prove that the following assertions are equivalent.
   (1) The fields $F(u)$ and $F(v)$ are F-isomorphic.
   (2) There exists $c \in F$ and $j \in \mathbb{N}$ with $\gcd(j, n) = 1$ and $a = b^j c^n$.

   (*Hint*: Use Lemma 6.1.6 and Exercise 17, Chapter 4.)

12. Let F be an arbitrary field, and let $n \in \mathbb{N}^*$ be such that $\zeta_n \in F$ and $\gcd(n, e(F)) = 1$. Let $X^n - a$, $X^n - b$ be irreducible polynomials in $F[X]$ with roots $u, v \in \Omega$, respectively, where Ω is an algebraically closed overfield of F. Prove that the following statements are equivalent.
   (1) $F(u) = F(v)$.
   (2) There exists $c \in F$ and $j \in \mathbb{N}$ with $\gcd(j, n) = 1$ and $a = b^j c^n$.

   (*Hint*: Use Exercise 11 and Theorem 7.1.6 (1).)

8.4. Bibliographical comments to Chapter 8

Section 8.1. The results of this section are due to Albu and Nicolae [**20**].

Section 8.2. The results of this section are due to Albu [**3**], [**7**], [**12**], [**13**], Albu and Nicolae [**20**], Albu, Nicolae, and Ţena [**23**]. Some of the

original results of Albu [3] and Albu, Nicolae, and Ţena [23] presented in
this section were either incorrect or had incomplete proofs and we have
made the necessary changes.

To the best of our knowledge, Proposition 8.2.1, Corollary 8.2.2, Corollary 8.2.3, as well as well as a similar result for finite classical Kummer
extensions, were not known before 1995, when the paper of Albu and Nicolae [20] appeared.

Corollary 8.2.5 is folklore, but without using Proposition 8.2.1, its proof
is not at all as brief as ours. Corollaries 8.2.15, 8.2.16, 8.2.17 appear in Kaplansky [74], but were originally proved in a different manner. Proposition
8.2.19 is essentially due to Baker and Stark [28], who proved it only for algebraic number fields. Our proof, based on Lemma 8.2.18 and taken from
Albu [7], is valid for more than just algebraic number fields.

CHAPTER 9

APPLICATIONS TO ALGEBRAIC NUMBER FIELDS

Cogalois Theory has nice applications to algebraic number fields. This chapter presents some of them. After reviewing, in Section 9.1, the standard notation, notions, and facts on algebraic number fields which will be used in the sequel, we show, in Section 9.2, that some classical results due to Hasse, Besicovitch, Mordell, Siegel are immediate consequences of basic facts in Cogalois Theory.

In Section 9.3 we present a surprising application of the Kneser Criterion in proving a very classical result in Algebraic Number Theory claimed by Hecke (but not proved) in his book [**71**] and related to the so-called *Hecke systems of ideal numbers*.

9.1. Number theoretic preliminaries

Below, we present some of the concepts and results from Algebraic Number Theory which will be used in the next sections of this chapter.

All rings considered in this section are supposed to be commutative with nonzero identity element. Throughout this section R will always denote such a ring.

9.1.1. INTEGRAL EXTENSION

Integral element. Let R be any ring, let S be an overring of R (that is, a ring containing R as a subring), and let $u \in S$. Then u is said to be an *integral element over* R if there exist $n \in \mathbb{N}^*$ and elements $a_0, \ldots, a_{n-1}$ in R such that u satisfies the equation

$$u^n + a_{n-1}u^{n-1} + \ldots + a_0 = 0,$$

in other words, if u is a root of the monic polynomial

$$X^n + a_{n-1}X^{n-1} + \ldots + a_0 \in R[X].$$

If E/F is a field extension, then $u \in E$ is integral over F if and only if u is algebraic over E.

Ring extension. If S is an overring of a ring R, then the pair (R, S) is called a *ring extension*, and in this case we shall write S/R.

If S/R is a ring extension, then the underlying additive group $(S, +)$ of the ring S may be viewed as a module $_RS$ via the scalar multiplication $R \times S \longrightarrow S$, $(a, x) \mapsto ax$, where ax is the product of the elements $a \in R$ and $x \in S$ under the given multiplication on S. In fact, this structure of R-module on S together with the given ring structure on S endow S in a natural way with a structure of an R-algebra.

A ring extension S/R is said to be *finite* if $_RS$ is a finitely generated module.

Integral extension. We say that a ring extension S/R is an *integral extension* if every element of S is integral over R.

A field extension is integral if and only if it is algebraic. As is known, any finite field extension E/F is necessarily an algebraic extension. The corresponding result for ring extensions is the following: any finite ring extension is an integral extension.

Integral closure. If S/R is a ring extension, then the set R'_S of all elements of S which are integral over R is called the *integral closure* of R in S. The integral closure R'_S of R in S is a subring of S that includes R. The ring R is said to be *integrally closed* in S if $R'_S = R$. The integral closure R'_S of R in S is integrally closed in S, and R'_S/R is an integral extension.

A *domain* is a ring without zero divisors, that is, a ring having the following property: whenever $a, b \in R$ and $ab = 0$, it follows that $a = 0$ or $b = 0$.

An *integrally closed domain* is a domain which is integrally closed in its field of quotients.

Algebraic integer. An *algebraic number field*, or simply, a *number field* is any subfield K (which automatically contains $\mathbb{Q}$ as a subfield) of the field $\mathbb{C}$ of complex numbers, such that the extension $K/\mathbb{Q}$ is finite. The positive integer $[K : \mathbb{Q}]$ is called the *degree of the number field K*.

Let K be an algebraic number field. The integral closure $\mathbb{Z}'_K$ of $\mathbb{Z}$, which is a subring of K, is called the *ring of algebraic integers of K*, and is denoted by $\mathfrak{O}_K$. Observe that $\mathfrak{O}_\mathbb{Q} = \mathbb{Z}$, and this is the reason why the elements of $\mathbb{Z}$ are called *rational integers*.

An *algebraic number* (resp. *algebraic integer*) is any $z \in \mathbb{C}$ which is algebraic (resp. integral) over $\mathbb{Q}$ (resp. over $\mathbb{Z}$).

Quadratic integer. A number field K is said to be *quadratic* if $[K : \mathbb{Q}] = 2$. A subfield K of $\mathbb{C}$ is a quadratic number field if and only if there exists a square-free integer $d \in \mathbb{Z} \setminus \{1\}$, which is uniquely determined, such that $K = \mathbb{Q}(\sqrt{d})$, that is, $K = \{a + b\sqrt{d} \mid a, b \in \mathbb{Q}\}$. A number $s \in \mathbb{Z}$ is said to be *square-free* if $p^2 \nmid s$ for every $p \in \mathbb{P}$.

For any square-free integer $d \in \mathbb{Z} \setminus \{1\}$, the ring $\mathfrak{O}_{\mathbb{Q}(\sqrt{d})}$ of integers of the quadratic field $\mathbb{Q}(\sqrt{d})$ has the following form:

$$\mathfrak{O}_{\mathbb{Q}(\sqrt{d})} = \begin{cases} \mathbb{Z}[\sqrt{d}] & \text{if } d \equiv 2 \ (\text{mod } 4) \ \text{ or } \ d \equiv 3 \ (\text{mod } 4) \\ \mathbb{Z}[\frac{1+\sqrt{d}}{2}] & \text{if } d \equiv 1 \ (\text{mod } 4). \end{cases}$$

9.1.2. FRACTIONAL IDEAL

Definition and examples. Let R be a domain with field of quotients K. A *fractional ideal* of R is any R-submodule I of $_R K$ having the property that there exists an $a \in R$, $a \neq 0$ such that $aI \subseteq R$, where $aI = \{ax \mid x \in I\}$.

If I is a fractional ideal of R, then $I \subseteq R$ if and only if I is an ideal of the ring R, and we will then say that I is an *integral ideal* of R. Any cyclic submodule Rx of $_R K$, with $x \in K$, is a fractional ideal of R, and we call it a *principal fractional ideal* of R. More generally, any finitely generated R-submodule of $_R K$ is a fractional ideal of R.

Operations on fractional ideals. If I and J are fractional ideals of R, then $I + J$, $I \cap J$, and $I \cdot J$ are also fractional ideals of R, where

$$I \cdot J := \left\{ \sum_{1 \leqslant k \leqslant n} x_k y_k \mid n \in \mathbb{N}^*, \, x_k \in I, \, y_k \in J, \, 1 \leqslant k \leqslant n \right\}$$

is the product of I and J.

A fractional ideal I of R is said to be *invertible* if there exists a fractional ideal J of R such that $I \cdot J = R$.

The set $\mathcal{F}_R$ of all nonzero fractional ideals of a domain R is a commutative monoid with respect to the multiplication of fractional ideals, having R as identity element. We will see in the next subsection that the monoid $(\mathcal{F}_R, \cdot)$ is a group precisely when R is a Dedekind domain.

9.1.3. DEDEKIND DOMAIN

PID. A *principal ideal domain*, abbreviated *PID*, is a domain R such that every ideal I of R is a principal ideal, i.e., $I = Ra$ for some $a \in R$.

Prime ideal. Let R be a ring. An ideal P of R is said to be *prime* if $P \neq R$ and for any $a, b \in R$ such that $ab \in P$ it follows that $a \in P$ or $b \in P$. The zero ideal 0 of R is a prime ideal if and only if R is a domain.

Prime element. Let R be a ring. An element $p \in R$ is said to be *prime* if $p \notin U(R)$ and for any $a, b \in R$ such that $p \mid ab$ it follows that $p \mid a$ or $p \mid b$, or equivalently, if the principal ideal Rp generated by p is a prime ideal.

UFD. A domain R is said to be a *unique factorization domain*, abbreviated *UFD*, if every element $a \in R \setminus (U(R) \cup \{0\})$ factors into a product of prime elements. The factorization is necessarily unique up to unit multiples and the order of the factors. Any PID is a UFD.

Maximal ideal. An ideal M of a ring R is said to be *maximal* if $M \neq R$ and whenever I is an ideal of R such that $M \subseteq I$, then $I = M$ or $I = R$.

Noetherian ring. A ring R is *Noetherian* if every ideal of R is finitely generated, or equivalently, if every ascending chain of ideals

$$I_1 \subseteq I_2 \subseteq I_3 \subseteq \dots$$

of R is eventually constant, i.e., there exists an $n_0 \in \mathbb{N}^*$ such that $I_n = I_{n+1}$ for all $n \geqslant n_0$.

Characterization. The following statements are equivalent for a domain R.

(a) R is a Noetherian integrally closed domain and every nonzero prime ideal of R is maximal.

(b) Every nonzero fractional ideal of R is invertible, in other words, the monoid $(\mathcal{F}_R, \cdot)$ of nonzero fractional ideals of R is a group.

(c) Every integral ideal I of R, $I \neq R$, is representable as a product of finitely many prime ideals of R.

A domain R satisfying one of the equivalent conditions above is called a *Dedekind domain*.

Examples.

(1) Any field is a Dedekind domain.
(2) Any PID is a Dedekind domain. In particular, the ring $\mathbb{Z}$ of rational integers, and the ring $F[X]$ of polynomials in one indeterminate over any field F are Dedekind domains.
(3) A Dedekind domain R is a UFD if and only if R is a PID.
(4) Let R be a Dedekind domain, let K be field of quotients of R, and let L be a finite separable extension of K. Then, the integral closure R'_L of R in L is a Dedekind domain. In particular, for any algebraic number field K, the ring $\mathcal{O} = \mathbb{Z}'_K$ of algebraic integers of K is a Dedekind domain.

Ideal class group. Let R be a Dedekind domain with field of quotients K. The set $\mathcal{P}_R = \{\, Rx \,|\, x \in K^* \,\}$ of all nonzero principal fractional ideals of R is a subgroup of the group $\mathcal{F}_R$ of all nonzero fractional ideals of R, so it makes sense to consider the quotient group

$$\mathcal{C}\ell_R := \mathcal{F}_R/\mathcal{P}_R$$

which is called the *ideal class group of R*. The order $h_R \in \mathbb{N}^* \cup \{\infty\}$ of the group $\mathcal{C}\ell_R$ is called the *class number of R*. The Dedekind domain R is a PID if and only if $h_R = 1$.

9.1.4. RING OF ALGEBRAIC INTEGERS

Throughout this subsection K will denote an algebraic number field and $\mathcal{O}_K$ the ring of integers of K.

Group of units. The group of units $U(\mathcal{O}_K)$ is called, by abuse of terminology, the *group of units of K*, and is denoted by U_K.

Root of unity. Since $K \subseteq \mathbb{C}$, for any $n \in \mathbb{N}^*$ we can choose as a primitive n-th root of unity over K the complex number

$$\zeta_n = \cos(2\pi/n) + i\sin(2\pi/n).$$

The group $\mu(K)$ of all roots of unity in K is a finite cyclic group.

Decomposition in prime ideals. Since $\mathcal{O}_K$ is a Dedekind domain, any ideal I of $\mathcal{O}_K$, $0 \neq I \neq \mathcal{O}_K$, is representable as a finite product of prime ideals of $\mathcal{O}_K$. Moreover, this representation is unique up to the order of prime ideals in the representation.

Ideal class group. The group $\mathcal{F}_{\mathfrak{O}_K}$ (resp. $\mathcal{P}_{\mathfrak{O}_K}$) of all nonzero fractional ideals (resp. principal fractional ideals) of $\mathfrak{O}_K$ is called, by abuse of terminology, the *group of nonzero fractional ideals* (resp. the *group of nonzero principal fractional ideals*) *of* K, and is denoted by $\mathcal{F}_K$ (resp. $\mathcal{P}_K$).

The ideal class group $\mathcal{Cl}_{\mathfrak{O}_K} := \mathcal{F}_K/\mathcal{P}_K$ is called, also by abuse of terminology, the *ideal class group of* K, and is denoted by $\mathcal{Cl}_K$. The elements of $\mathcal{Cl}_K$ are called *ideal classes of* K. A famous theorem due to Dirichlet states that $\mathcal{Cl}_K$ is a finite group. The order of $\mathcal{Cl}_K$ is denoted by h_K, or simply by h when there is no danger of confusion, and is called, by abuse of terminology, the *class number of* K.

Extension of ideals. Let L/K be a field extension of algebraic number fields, and let $I \in \mathcal{F}_L$. Then $\mathfrak{O}_L/\mathfrak{O}_K$ is a ring extension, and $\mathfrak{O}_L$ is the integral closure of $\mathfrak{O}_K$ in L. For any $I \in \mathcal{F}_K$ set

$$I\mathfrak{O}_L := \Big\{ \sum_{1 \leqslant k \leqslant n} x_k y_k \mid n \in \mathbb{N}^*, \, x_k \in I, \, y_k \in \mathfrak{O}_L, \, 1 \leqslant k \leqslant n \Big\}.$$

Then $I\mathfrak{O}_L \in \mathcal{F}_L$ and

$$I = I\mathfrak{O}_L \cap \mathfrak{O}_K = I\mathfrak{O}_L \cap K.$$

9.2. Some classical results via Cogalois Theory

The aim of this section is to present some classical results due to Hasse, Besicovitch, Mordell, and Siegel as easy consequences of basic facts from Cogalois Theory.

Throughout this section K denotes a fixed algebraic number field, and for any $a \in K^*$ and $n \in \mathbb{N}^*$, $\sqrt[n]{a}$ designates a root (which in general is not specified) in $\mathbb{C}$ of the polynomial $X^n - a \in K[X]$.

However, if K is a subfield of the field $\mathbb{R}$ of real numbers and $a > 0$, then $\sqrt[n]{a}$ will always mean the unique positive root in $\mathbb{R}$ of the polynomial $X^n - a$.

THEOREM 9.2.1 (HASSE, 1930). *Let K be an algebraic number field containing ζ_n for some $n \in \mathbb{N}^*$, let $r \in \mathbb{N}^*$, and let $x_1, \ldots, x_r \in \mathbb{C}^*$ be such that $x_k^n \in K$ for all k, $1 \leqslant k \leqslant r$. Assume that the following condition is satisfied:*

(†) $m_1, \ldots, m_r \in \mathbb{N}$ *and* $x_1^{m_1} \cdot \ldots \cdot x_r^{m_r} \in K \implies x_k^{m_k} \in K$, $k = 1, \ldots, r$.

Then

$$[K(x_1,\ldots,x_r):K] = \prod_{1\leqslant k\leqslant r} |K^*\langle x_k\rangle/K^*|.$$

PROOF. Clearly, $K(x_1,\ldots,x_r)/K$ is a classical n-Kummer extension, so it is a G-Cogalois extension by Theorem 7.1.6 (1), where $G = K\langle x_1,\ldots,x_r\rangle$. In particular, $K(x_1,\ldots,x_r)/K$ is a G-Kneser extension, hence

$$[K(x_1,\ldots,x_r):K] = |G/K^*| = |K^*\langle x_1,\ldots,x_r\rangle/K^*|.$$

Now apply Lemma 8.1.5 to deduce that

$$|K^*\langle x_1,\ldots,x_r\rangle/K^*| = \prod_{1\leqslant k\leqslant r} |K^*\langle x_k\rangle/K^*|.$$

The fact that $K(x_1,\ldots,x_r)/K$ is a G-Kneser extension can be shown more directly as follows, using only the Kneser Criterion.

Let $p \in \mathcal{P}$. If $p\,|\,n$, then we have $\zeta_p \in K$, since $\zeta_n \in K$ by hypothesis. Now assume that $p\nmid n$. Then $ap+bn = 1$ for some $a,b \in \mathbb{Z}$. Observe that $G^n \subseteq K^*$, so, if $\zeta_p \in G$ for an odd prime p, then $\zeta_p^n \in K$. It follows that $\zeta_p = \zeta_p^{ap+bn} = (\zeta_p^n)^b \in K$, as desired.

Similarly, $1 + \zeta_4 \in G$ implies that $(1 + \zeta_4)^n \in K$. Since $(1 + \zeta_4)^4 = -4 \in K$, and we are assuming that $4\nmid n$, we deduce as in the proof of the Kneser Criterion (Theorem 2.2.1) that the element $\widehat{1+\zeta_4} \in G/K^*$ has order 1 or 2, which implies that $\zeta_4 \in K$. By the Kneser Criterion, it follows that $K(x_1,\ldots,x_r)/K$ is a G-Kneser extension. $\qquad\square$

THEOREM 9.2.2 (SIEGEL, 1972). *Let K be an algebraic number field, let $r \in \mathbb{N}^*$, let $n_1,\ldots,n_r \in \mathbb{N}^*$, and let $x_1,\ldots,x_r \in \mathbb{C}$ be such that $x_k^{n_k} \in K$ for all $k, 1 \leqslant k \leqslant r$. Assume that either $\zeta_{n_k} \in K$ for all $k, 1 \leqslant k \leqslant r$, or $K \subseteq \mathbb{R}$ and $x_k \in \mathbb{R}_+^*$ for all $k, 1 \leqslant k \leqslant r$. Then*

$$[K(x_1,\ldots,x_r):K] = |K^*\langle x_1,\ldots,x_r\rangle/K^*|.$$

PROOF. Set $G = K\langle x_1,\ldots,x_r\rangle$ and $L = K(x_1,\ldots,x_r)$. Observe that L/K is a separable G-radical extension, and so, the desired equality means precisely that L/K is a G-Kneser extension.

If $K \subseteq \mathbb{R}$ and $x_k \in \mathbb{R}_+^*$ for all $k, 1 \leqslant k \leqslant r$, then $L \subseteq \mathbb{R}$, hence L/K is clearly a pure extension (see also Examples 3.1.4 (1)), and thus, it is G-Kneser by Lemma 3.1.6.

Now, assume that $\zeta_{n_k} \in K$ for all $k, 1 \leqslant k \leqslant r$, and let $n = \mathrm{lcm}(n_1,\ldots,n_r)$. Then $n = n_k b_k$ for some $b_k \in \mathbb{Z}, 1 \leqslant k \leqslant r$. Clearly,

$\gcd(b_1,\ldots,b_r) = 1$, hence $1 = \sum_{1 \leqslant k \leqslant r} b_k c_k$ for some $c_1,\ldots,c_r \in \mathbb{Z}$. It follows that

$$\zeta_n = \zeta_n^{b_1 c_1 + \cdots + b_r c_r} = (\zeta_n^{b_1})^{c_1} \cdot \ldots \cdot (\zeta_n^{b_r})^{c_r} \in K,$$

since $\zeta_{n_k} \in K$ and $\zeta_n^{b_k}$ is a primitive n_k-th root of unity for all $k = 1,\ldots,r$. By the proof of Theorem 9.2.1, L/K is a G-Kneser extension. $\quad\square$

THEOREM 9.2.3 (MORDELL, 1953). *Let K be an algebraic number field, let $r \in \mathbb{N}^*$, let $n_1,\ldots,n_r \in \mathbb{N}^*$, and let $x_1,\ldots,x_r \in \mathbb{C}$ be such that $x_k^{n_k} \in K$ for all k, $1 \leqslant k \leqslant r$, and*

$(\dagger\dagger)\quad m_1,\ldots,m_r \in \mathbb{N}$ *and* $x_1^{m_1} \cdot \ldots \cdot x_r^{m_r} \in K \Longrightarrow n_k \mid m_k,\ k = 1,\ldots,r.$

Assume that either $\zeta_{n_k} \in K$ for all k, $1 \leqslant k \leqslant r$, or $K \subseteq \mathbb{R}$ and $x_k \in \mathbb{R}_+^*$ for all k, $1 \leqslant k \leqslant r$. Then

$$[\,K(x_1,\ldots,x_r) : K\,] = n_1 \cdot \ldots \cdot n_r.$$

PROOF. The extension $K(x_1,\ldots,x_r)/K$ is $K\langle x_1,\ldots,x_r\rangle$-Kneser by Theorem 9.2.2. To conclude, observe that condition $(\dagger\dagger)$ is the same as condition $(\dagger)$ in Theorem 9.2.1. Now, apply Lemma 8.1.5 to obtain the desired equality. $\quad\square$

LEMMA 9.2.4. *Let $r, n, k_1,\ldots,k_r \in \mathbb{N}^*$, and let $p_1,\ldots,p_r$ be different positive prime integers. Then, the following statements are equivalent.*

(1) $\sqrt[n]{p_1^{k_1} \cdot \ldots \cdot p_r^{k_r}} \in \mathbb{Q}.$

(2) $\sqrt[n]{p_1^{k_1} \cdot \ldots \cdot p_r^{k_r}} \in \mathbb{N}.$

(3) $n \mid k_j$ *for all* $j = 1,\ldots,r.$

PROOF. For simplicity, denote $a = p_1^{k_1} \cdot \ldots \cdot p_r^{k_r}$ and $x = \sqrt[n]{a}$.

$(1) \Longrightarrow (2)$: Assume that $x \in \mathbb{Q}$. Observe that x is a root of the monic polynomial $X^n - a \in \mathbb{Z}[X]$, i.e., x is an integral element over $\mathbb{Z}$. It follows that $x \in \mathbb{Z}$ since $\mathbb{Z}$ is integrally closed. But $x > 0$, hence necessarily $x \in \mathbb{N}$.

$(2) \Longrightarrow (3)$: Assume that $x \in \mathbb{N}$. Since $x \geqslant 2$ and $\mathbb{Z}$ is a UFD, we can factor x as

$$x = q_1^{i_1} \cdot \ldots \cdot q_s^{i_s},$$

where $s, i_1,\ldots,i_s \in \mathbb{N}^*$ and $q_1,\ldots,q_s$ are different positive prime integers. This implies that

$$q_1^{n i_1} \cdot \ldots \cdot q_s^{n i_s} = x^n = a = p_1^{k_1} \cdot \ldots \cdot p_r^{k_r}.$$

Since the decomposition of a into a product of positive prime integers is unique up to the order of the factors, we deduce that $r = s$ and there exists a permutation $\sigma \in \mathfrak{S}_r$ such that $p_j = q_{\sigma(j)}$ and $k_j = ni_{\sigma(j)}$ for all $j = 1, \ldots, r$, hence $n \mid k_j$ for all $j = 1, \ldots, r$.

(3) $\Longrightarrow$ (2) and (2) $\Longrightarrow$ (1) are obvious. $\qquad\qquad\qquad\square$

COROLLARY 9.2.5. *Let* $r \in \mathbb{N}$, $r \geqslant 2$, *and let* $a_1, \ldots, a_r, n_1, \ldots, n_r \in \mathbb{N}^*$ *be such that for every* k, $1 \leqslant k \leqslant r$, *and every* l_k, $1 \leqslant l_k < n_k$ *one has* $(\sqrt[n_k]{a_k})^{l_k} \notin \mathbb{N}$. *Assume that the following condition is satisfied.*

(*) $\qquad\qquad a_1, \ldots, a_k$ *are relatively prime in pairs.*

Then
$$[\mathbb{Q}(\sqrt[n_1]{a_1}, \ldots, \sqrt[n_r]{a_r}) : \mathbb{Q}] = n_1 \cdot \ldots \cdot n_r.$$

PROOF. We are going to show that condition (*) implies condition (††) in Theorem 9.2.3. Let $m_1, \ldots, m_r \in \mathbb{N}$ be such that
$$\sqrt[n_1]{a_1}^{\,m_1} \cdot \ldots \cdot \sqrt[n_r]{a_r}^{\,m_r} \in \mathbb{Q}.$$
If $n = \mathrm{lcm}(n_1, \ldots, n_r)$, then $n = n_i k_i$ for all $i = 1, \ldots, r$, hence the relation above can be written as
$$\sqrt[n]{a_1^{k_1 m_1} \cdot \ldots \cdot a_r^{k_r m_r}} \in \mathbb{Q}.$$

Since $a_i \geqslant 2$ for all $i = 1, \ldots, r$, we can factor each a_i into a product of prime numbers. Let $a_1 = q_1^{j_1} \cdot \ldots \cdot q_s^{j_s}$, with $s, j_1, \ldots, j_s \in \mathbb{N}^*$ and $q_1, \ldots, q_s$ different positive prime integers. According to condition (*), any of the prime numbers $q_1, \ldots, q_s$ does not occur in the factorizations of $a_2, \ldots, a_r$ into products of prime numbers, so, by Lemma 9.2.4, we deduce that $n \mid k_1 m_1 j_i$, hence $n_1 \mid m_1 j_i$ for all $i = 1, \ldots, s$. Consequently, $a_1^{m_1} = q_1^{m_1 j_1} \cdot \ldots \cdot q_s^{m_1 j_s}$ is an n_1-th power of some natural number, i.e., $(\sqrt[n_1]{a_1})^{m_1} \in \mathbb{N}$. But, by Lemma 9.2.4, the condition $(\sqrt[n_1]{a_1})^{l_1} \notin \mathbb{N}$ for every l_1, $1 \leqslant l_1 < n_1$ means exactly that $\mathrm{ord}(\widehat{\sqrt[n_1]{a_1}}) = n_1$ in the group $\mathbb{R}^*/\mathbb{Q}^*$. Thus, we must have $n_1 \mid m_1$. In a similar way, one shows that $n_i \mid m_i$ for all $i = 2, \ldots, r$, and we are done. $\qquad\qquad\square$

Another consequence of Theorem 9.2.3 is the following classical result.

THEOREM 9.2.6 (BESICOVITCH, 1940). *Let* $r \in \mathbb{N}^*$, *let* $p_1, \ldots, p_r$ *be different positive prime integers, let* $b_1, \ldots, b_r \in \mathbb{N}^*$ *be not divisible by any of these primes, and let* $a_1 = b_1 p_1, \ldots, a_r = b_r p_r$. *Then, for any* $n_1, \ldots, n_r \in \mathbb{N}^*$ *one has*
$$[\mathbb{Q}(\sqrt[n_1]{a_1}, \ldots, \sqrt[n_r]{a_r}) : \mathbb{Q}] = n_1 \cdot \ldots \cdot n_r.$$

PROOF. We will show that condition (††) in Theorem 9.2.3 is satisfied. Let $m_1, \ldots, m_r \in \mathbb{N}$ be such that

$$\sqrt[n_1]{a_1}^{\,m_1} \cdot \ldots \cdot \sqrt[n_r]{a_r}^{\,m_r} \in \mathbb{Q}.$$

If $n = \operatorname{lcm}(n_1, \ldots, n_r)$, then $n = n_i k_i$ for all $i = 1, \ldots, r$, hence the relation above can be written as

$$\sqrt[n]{b_1^{k_1 m_1} p_1^{k_1 m_1} \cdot \ldots \cdot b_r^{k_r m_r} p_r^{k_r m_r}} \in \mathbb{Q}.$$

By hypotheses, $p_1, \ldots, p_r$ are different positive prime integers and $b_1, \ldots, b_r$ are not divisible by any of these primes. By Lemma 9.2.4, we deduce that $n \mid k_i m_i$, hence $n_i \mid m_i$ for all $i = 1, \ldots, r$, as desired. $\qquad\square$

COROLLARY 9.2.7. *Let $r \in \mathbb{N}^*$, let $p_1, \ldots, p_r$ be different positive prime integers, and let $n_1, \ldots, n_r \in \mathbb{N}^*$ be arbitrary. Then*

$$[\mathbb{Q}(\sqrt[n_1]{p_1}, \ldots, \sqrt[n_r]{p_r}) : \mathbb{Q}] = n_1 \cdot \ldots \cdot n_r.$$

PROOF. Apply Theorem 9.2.6 for $b_1 = \ldots = b_r = 1$. $\qquad\square$

REMARKS 9.2.8. (1) In view of Theorem 7.1.6 and Lemma 8.1.5, Theorem 9.2.1 holds not only for algebraic number fields K, but also for any field K and any $n \in \mathbb{N}^*$ such that $\gcd(n, e(F)) = 1$. Of course, the field $\mathbb{C}$ in the statement of Theorem 9.2.1 should be replaced by an algebraically closed field Ω containing K as a subfield. Also, according to Theorem 7.2.3 and Lemma 8.1.5, Theorems 9.2.2 and 9.2.3 are valid for any field K, any $n_1, \ldots, n_r \in \mathbb{N}^*$, and any $x_1, \ldots, x_r \in \Omega$ with $x_k^{n_k} \in K$ for all k, $1 \leqslant k \leqslant r$, such that $\gcd(n, e(K)) = 1$ and $\mu_n(K(x_1, \ldots, x_r)) \subseteq K$, where $n = \operatorname{lcm}(n_1, \ldots, n_r)$.

(2) Corollary 9.2.5 and Theorem 9.2.6 can be extended from $\mathbb{Q}$ to the field of quotients of any UFD (see Exercises 1 and 2).

(3) We have seen that the conditions in Theorem 9.2.6, as well as that in Corollary 9.2.5 imply for $K = \mathbb{Q}$ the condition (††) in Theorem 9.2.3. Observe that the extension $\mathbb{Q}(\sqrt{2}, \sqrt{6})/\mathbb{Q}$ satisfies the condition (††) in Theorem 9.2.3 with $r = 2$, $n_1 = n_2 = 2$, $x_1 = \sqrt{2}$, $x_2 = \sqrt{6}$, but satisfies neither the conditions in Corollary 9.2.5 nor the conditions in Theorem 9.2.6.

The extension $\mathbb{Q}(\sqrt{8}, \sqrt{3})/\mathbb{Q}$ satisfies the conditions in Corollary 9.2.5, but does not satisfy the conditions in Theorem 9.2.6. Also, the extension $\mathbb{Q}(\sqrt{6}, \sqrt{10})/\mathbb{Q}$ satisfies the conditions in Theorem 9.2.6, but does not satisfy the ones in Corollary 9.2.5. This shows that we cannot deduce Corollary 9.2.5 from Theorem 9.2.6, and conversely.

(4) Necessary and sufficient conditions for an arbitrary finite radical extension $F(x_1, \ldots, x_r)/F$ to have degree $n_1 \cdot \ldots \cdot n_r$, where $x_1, \ldots, x_r \in \Omega^*$ are such that $x_i^{n_i} \in F^*$ for every i, $1 \leqslant i \leqslant r$, are provided in Exercise 8. $\square$

EXAMPLE 9.2.9. We are going to calculate the degree

$$[\mathbb{Q}(\sqrt[4]{12} + \sqrt[6]{108}) : \mathbb{Q}]$$

and to exhibit a vector space basis of the extension $\mathbb{Q}(\sqrt[4]{12} + \sqrt[6]{108})/\mathbb{Q}$. To do this, we first apply Corollary 8.2.3 to deduce that

$$\mathbb{Q}(\sqrt[4]{12} + \sqrt[6]{108}) = \mathbb{Q}(\sqrt[4]{12}, \sqrt[6]{108}).$$

Clearly

$$\sqrt[4]{12} = \sqrt[4]{2^2 \cdot 3} = \sqrt[12]{2^6 \cdot 3^3} \text{ and } \sqrt[6]{108} = \sqrt[6]{2^2 \cdot 3^3} = \sqrt[12]{2^4 \cdot 3^6},$$

hence

$$\mathbb{Q}(\sqrt[4]{12}, \sqrt[6]{108}) = \mathbb{Q}(\sqrt[12]{2^6 \cdot 3^3}, \sqrt[12]{2^4 \cdot 3^6}).$$

Set $E = \mathbb{Q}(\sqrt[12]{2^6 \cdot 3^3}, \sqrt[12]{2^4 \cdot 3^6})$, and observe that $E/\mathbb{Q}$ is a 12-Kummer extension with few roots of unity. Then, by Theorem 7.3.2, we have

$$[E : \mathbb{Q}] = |\mathbb{Q}^* \langle \sqrt[12]{a}, \sqrt[12]{b} \rangle / \mathbb{Q}^*| = |\langle \widehat{\sqrt[12]{a}}, \widehat{\sqrt[12]{b}} \rangle| = |\langle \widehat{a}, \widehat{b} \rangle|,$$

where $a = 2^4 \cdot 3^6$, $b = 2^6 \cdot 3^3$, and $\widehat{x}$ denotes for any $x \in \mathbb{Q}^*$ its coset $x\mathbb{Q}^{*12}$ in the quotient group $\mathbb{Q}^*/\mathbb{Q}^{*12}$.

Now we shall describe explicitly the group $\langle \widehat{a}, \widehat{b} \rangle$. Observe that $a^6 = (2^4 \cdot 3^6)^6 \in \mathbb{Q}^{*12}$, hence $\widehat{a}^6 = \widehat{1}$, and $\widehat{a}^i \neq \widehat{1}$ for every $i, 1 \leqslant i < 6$, in other words $\operatorname{ord}(\widehat{a}) = 6$. Also, $b^4 = (2^6 \cdot 3^3)^4 \in \mathbb{Q}^{*12}$, hence $\widehat{b}^4 = \widehat{1}$, and $\widehat{b}^i \neq \widehat{1}$ for every $i, 1 \leqslant i < 4$, in other words $\operatorname{ord}(\widehat{b}) = 4$. Moreover, $\widehat{b}^2 = \widehat{a}^3 = \widehat{3^6}$, hence

$$\langle \widehat{a}, \widehat{b} \rangle = \{ \widehat{a}^i \cdot \widehat{b}^j \mid 0 \leqslant i \leqslant 5, \ 0 \leqslant j \leqslant 3 \} = \{ \widehat{a}^i \cdot \widehat{b}^j \mid 0 \leqslant i \leqslant 5, \ 0 \leqslant j \leqslant 1 \}$$

$$= \{ \widehat{1}, \widehat{a}, \widehat{a}^2, \widehat{a}^3, \widehat{a}^4, \widehat{a}^5, \widehat{b}, \widehat{a} \cdot \widehat{b}, \widehat{a}^2 \cdot \widehat{b}, \widehat{a}^3 \cdot \widehat{b}, \widehat{a}^4 \cdot \widehat{b}, \widehat{a}^5 \cdot \widehat{b} \}.$$

Note that $\widehat{b} \notin \langle \widehat{a} \rangle$, and consequently $|\langle \widehat{a}, \widehat{b} \rangle| = 12$. Thus $[E : \mathbb{Q}] = 12$, and, by Corollary 2.1.10, a basis of the extension $\mathbb{Q}(\sqrt[4]{12} + \sqrt[6]{108})/\mathbb{Q}$ is the set

$$\{ \sqrt[4]{12}^{\,i} \cdot \sqrt[6]{108}^{\,j} \mid 0 \leqslant i \leqslant 5, 0 \leqslant j \leqslant 1 \}.$$

$\square$

9.3. Hecke systems of ideal numbers

The Kneser Criterion (Theorem 2.2.1) has nice applications not only in investigating field extensions with Cogalois correspondence, but also in proving some very classical results from Algebraic Number Theory. In this section we present such applications.

A classical construction from the Algebraic Number Theory is the following one: to every algebraic number K one can associate a so-called *system of ideal numbers* S which is a subgroup of the multiplicative group $\mathbb{C}^*$ of complex numbers such that $K^* \leqslant S$ and the quotient group S/K^* is canonically isomorphic to the ideal class group $\mathcal{C}\ell_K$ of K. This construction, originating with Hecke [70], has the following important property, that the Hilbert class field also possesses: every ideal of K becomes a principal ideal in the algebraic number field $K(S)$. The equality $[K(S) : K] = |\mathcal{C}\ell_K|$ was claimed by Hecke, but never proved by him. The main aim of this section is to provide a short proof of this equality by using the Kneser Criterion, and to discuss some other related questions.

First let us fix the notation and terminology needed to prove the equality mentioned above. Throughout this section K will denote a fixed algebraic number field. We will denote by $\mathfrak{O}_K$ the ring of algebraic integers of K, by $\mathcal{F}_K$ the group of nonzero fractional ideals of K, by $\mathcal{P}_K$ the group of nonzero principal fractional ideals of K, by $\mathcal{C}\ell_K = \mathcal{F}_K/\mathcal{P}_K$ the ideal class group of K, and by $h = |\mathcal{C}\ell_K|$ the class number of K. For any $a \in K^*$, (a) will denote the principal fractional ideal $a\mathfrak{O}_K$ of K.

The following basic facts from the Algebraic Number Theory will be used several times in what follows without further comment. Let L/K be an extension of algebraic number fields. Then

- $I = I\mathfrak{O}_L \cap \mathfrak{O}_K = I\mathfrak{O}_L \cap K$ for any $I \in \mathcal{F}_K$.
- If $I, J \in \mathcal{F}_K$ are such that $I^n = J^n$ for some $n \in \mathbb{N}^*$, then $I = J$.

As is well-known, the finite abelian group $\mathcal{C}\ell_K$ is an internal direct sum of finitely many cyclic subgroups. This means that there exist ideal classes $\mathcal{C}_1, \ldots, \mathcal{C}_s$ in $\mathcal{C}\ell_K$, $s \geqslant 1$, such that every ideal class $\mathcal{C} \in \mathcal{C}\ell_K$ has a unique decomposition

$$\mathcal{C} = \mathcal{C}_1^{r_1} \cdot \ldots \cdot \mathcal{C}_s^{r_s},$$

where $0 \leqslant r_k < h_k$, $h_k > 1$ is the order of the ideal class $\mathcal{C}_k$ in $\mathcal{C}\ell_K$, $k = 1, \ldots, s$, and $h = h_1 \cdot \ldots \cdot h_s$. For any $k = 1, \ldots, s$ let I_k be an integral ideal from the ideal class $\mathcal{C}_k$. Then, every fractional ideal $I \in \mathcal{F}_K$

has a unique decomposition

$$I = (a)I_1^{r_1} \cdot \ldots \cdot I_s^{r_s},$$

where $a \in K^*$, $0 \leqslant r_k < h_k$, $k = 1, \ldots, s$, and the exponents r_k are uniquely determined.

Since $C_k^{h_k} = 1$, one deduces that $I_k^{h_k} = (c_k) \in \mathcal{P}_K$ for suitable numbers $c_k \in K^*$, $k = 1, \ldots, s$, which are uniquely determined up to units of K. Assume that we have fixed the numbers c_k and consider the number field $K(\gamma_1, \ldots, \gamma_s)$, where γ_k is a complex root of the polynomial $X^{h_k} - c_k \in K[X]$, i.e., $\gamma_k^{h_k} = c_k$.

DEFINITION 9.3.1. *With the notation above, the group* $K^*\langle \gamma_1, \ldots, \gamma_s \rangle$ *is called a* Hecke system of ideal numbers *of* K, *and the field* $K(\gamma_1, \ldots, \gamma_s)$, *denoted by* H_K, *is called the* Hecke field *of* K *associated with the Hecke system of ideal numbers* $K^*\langle \gamma_1, \ldots, \gamma_s \rangle$ *of* K. $\square$

By Exercise 10, the Hecke field of any algebraic number field K, which clearly depends on the chosen Hecke system of ideal numbers of K, is uniquely determined up to a K-isomorphism. Therefore, we will just call it the *Hecke field* of K.

The group morphism

$$\mathcal{F}_K \longrightarrow K^*\langle \gamma_1, \ldots, \gamma_s \rangle / K^*, \quad (a)I_1^{r_1} \cdot \ldots \cdot I_s^{r_s} \longmapsto \widehat{\gamma_1^{r_1} \cdot \ldots \cdot \gamma_s^{r_s}},$$

with $a \in K^*, 0 \leqslant r_k < h_k$, $k = 1, \ldots s$, clearly induces a surjective group morphism

$$\psi_K : \mathcal{F}_K / \mathcal{P}_K \longrightarrow K^*\langle \gamma_1, \ldots, \gamma_s \rangle / K^*.$$

LEMMA 9.3.2. *With the notation above, the map*

$$\psi_K : \mathcal{C\ell}_K \longrightarrow K^*\langle \gamma_1, \ldots, \gamma_s \rangle / K^*$$

is a group isomorphism. In particular, one has

$$|K^*\langle \gamma_1, \ldots, \gamma_s \rangle / K^*| = |\mathcal{C\ell}_K| = h.$$

PROOF. We are going to show that ψ_K is injective. To do that, we have to prove that $r_1 = \ldots = r_s = 0$ whenever $\gamma := \gamma_1^{r_1} \cdot \ldots \cdot \gamma_s^{r_s} \in K^*$, with $0 \leqslant r_k < h_k$, $k = 1, \ldots, s$. Let us consider the integral ideal

$$I := I_1^{r_1} \cdot \ldots \cdot I_s^{r_s}.$$

Since

$$I_k^{h_k} = (\gamma_k)^{h_k} = (c_k),$$

it follows that
$$(I_k \mathfrak{O}_{H_K})^{h_k} = (\gamma_k \mathfrak{O}_{H_K})^{h_k}$$
for all $k = 1, \ldots, s$, hence
$$I_k \mathfrak{O}_{H_K} = \gamma_k \mathfrak{O}_{H_K}$$
for all $k = 1, \ldots, s$. Thus,
$$I \mathfrak{O}_{H_K} = (\gamma_1^{r_1} \cdot \ldots \cdot \gamma_s^{r_s}) \mathfrak{O}_{H_K} = \gamma \mathfrak{O}_{H_K},$$
and consequently
$$I = I \mathfrak{O}_{H_K} \cap \mathfrak{O}_K = \gamma \mathfrak{O}_{H_K} \cap \mathfrak{O}_K = \gamma \mathfrak{O}_K.$$
This shows that the ideal class $\mathcal{C}$ of I in $\mathcal{C}\ell_K$ is the identity class. Since
$$\mathcal{C} = \mathcal{C}_1^{r_1} \cdot \ldots \cdot \mathcal{C}_s^{r_s},$$
we deduce that $r_1 = \ldots = r_s = 0$, as desired. $\square$

LEMMA 9.3.3. *If $\varepsilon \in K^* \langle \gamma_1, \ldots, \gamma_s \rangle$ is a unit of H_K, then $\varepsilon \in K$.*

PROOF. Since $\varepsilon \in K^* \langle \gamma_1, \ldots, \gamma_s \rangle$, there exist $a \in K^*$, $0 \leqslant r_k < h_k$ $k = 1, \ldots, s$, such that $\varepsilon = a \gamma_1^{r_1} \cdot \ldots \cdot \gamma_s^{r_s}$. From the proof of Lemma 9.3.2 we have
$$I_k \mathfrak{O}_{H_K} = \gamma_k \mathfrak{O}_{H_K}$$
for all $k = 1, \ldots, s$. Therefore, the fractional ideal $I = (a) I_1^{r_1} \cdot \ldots \cdot I_s^{r_s}$ of K has the following property:
$$I \mathfrak{O}_{H_K} = (a \gamma_1^{r_1} \cdot \ldots \cdot \gamma_s^{r_s}) \mathfrak{O}_{H_K} = (\varepsilon) \mathfrak{O}_{H_K} = \mathfrak{O}_{H_K}.$$
We deduce that
$$I = I \mathfrak{O}_{H_K} \cap \mathfrak{O}_K = \mathfrak{O}_{H_K} \cap \mathfrak{O}_K = \mathfrak{O}_K = (1).$$
Consequently, the ideal class $\mathcal{C}$ of I in $\mathcal{C}\ell_K$ is the identity class. Since
$$\mathcal{C} = \mathcal{C}_1^{r_1} \cdot \ldots \cdot \mathcal{C}_s^{r_s},$$
it follows that $r_1 = \ldots = r_s = 0$. Thus, $\varepsilon = a \in K$, and we are done. $\square$

THEOREM 9.3.4. *Let K be an algebraic number field, and let H_K be its Hecke field. Then*
$$[H_K : K] = h.$$

PROOF. With the notation above, by Lemma 9.3.2, we have to prove that

$$[K\langle\gamma_1,\ldots,\gamma_s\rangle : K] = |K^*\langle\gamma_1,\ldots,\gamma_s\rangle/K^*| = h.$$

Since $K\langle\gamma_1,\ldots,\gamma_s\rangle = K(K^*\langle\gamma_1,\ldots,\gamma_s\rangle)$ and $\gamma_i^{h_i} = c_i \in K^*$ for all $i = 1,\ldots,s$, it follows that the extension $K\langle\gamma_1,\ldots,\gamma_s\rangle/K$ is $K^*\langle\gamma_1,\ldots,\gamma_s\rangle$-radical. Consequently, the desired equality means precisely that the extension $K\langle\gamma_1,\ldots,\gamma_s\rangle/K$ is $K^*\langle\gamma_1,\ldots,\gamma_s\rangle$-Kneser. To prove that, we will check that the conditions from the Kneser Criterion are satisfied. Let $p \in \mathbb{P}$ be an odd prime such that $\zeta_p \in K^*\langle\gamma_1,\ldots,\gamma_s\rangle$. Then $\zeta_p \in K$ by Lemma 9.3.3. Now, assume that $1 + \zeta_4 \in K^*\langle\gamma_1,\ldots,\gamma_s\rangle$. Then $(1 + \zeta_4)^2 = 2\zeta_4 \in K^*\langle\gamma_1,\ldots,\gamma_s\rangle$, so $\zeta_4 \in K^*\langle\gamma_1,\ldots,\gamma_s\rangle$. Thus, $\zeta_4 \in K$ again by Lemma 9.3.3. Therefore, H_K/K is a $K^*\langle\gamma_1,\ldots,\gamma_s\rangle$-Kneser extension, which finishes the proof. □

COROLLARY 9.3.5. *Let $\mathcal{C}$ be an ideal class of K, and let $\widehat{\gamma}$ be the associated element in the group $K^*\langle\gamma_1,\ldots,\gamma_s\rangle/K^*$ via the isomorphism ψ_K from Lemma 9.3.2. If $m = \mathrm{ord}(\mathcal{C})$ in $\mathcal{C\ell}_K$, then $[K(\gamma) : K] = m$, and every ideal of $\mathcal{C}$ becomes a principal ideal in $K(\gamma)$.*

PROOF. Since $K^*\langle\gamma\rangle \leqslant K^*\langle\gamma_1,\ldots,\gamma_s\rangle$ and $K\langle\gamma_1,\ldots,\gamma_s\rangle/K$ is a $K^*\langle\gamma_1,\ldots,\gamma_s\rangle$-Kneser extension by the proof of Theorem 9.3.4, it follows that $K(K^*\langle\gamma\rangle)/K$ is a $K^*\langle\gamma\rangle$-Kneser extension by Corollary 2.1.10. But $K(K^*\langle\gamma\rangle) = K(\gamma)$, hence

$$[K(\gamma) : K] = |K^*\langle\gamma\rangle/K^*| = \mathrm{Ord}(\widehat{\gamma}) = \mathrm{Ord}(\mathcal{C}) = m.$$

By the proof of Lemma 9.3.2 we have

$$I\mathfrak{O}_{H_K} = \gamma\mathfrak{O}_{H_K}$$

for any integral ideal I of the ideal class $\mathcal{C}$. This implies that

$$I\mathfrak{O}_{K(\gamma)} = \gamma\mathfrak{O}_{K(\gamma)},$$

and we are done. □

REMARK 9.3.6. The Hecke field H_K of a number field K has two of the basic properties of the *Hilbert class field* HCF_K of K, namely:

- $[H_K : K] = h$.
- Every ideal of K becomes a principal ideal in H_K.

However, the fields H_K and HCF_K are different since the extension HCF_K/K is always Galois, while, in general, the extension H_K/K is not necessarily Galois. Corollary 9.3.5 shows that every ideal class $\mathcal{C} \in \mathcal{C\ell}_K$ of order m becomes a principal ideal class in a suitable intermediate field of H_K/K of degree m over K. As is known, the question of whether HCF_K has the same property, was answered in the negative by Artin and Furtwängler (see Hasse [**68**, pp. 173-174]).   □

PROPOSITION 9.3.7. *Let* $I \in \mathcal{F}_K$, *let* $\mathcal{C}$ *the class of* I *in* $\mathcal{Cl}_K$, *let* $m = \mathrm{ord}(\mathcal{C})$ *in* $\mathcal{Cl}_K$, *and let* $c \in K^*$ *with* $I^m = (c)$. *Then, the polynomial* $X^m - c$ *is irreducible in* $K[X]$.

PROOF. Let γ denote a root in $\mathbb{C}$ of the polynomial $X^m - c$. Then $(I\mathfrak{O}_{K(\gamma)})^m = c\mathfrak{O}_{K(\gamma)} = (\gamma\mathfrak{O}_{K(\gamma)})^m$, hence $I\mathfrak{O}_{K(\gamma)} = \gamma\mathfrak{O}_{K(\gamma)}$.

Let ε be a unit of $\mathfrak{O}_{K(\gamma)}$ with $\varepsilon \in K^*\langle\gamma\rangle$. Then $\varepsilon = a\gamma^r$ for some $a \in K^*$ and $0 \leqslant r < m$. Hence

$$I^r = (I\mathfrak{O}_{K(\gamma)})^r \cap K = (\gamma)^r\mathfrak{O}_{K(\gamma)} \cap K = (\varepsilon a^{-1})\mathfrak{O}_{K(\gamma)} \cap K$$

$$= (a^{-1})\mathfrak{O}_{K(\gamma)} \cap K = a^{-1}\mathfrak{O}_K = (a^{-1}).$$

Thus, I^r is a principal fractional ideal, hence its ideal class $\mathcal{C}^r$ is the identity class. Since $\mathrm{ord}(\mathcal{C}) = m$, it follows that $m \,|\, r$, hence necessarily $r = 0$, and so, $\varepsilon = a \in K^*$. Now, proceed as in the proof of Theorem 9.3.4 to deduce that $K(\gamma)/K$ is a $K^*\langle\gamma\rangle$-Kneser extension. Therefore

$$[K(\gamma) : K] = |K^*\langle\gamma\rangle/K^*| - \mathrm{Ord}(\widehat{\gamma}).$$

If $n = \mathrm{Ord}(\widehat{\gamma})$, then $n \,|\, m$, hence $m = nt$ for some $t \in \mathbb{N}^*$. On the other hand, we have

$$(c) = (\gamma)^m = (\gamma)^{nt} = ((\gamma)^n)^t = (\gamma^n)^t = I^m = I^{nt} = (I^n)^t,$$

which implies that $I^n = (\gamma^n)$, hence I^n is a principal fractional ideal. As above, we deduce that $m \,|\, n$. Then $m = n$, and so $[K(\gamma) : K] = m$. We conclude that the polynomial $X^m - c$ is irreducible in $K[X]$.   □

COROLLARY 9.3.8. *With the notation above, the following statements hold.*

(1) *The polynomials* $X^{h_k} - c_k$ *are irreducible in* $K[X]$ *for every* k, $1 \leqslant k \leqslant s$.
(2) *The fields* $K(\gamma_1), \ldots, K(\gamma_s)$ *are linearly disjoint over* K.
(3) *There exists a canonical group isomorphism*

$$K^*\langle\gamma_1, \ldots, \gamma_s\rangle/K^* \cong \prod_{1 \leqslant i \leqslant s} (K^*\langle\gamma_i\rangle/K^*).$$

PROOF. (1) follows immediately from Proposition 9.3.7.

(2) By (1), we have $[K(\gamma_j) : K] = h_j$ for $j = 1, \ldots, s$, and so

$$[K(\gamma_1, \ldots, \gamma_s) : K] = h = h_1 \cdot \ldots \cdot h_s = [K(\gamma_1) : K] \cdot \ldots \cdot [K(\gamma_s) : K].$$

This shows that the fields $K(\gamma_1), \ldots, K(\gamma_s)$ are linearly disjoint over K.

(3) follows from (2) and Lemma 8.1.5.   □

PROPOSITION 9.3.9. *Let K be an algebraic number field, let $S_K = K^*\langle\gamma_1, \ldots, \gamma_s\rangle$ be any Hecke system of ideal numbers of K, and let $H_K = K^*(\gamma_1, \ldots, \gamma_s)$ be its associated Hecke field. Then H_K/K is an S_K-Kneser extension.*

PROOF. The result follows at once from the proof of Theorem 9.3.4.   □

The next example shows that, in general, the S_K-Kneser extension H_K/K is not necessarily S_K-Cogalois.

EXAMPLE 9.3.10. Let $K = \mathbb{Q}(\sqrt{-87})$. It is known that the class number h of K is 6 (see e.g., the Table in Borevitch and Shafarevitch [37]). Since $-87 \equiv 1 \pmod 4$, it follows that $\mathfrak{O}_K = \mathbb{Z}[\frac{1+\sqrt{-87}}{2}]$. The decomposition of 3 as a product of prime ideals in the ring $\mathfrak{O}_K$ is

$$3\mathfrak{O}_K = (3, \sqrt{-87})^2.$$

We claim that the prime ideal $I = (3, \sqrt{-87})$ is not a principal ideal of $\mathfrak{O}_K$. Indeed, assume that $I = (\delta)$ for some

$$\delta = a + b \, \frac{1 + \sqrt{-87}}{2} \in \mathfrak{O}_K,$$

and consider the norm $N = N_{K/\mathbb{Q}}$ of the extension $K/\mathbb{Q}$. Since $(\delta) = (3, \sqrt{-87})$, we deduce that $N(\delta) \mid N(3) = 9$ and $N(\delta) \mid N(\sqrt{-87}) = 87$, hence

$$N(\delta) = \left(a + \frac{b}{2}\right)^2 + 87\, \frac{b^2}{4} = a^2 + ab + 22b^2 = 3.$$

This equation is not solvable in $\mathbb{Z}$. Indeed, this is clear if $ab \geqslant 0$. If $ab < 0$ then $a^2 + b^2 > -ab$ and $21b^2 \geqslant 21 > 3$, hence we obtain $a^2 + 22b^2 > 3 - ab$, and so, $a^2 + ab + 22b^2 > 3$.

Thus, the ideal class $\mathcal{C}_1$ of I in $\mathcal{C}\ell_K$ has order 2, and $I^2 = (3) = (-3)$. We choose $\gamma_1 = \sqrt{-3}$ as an ideal number of I. Since the group $\mathcal{C}\ell_K$ is cyclic of order 6, it contains an ideal class $\mathcal{C}_2$ of order 3. Then, we have necessarily

$$\mathcal{C}\ell_K = \langle\mathcal{C}_1\rangle \oplus \langle\mathcal{C}_2\rangle.$$

Let γ_2 be an ideal number of an integral ideal I_2 from the ideal class $\mathcal{C}_2$, that is, γ_2 is one of the complex roots of the polynomial $X^3 - c_2 \in K[X]$, where $c_2 \in K^*$ is such that $I_2^3 = (c_2)$.

Thus, the group $K^*\langle\sqrt{-3}, \gamma_2\rangle$ is a Hecke system of ideal numbers of K, and $H_K = K(\sqrt{-3}, \gamma_2)$ is its associated Hecke field of K. By Proposition 9.3.9, the extension H_K/K is a $K^*\langle\sqrt{-3}, \gamma_2\rangle$-Kneser extension. Observe that $\exp(K^*\langle\sqrt{-3}, \gamma_2\rangle/K^*) = 6$, $3\,|\,6$ and $\zeta_3 = \frac{-1+\sqrt{-3}}{2} \in H_K \setminus K$, so the extension H_K/K is not 6-pure. By the n-Purity Criterion, it follows that H_K/K is not a $K^*\langle\sqrt{-3}, \gamma_2\rangle$-Cogalois extension.   □

The next result provides two cases when the extension H_K/K is S_K-Cogalois.

PROPOSITION 9.3.11. *The following statements hold for an algebraic number field K with class number h.*

(1) *If $\zeta_h \in K$, then the extension $K^*(\gamma_1, \ldots, \gamma_s)/K$ is $K^*\langle\gamma_1, \ldots, \gamma_s\rangle$-Cogalois for any choice of the Hecke system $K^*\langle\gamma_1, \ldots, \gamma_s\rangle$ of ideal numbers of K.*

(2) *If K can be embedded into the field $\mathbb{R}$ of real numbers, then there exists a Hecke system $K^*\langle\gamma_1, \ldots, \gamma_s\rangle$ of ideal numbers of K such that the extension $K^*(\gamma_1, \ldots, \gamma_s)/K$ is $K^*\langle\gamma_1, \ldots, \gamma_s\rangle$-Cogalois.*

PROOF. (1) If $\zeta_h \in K$, then clearly $\mu_h(K^*(\gamma_1, \ldots, \gamma_s)) \subseteq K$. Since $h = h_1 \cdot \ldots \cdot h_s$ and $\gamma_k^{h_k} = c_k \in K$ for all k, $1 \leqslant k \leqslant s$, we deduce that $K^*(\gamma_1, \ldots, \gamma_s)/K$ is a generalized h-Kummer extension. Now apply Theorem 7.2.3 to obtain the desired result.

(2) Without loss of generality, we may assume that K is a subfield of the field $\mathbb{R}$ of real numbers. In the construction of ideal numbers we may choose $c_1 > 0, \ldots, c_s > 0$. For γ_k we choose the positive real root of the polynomial $X^{h_k} - c_k$, $1 \leqslant k \leqslant s$. Then, $H_K = K(\gamma_1, \ldots, \gamma_s)$ is a subfield of $\mathbb{R}$. Thus, H_K/K is a pure extension, and, a fortiori, an n-pure extension, where $n = \exp(K^*\langle\gamma_1, \ldots, \gamma_s\rangle/K^*)$. By the n-Purity Criterion, we deduce that H_K/K is a $K^*\langle\gamma_1, \ldots, \gamma_s\rangle$-Cogalois extension.   □

REMARK 9.3.12. Let A be a Dedekind ring having a finite ideal class group $\mathcal{C}\ell_A$ of order h, and denote by L its quotient field. Assume that the characteristic of L is not 2 and is relatively prime with h. Then, the main part of the results presented in this section can be extended from algebraic number fields K to such more general fields L (see Exercises 11 and 12). □

9.4. Exercises to Chapter 9

1. Let R be a UFD which is not a field, let K be its field of quotients, and let Ω be an algebraically closed field containing K as a subfield. Denote by $\mathbb{P}_R$ a representative set of mutually nonassociated in divisibility nonzero prime elements of R, and by $[\mathbb{P}_R]$ the subsemigroup of the multiplicative monoid $(R, \cdot)$ generated by $\mathbb{P}_R$, i.e.,

$$[\mathbb{P}_R] = \{ p_1 \cdot \ldots \cdot p_n \mid n \in \mathbb{N}^*, \, p_k \in \mathbb{P}_R, \, k = 1, \ldots, n \}.$$

Let $r \in \mathbb{N}$, $r \geqslant 2$, let $n_1, \ldots, n_r \in \mathbb{N}^*$, let $a_1, \ldots, a_r \in [\mathbb{P}_R]$, and let $x_1, \ldots, x_r \in \Omega$ be such that $x_k^{n_k} = a_k$ and $\operatorname{ord}(\widehat{x_k}) = n_k$ in Ω/K^* for every k, $1 \leqslant k \leqslant r$. Assume that $\gcd(n, e(K)) = 1$, where $n = \operatorname{lcm}(n_1, \ldots, n_r)$, and $\mu_n(K(x_1, \ldots, x_r)) \subseteq K$. Then, prove that

$$[K (\sqrt[n_1]{a_1}, \ldots, \sqrt[n_r]{a_r}) : K] = n_1 \cdot \ldots \cdot n_r,$$

provided the following condition is satisfied:

$(*)$ $a_1, \ldots, a_k$ are relatively prime in pairs.

2. Show that Theorem 9.2.6 can be extended from $\mathbb{Z}$ to a UFD satisfying the conditions in Exercise 1.

3. Show that $[\mathbb{Q}(\sqrt[3]{12}, \sqrt[4]{3}) : \mathbb{Q}] = 8$, and exhibit a vector space basis of the extension $\mathbb{Q}(\sqrt[3]{12}, \sqrt[4]{3})/\mathbb{Q}$.

4. Show that $[\mathbb{Q}(\sqrt[9]{18}, \sqrt[6]{162}) : \mathbb{Q}] = 18$, and exhibit a vector space basis of the extension $\mathbb{Q}(\sqrt[9]{18}, \sqrt[6]{162})/\mathbb{Q}$.

5. Find $[\mathbb{Q}(\sqrt{2}, \sqrt[3]{6}, \sqrt[5]{60}) : \mathbb{Q}]$.

6. Let Ω be an algebraic closure of the field $F := \mathbb{Q}(X)$ of rational fractions in the indeterminate X, and consider the elements

$$x_1 = \sqrt[6]{X^2 + 1} \quad \text{and} \quad x_2 = \sqrt[6]{X^4 - 2X^3 + 2X^2 - 2X + 1}$$

of Ω. Find $[F(x_1, x_2) : F]$.

7. Let $r \in \mathbb{N}^*$ and $a_1, \ldots, a_r \in \mathbb{Z}^*$. Prove that

$$[\mathbb{Q}(\sqrt{a_1}, \ldots, \sqrt{a_r}) : \mathbb{Q}] = | \mathbb{Q}^* \langle \sqrt{a_1}, \ldots, \sqrt{a_r} \rangle/\mathbb{Q}^* |.$$

8. (*Schinzel* [**93**]). Let F be any field, let $r \in \mathbb{N}^*$, let $n_1, \dots, n_r \in \mathbb{N}^*$ be positive integers at most one of them divisible by the characteristic of F, let $a_1, \dots, a_r \in F^*$, and let $x_1, \dots, x_r \in \Omega$ be such that $x_i^{n_i} = a_i$ for every $i = 1, \dots, r$. Prove that

$$[F(x_1, \dots, x_r) : F] = n_1 \cdot \dots \cdot n_r$$

if and only if the following two conditions are satisfied.
  (a) Whenever $p \in \mathbb{P}$ and $k_1, \dots, k_r \in \mathbb{N}$ are such that $p \mid n_i k_i$ for every $i = 1, \dots, r$ and $a_1^{k_1} \cdot \dots \cdot a_r^{k_r} \in F^p$, then $p \mid k_i$ for every $i = 1, \dots, r$.
  (b) Whenever $k_1, \dots, k_r \in \mathbb{N}$ are such that $4 \mid n_i k_i$ for every $i = 1, \dots, r$ and $a_1^{k_1} \cdot \dots \cdot a_r^{k_r} \in -4F^4$, then $p \mid k_i$ for every $i = 1, \dots, r$.

9. With the notation of Section 9.3, show that for every $a \in K^*$ one has $a \gamma_1^{r_1} \cdot \dots \cdot \gamma_s^{r_s} \mathfrak{O}_{H_K} \cap K = (a) I_1^{r_1} \cdot \dots \cdot I_s^{r_s}$.

10. Prove that the Hecke field of any algebraic number field K is uniquely determined up to a K-isomorphism. More precisely, with the notation of Section 9.3, show that if γ_k' is another root of the polynomials $X^{h_k} - c_k$, $k = 1, \dots, s$, then the fields $K(\gamma_1, \dots, \gamma_s)$ and $K(\gamma_1', \dots, \gamma_s')$ are K-isomorphic via a K-isomorphism sending γ_k to γ_k', $k = 1, \dots, s$.

11. Let R be a Dedekind ring having a finite ideal class group $\mathcal{C}\ell_A$ of order h, and denote by L its field of quotients. Assume that the characteristic of L is not 2 and is relatively prime with h. Show that we can perform mutatis-mutandis the construction presented in Section 9.3 to define a *Hecke system* $L^* \langle \gamma_1, \dots, \gamma_s \rangle$ *of ideal elements* of L and a *Hecke field* H_L of L.

12. Prove that if R is a Dedekind domain as in Exercise 11, and L is its field of quotients, then Theorem 9.3.4, Corollary 9.3.5, Proposition 9.3.7 and Corollary 9.3.8 also hold for the field L.

13. Let K_1 and K_2 be algebraic number fields such that their discriminants δ_{K_1} and δ_{K_2} are relatively prime. Prove that

$$[K_1 K_2 : \mathbb{Q}] = [K_1 : \mathbb{Q}] \cdot [K_2 : \mathbb{Q}].$$

9.5. Bibliographical comments to Chapter 9

Section 9.1. The short review presented in this section is mainly based on Ribenboim [89].

Section 9.2. The evaluation of the degree of a radical extension of type $K(\sqrt[n]{a_1}, \ldots, \sqrt[n]{a_r})/K$ with K an algebraic number field containing ζ_n, $n, r \in \mathbb{N}^*$, and $a_1, \ldots, a_r \in K^*$, as the order of the quotient group $K^* \langle \sqrt[n]{a_1}, \ldots, \sqrt[n]{a_r} \rangle / K^*$ has been known for many years by the classical Kummer Theory. The problem "When is $[K(\sqrt[n]{a_1}, \ldots, \sqrt[n]{a_r}) : K] = \prod_{1 \leqslant i \leqslant r} [K(\sqrt[n]{a_i}) : K]$?" was first answered by Hasse in 1930 (see the second edition [68] of his mimeographed lectures on Class Field Theory).

The case when the algebraic number field K does not necessarily contain a primitive n-th root ζ_n of unity was, surprisingly, first discussed fairly late, only in 1940 by Besicovitch [35] for $K = \mathbb{Q}$ and $\sqrt[n]{a_1}, \ldots, \sqrt[n]{a_r}$ real roots of positive integers $a_1, \ldots a_r$, and then, by Mordell [83] in 1953 for K any real number field and $\sqrt[n]{a_1}, \ldots, \sqrt[n]{a_r} \in \mathbb{R}$. Later, in 1972, Siegel [98] shows that the degree $[K(\sqrt[n]{a_1}, \ldots, \sqrt[n]{a_r}) : K]$ is the order of the quotient group $K^* \langle \sqrt[n]{a_1}, \ldots, \sqrt[n]{a_r} \rangle / K^*$ for any real number field K and any real roots $\sqrt[n]{a_1}, \ldots, \sqrt[n]{a_r}$.

A particular case of Corollary 9.2.7 was proved by Richards [90] (see also Gaal's book [58], where Richards' proof is reproduced). Corollary 9.2.5 is due to Ursell [106], but our proof is different from his.

The presentation of this section is based on Schinzel [93] and Albu [3].

Section 9.3. Hecke systems of ideal numbers $K^* \langle \gamma_1, \ldots, \gamma_s \rangle$ associated with any algebraic number field K were introduced in 1920 by Hecke [70]. In his monograph [71] published in 1948, Hecke claims on page 122 that the degree $[H_K : K]$ is precisely the class number h of K, where H_K is the field $K(\gamma_1, \ldots, \gamma_s)$ associated with the system $K^* \langle \gamma_1, \ldots, \gamma_s \rangle$ of ideal numbers.

To the best of our knowledge, no proof of this assertion excepting the one in Albu and Nicolae [21] is available in the literature. Note that Ribenboim gives on page 124 of his monograph [89] only the inequality $[H_K : K] \leqslant h$.

Related to Hecke systems of ideal numbers a natural question arose: are the polynomials $X^{h_k} - c_k$ irreducible in $K[X]$, where $c_k = \gamma_k^{h_k} \in K$? This problem was only mentioned (but not settled) by Hasse [67, p. 544] as follows: "Auf die Frage nach der Irreduzibilität der Polynome $X^{h_k} - c_k$ über

K wollen wir hier nicht eingehen." The positive answer to this question was given by Albu and Nicolae [21].

The results of this section are taken from Albu and Nicolae [21].

CHAPTER 10

CONNECTIONS WITH GRADED ALGEBRAS AND HOPF ALGEBRAS

The aim of this chapter is to present some interesting connections of Cogalois Theory with graded algebras and Hopf algebras.

In Section 10.1 we analyze the concepts of G-radical, G-Kneser, and G-Cogalois extension in terms of graded algebras.

After reviewing some basic notions and facts on Hopf algebras, we describe in Section 10.2 the G-Kneser extensions and Cogalois extensions in terms of Galois H-objects appearing in Hopf algebras.

10.1. G-Cogalois extensions via strongly graded fields

In this section we describe the concepts of G-radical, G-Kneser, and G-Cogalois extension in terms of graded ring theory.

Throughout this section all algebras are assumed to be associative with unit, and K will denote a fixed commutative ring with nonzero identity element. If A is a K-algebra and X, Y are subsets of A, then XY will denote the K-submodule of the underlying K-module of the algebra A which is generated by the set $\{\, xy \,|\, x \in X, y \in Y \,\}$. Any ring can be viewed in a canonical way as a $\mathbb{Z}$-algebra.

For a K-module M and a family $(M_i)_{i \in I}$ of submodules of M, the notation $M = \bigoplus_{i \in I} M_i$ will mean throughout this chapter that M is the "internal direct sum" of the independent family $(M_i)_{i \in I}$ of its submodules, that is, any element $x \in M$ can be uniquely expressed as $x = \sum_{i \in I} x_i$, where $(x_i)_{i \in I}$ is a family of finite support, with $x_i \in M_i$ for every $i \in I$.

DEFINITION 10.1.1. *Let Γ be a multiplicative group with identity element e. A K-algebra A is said o be a Γ-graded algebra if $A = \bigoplus_{\gamma \in \Gamma} A_\gamma$*

is a direct sum of K-submodules A_γ of A, with $A_\gamma A_\delta \subseteq A_{\gamma\delta}$ for every γ, $\delta \in \Gamma$. A Γ-graded algebra $A = \bigoplus_{\gamma\in\Gamma} A_\gamma$ is said to be strongly graded *if $A_\gamma A_\delta = A_{\gamma\delta}$ for every γ, $\delta \in \Gamma$. A (strongly) Γ-graded ring is a (strongly) Γ-graded algebra over the ring $\mathbb{Z}$ of rational integers.*

A left module M over the Γ-graded algebra $A = \bigoplus_{\gamma\in\Gamma} A_\gamma$ is said to be a graded module (*resp. a* strongly graded module) *if $M = \bigoplus_{\gamma\in\Gamma} M_\gamma$ is a direct sum of K-submodules M_γ of M, with $A_\gamma M_\delta \subseteq M_{\gamma\delta}$ (resp. $A_\gamma M_\delta = M_{\gamma\delta}$) for every γ, $\delta \in \Gamma$. The elements of $h(M) = \bigcup_{\gamma\in\Gamma} M_\gamma$ are called the* homogeneous elements *of M. Any element $x \in M$ has a unique decomposition $x = \sum_{\gamma\in\Gamma} x_\gamma$, with $x_\gamma \in M_\gamma$, $\gamma \in \Gamma$, and all but a finite number of the x_γ are zero; the elements x_γ are called the* homogeneous components *of x.*   □

DEFINITION 10.1.2. *Let E/F be a field extension, and let Γ be a multiplicative group with identity element e. One says that E/F is a Γ-Clifford extension (resp. a* strongly Γ-graded extension) *if there exists a family $(E_\gamma)_{\gamma\in\Gamma}$ of F-subspaces of ${}_F E$ indexed by the group Γ, satisfying the following conditions.*

(1) $E = \sum_{\gamma\in\Gamma} E_\gamma$ (*resp.* $E = \bigoplus_{\gamma\in\Gamma} E_\gamma$).
(2) $E_\gamma E_\delta = E_{\gamma\delta}$ *for every* γ, $\delta \in \Gamma$.
(3) $E_\gamma = F \Longleftrightarrow \gamma = e$.

An element $x \in E$ is said to be homogeneous of degree γ *if $x \in E_\gamma$. The set of all nonzero homogeneous elements of E will be denoted by $U^h(E)$.* □

Note that we do not assume any finiteness condition on a Γ-Clifford extension, though we are mainly interested in finite extensions.

LEMMA 10.1.3. *The following statements hold for a Γ-Clifford extension E/F.*

(1) *If γ, $\delta \in \Gamma$, then $E_\gamma = E_\delta \Longleftrightarrow \gamma = \delta$.*
(2) $\dim_F(E_\gamma) = 1$ *for all $\gamma \in \Gamma$. In particular, $E_\gamma = Fx$ for every $x \in E_\gamma^*$.*
(3) *For every $\gamma \in \Gamma$ let $x_\gamma \in E_\gamma^*$ be arbitrary. Then $X = \{ x_\gamma \mid \gamma \in \Gamma \}$ is a set of generators of the vector space ${}_F E$, and $[E : F] \leqslant |\Gamma|$. If E/F is a strongly Γ-graded extension, then X is a basis of the vector space ${}_F E$, and*

$$[E : F] = |\Gamma|.$$

Conversely, if there exists a family $(x_\gamma)_{\gamma \in \Gamma} \in \prod_{\gamma \in \Gamma} E_\gamma^*$ which is a basis of the vector space $_F E$, then E/F is a strongly Γ-graded extension.

(4) *If E/F is a finite extension, then E/F is a strongly Γ-graded extension if and only if*

$$[E : F] = |\Gamma|.$$

(5) *The group Γ is Abelian.*

(6) *If Γ is a torsion group, then E/F is an algebraic extension.*

PROOF. (1) If $E_\gamma = E_\delta$, then

$$F = E_e = E_{\gamma\gamma^{-1}} = E_\gamma E_{\gamma^{-1}} = E_\delta E_{\gamma^{-1}} = E_{\delta\gamma^{-1}}$$

by Definition 10.1.2 (2), hence $\delta\gamma^{-1} = e$, i.e., $\gamma = \delta$ by Definition 10.1.2 (3).

(2) Again by Definition 10.1.2 (2), we have $E_{\gamma\gamma^{-1}} = E_e = F$, hence $E_{\gamma^{-1}} \neq 0$. Take an arbitrary $y \in E_{\gamma^{-1}}^*$. Then, the map

$$E_\gamma \longrightarrow F, \; x \mapsto xy, \; x \in E_\gamma$$

yields an isomorphism of F-vector spaces.

(3) By (2), $E_\gamma = Fx_\gamma$ for every $\gamma \in \Gamma$, hence

$$E = \sum_{\gamma \in \Gamma} E_\gamma = \sum_{\gamma \in \Gamma} Fx_\gamma,$$

i.e., X is a set of generators of the vector space $_F E$. This clearly implies that

$$[E : F] \leqslant |X| \leqslant |\Gamma|.$$

By definition, the Γ-Clifford extension E/F is strongly Γ-graded if and only if

$$E = \bigoplus_{\gamma \in \Gamma} E_\gamma = \bigoplus_{\gamma \in \Gamma} Fx_\gamma,$$

i.e., if and only if the set of generators X of the vector space $_F E$ is linearly independent over F, hence a basis of $_F E$, and in this case we clearly have $[E : F] = |X|$. Now, observe that if E/F is a strongly Γ-graded extension, then the map $\Gamma \longrightarrow X, \; \gamma \mapsto x_\gamma$, is bijective, hence $|\Gamma| = |X|$.

(4) follows from (3) since in any finite dimensional vector space $_F V$ of dimension n, any set of generators S of $_F V$ with $|S| = n$ is necessarily a basis of $_F V$.

(5) Since the multiplication on E is commutative, we have

$$E_{\gamma\delta} = E_\gamma E_\delta = E_\delta E_\gamma = E_{\delta\gamma},$$

hence $\gamma\delta = \delta\gamma$ for every $\gamma, \delta \in \Gamma$ by (1).

(6) Since Γ is torsion group, it follows that for every $\gamma \in \Gamma$ and every $x_\gamma \in E_\gamma$ there exists an $n_\gamma \in \mathbb{N}^*$ such that $\gamma^{n_\gamma} = e$, hence $x_\gamma^{n_\gamma} \in E_\gamma^{n_\gamma} = E_{\gamma^{n_\gamma}} = E_e = F$. Thus, every homogeneous element of E is algebraic over F, and since any element of E is a sum of homogeneous elements, it follows that E/F is an algebraic extension. $\qquad\square$

The next example shows that, in general, the converse implication in Lemma 10.1.3 (6) does not hold.

EXAMPLE 10.1.4. Let $\theta = \sqrt{1 + \sqrt{2}}$, $F = \mathbb{Q}$ and $E = \mathbb{Q}(\theta)$. Then E/F is a quartic extension, hence an algebraic extension, which is also a $\mathbb{Z}$-Clifford extension with

$$E = \sum_{n \in \mathbb{Z}} E_n, \ E_n = F\theta^n, \ n \in \mathbb{Z}.$$

Indeed, the only nontrivial fact is that $E_n = F \implies n = 0$. But $E_n = F$ clearly implies that $\theta^n \in \mathbb{Q}$, which can happen only if $n = 0$ by Proposition 3.2.6 (g). Thus, E/F is a $\mathbb{Z}$-Clifford extension, but $\mathbb{Z}$ is a torsion-free group. $\qquad\square$

For any Γ-Clifford extension E/F and any $\Delta \leqslant \Gamma$ we shall denote

$$E_\Delta := \sum_{\gamma \in \Delta} E_\gamma.$$

Obviously, E_Δ/F is a Δ-Clifford extension, which is strongly Δ-graded whenever the extension E/F is strongly Γ-graded.

LEMMA 10.1.5. *Let E/F be a Γ-Clifford extension, let $\varnothing \neq \Lambda \subseteq \Gamma$, and let $x_\lambda \in E_\lambda^*$, $\lambda \in \Lambda$. If E/F is an algebraic extension (in particular, if Γ is a torsion group), then*

$$E_{\langle \Lambda \rangle} = F(\{ x_\lambda \mid \lambda \in \Lambda \}).$$

PROOF. Set $M = \{ x_\lambda \mid \lambda \in \Lambda \}$ and $K = F(M)$. Since $\dim_F(E_\lambda) = 1$ and $x_\lambda \in E_\lambda \cap M$, it follows that $E_\lambda \subseteq K$ for every $\lambda \in \Lambda$. Further, if $\lambda \in \Lambda$ and $x \in E_{\lambda^{-1}}^*$, then $a = xx_\lambda \in E_{\lambda^{-1}}E_\lambda = E_e = F$, hence $x = ax_\lambda^{-1} \in K$. This shows that $E_{\lambda^{-1}} \subseteq K$. Now, let $\gamma \in \langle \Lambda \rangle$. Since Γ is Abelian by Lemma 10.1.3 (5), there exist $\lambda_1, \ldots, \lambda_r, \mu_1, \ldots, \mu_t \in \Lambda$ such that $\gamma = \lambda_1 \cdots \lambda_r \mu_1^{-1} \cdots \mu_t^{-1}$. But $E_\gamma = E_{\lambda_1} \cdots E_{\lambda_r} E_{\mu_1^{-1}} \cdots E_{\mu_t^{-1}}$, hence we deduce that $E_{\langle \Lambda \rangle} \subseteq K$.

For the opposite inclusion, let $y \in K$. Since E/F is an algebraic extension, we have $K = F(M) = F[M]$, hence there exists a finite subset $\{\lambda_1, \ldots, \lambda_n\}$ of Λ such that y can be written as a finite sum of monomials $a x_{\lambda_1}^{i_1} \cdots x_{\lambda_n}^{i_n}$, with $a \in F$ and $i_1, \ldots, i_n \in \mathbb{N}$. Since $a x_{\lambda_1}^{i_1} \cdots x_{\lambda_n}^{i_n} \in E_{\lambda_1^{i_1} \cdots \lambda_n^{i_n}} \subseteq E_{\langle \Lambda \rangle}$, it follows that $y \in E_{\langle \Lambda \rangle}$, which proves the inclusion $K \subseteq E_{\langle \Lambda \rangle}$. $\qquad\square$

PROPOSITION 10.1.6. *The following assertions hold for a field extension E/F.*

(1) *Let G be a group with $F^* \leqslant G \leqslant E^*$. If E/F is a G-radical extension, then E/F is a G/F^*-Clifford extension.*

(2) *Conversely, if E/F is a Γ-Clifford extension for some torsion group Γ, then there exists a group G such that $F^* \leqslant G \leqslant E^*$, $\Gamma \cong G/F^*$, E/F is G-radical, and $U^h(E) = G$.*

(3) *If E/F is a G/F^*-Clifford extension, where $F^* \leqslant G \leqslant T(E/F)$ and $U^h(E) \subseteq G$, then E/F is G-radical.*

(4) *Let G be a group with $F^* \leqslant G \leqslant E^*$. If E/F is a finite G-Kneser extension, then E/F is a strongly G/F^*-graded extension.*

(5) *Conversely, if E/F is a strongly Γ-graded extension for some finite group Γ, then there exists a group G such that $F^* \leqslant G \leqslant E^*$, $\Gamma \cong G/F^*$, E/F is G-Kneser, and $U^h(E) = G$.*

(6) *If E/F is a strongly G/F^*-graded extension, with $F^* \leqslant G \leqslant E^*$, G/F^* finite and $U^h(E) \subseteq G$, then E/F is G-Kneser.*

PROOF. (1) Assume that E/F is a G-radical extension. Then, by Remark 2.1.4 (3), we have $E = F(G) = F[G]$, hence G generates E as a vector space over F, i.e.,

$$E = \sum_{g \in G} Fg.$$

Observe that if $g_1, g_2 \in G$, then $Fg_1 = Fg_2 \Longleftrightarrow \widehat{g_1} = \widehat{g_2}$, hence it makes sense to set $E_{\widehat{g}} = Fg$ for every $g \in G$. Then clearly

$$E = \sum_{\widehat{g} \in G/F^*} E_{\widehat{g}}.$$

For $g, g_1, g_2 \in G$ we have

$$E_{\widehat{g}} = E_{\widehat{1}} = F \Longleftrightarrow Fg = F1 \Longleftrightarrow \widehat{g} = \widehat{1},$$

and
$$E_{\widehat{g_1}} E_{\widehat{g_2}} = F g_1 F g_2 = F g_1 g_2 = E_{\widehat{g_1 g_2}} = E_{\widehat{g_1}\, \widehat{g_2}},$$
which show that E/F is a G/F^*-Clifford extension.

(2) Assume that E/F is a Γ-Clifford extension with $E = \sum_{\gamma \in \Gamma} E_\gamma$. Pick for every $\gamma \in \Gamma$ an arbitrary element $x_\gamma \in E_\gamma^*$. Then $E_\gamma = F x_\gamma$ by Lemma 10.1.3 (2). For any $\gamma, \delta \in \Gamma$ we have $x_\gamma x_\delta \in E_\gamma E_\delta = E_{\gamma\delta}$, so there exists $a_{\gamma,\delta} \in F^*$ such that $x_\gamma x_\delta = a_{\gamma,\delta} x_{\gamma\delta}$. This shows that the map
$$\omega : \Gamma \longrightarrow E^*/F^*, \; \omega(\gamma) = \widehat{x_\gamma}, \; \gamma \in \Gamma,$$
is a group morphism. This morphism is injective since $\widehat{x_\gamma} = \widehat{1}$ implies that $x_\gamma \in F^*$, hence $E_\gamma = F x_\gamma = F$, and so, $\gamma = e$ by Definition 10.1.2 (3). Set $\overline{G} = \mathrm{Im}(\omega)$. Then $\overline{G} \cong \Gamma$, and there exists G with $F^* \leqslant G \leqslant E^*$ and $\overline{G} = G/F^*$. Observe that
$$G = \bigcup_{\gamma \in \Gamma} F^* x_\gamma = \bigcup_{\gamma \in \Gamma} E_\gamma^* = U^h(E)$$
hence
$$E = E_\Gamma = F(\{\, x_\gamma \mid \gamma \in \Gamma \,\}) = F(G)$$
by Lemma 10.1.5. We have $F^* x_e = F^* \leqslant G$. Since Γ is a torsion group, it follows that $G \leqslant T(E/F)$, hence E/F is a G-radical extension.

(3) For every $\widehat{g} \in G/F^*$ pick an $x_{\widehat{g}} \in E_{\widehat{g}}^*$. Since
$$\bigcup_{\widehat{g} \in G/F^*} E_{\widehat{g}}^* = U^h(E) \subseteq G$$
by hypothesis, we have $\{\, x_{\widehat{g}} \mid \widehat{g} \in G/F^* \,\} \subseteq G$. Now, applying Lemma 10.1.5, with $\Lambda = \Gamma = G/F^*$, we deduce that
$$E = E_{G/F^*} = F(\{\, x_{\widehat{g}} \mid \widehat{g} \in G/F^* \,\}) \subseteq F(G) \subseteq E,$$
i.e., $E = F(G)$. This shows that E/F is a G-radical extension.

(4) By (1), E/F is a G/F^*-Clifford extension, with
$$E = \sum_{\widehat{g} \in G/F^*} E_{\widehat{g}} \quad \text{and} \quad E_{\widehat{g}} = F g, \; \forall\, g \in G.$$
This implies that if S is any fixed set of representatives of G/F^*, then
$$E = \sum_{g \in S} F g.$$

On the other hand, since E/F is G-Kneser, S is a vector space basis of E over F by Proposition 2.1.8. Therefore

$$E = \bigoplus_{g \in S} Fg = \bigoplus_{\widehat{g} \in G/F^*} E_{\widehat{g}},$$

which shows that E/F is a strongly G/F^*-graded extension.

(5) According to (2), there exists a group G such that $F^* \leqslant G \leqslant E^*$, $\Gamma \cong G/F^*$, E/F is G-radical, and $U^h(E) = G$. Pick for every $\gamma \in \Gamma$ an arbitrary element $x_\gamma \in E_\gamma^*$. Then, the set $B = \{\, x_\gamma \mid \gamma \in \Gamma \,\}$ is a basis of the vector space $_FE$ by Lemma 10.1.3 (3), and clearly $B \subseteq U^h(E) = G$. By the proof of (2), $\widehat{x_\gamma} \neq \widehat{x_\delta}$ for every $\gamma \neq \delta$ in Γ, so B is a set of representatives of the group $\overline{G} = G/F^* = \{\, \widehat{x_\gamma} \mid \gamma \in \Gamma \,\}$. Now, apply Corollary 2.1.10, to deduce that E/F is a G-Kneser extension.

(6) Since E/F is a strongly G/F^*-graded extension, we have $E = \bigoplus_{\widehat{g} \in G/F^*} E_{\widehat{g}}$, hence $[\, E : F\,] = |G/F^*|$ since $\dim_F(E_{\widehat{g}}) = 1$ for all $\widehat{g} \in G/F^*$ by Lemma 10.1.3 (2). By (3), the extension E/F is G-radical, hence it is G-Kneser. $\qquad\square$

The graded version of the concept of G-Cogalois extension is that of Γ-Cogalois extension we are going to introduce below.

For any Γ-Clifford extension E/F with $E = \sum_{\gamma \in \Gamma} E_\gamma$, and any intermediate field K of E/F we shall use the following notation:

$$\Gamma_K = \{\, \gamma \in \Gamma \mid E_\gamma \subseteq K \,\}.$$

Then clearly Γ_K is a subgroup of Γ, hence it makes sense to consider the map

$$\Phi : \underline{\mathrm{Intermediate}}(E/F) \longrightarrow \underline{\mathrm{Subgroups}}(\Gamma), \quad \Phi(K) = \Gamma_K,$$

which is clearly order-preserving.

Also, we have an order-preserving map

$$\Psi : \underline{\mathrm{Subgroups}}(\Gamma) \longrightarrow \underline{\mathrm{Intermediate}}(E/F), \quad \Psi(\Delta) = E_\Delta = \sum_{\gamma \in \Delta} E_\gamma.$$

Clearly, the maps Φ and Φ define a Cogalois connection between the lattices $\underline{\mathrm{Intermediate}}(E/F)$ and $\underline{\mathrm{Subgroups}}(\Gamma)$.

LEMMA 10.1.7. *With the notation above, the following statements hold for any* Γ-*Clifford extension* E/F.

(1) E/K *is a* Γ/Γ_K-*Clifford extension for every intermediate field* K *of* E/F.

(2) *If E/F is a strongly Γ-graded extension, then $\Phi \circ \Psi = 1_{\underline{\text{Subgroups}}(\Gamma)}$.*

PROOF. (1) For any $\gamma \in \Gamma$ we denote by $\widehat{\gamma}$ its coset modulo the subgroup Γ_K of Γ. Let $\gamma_1, \gamma_2 \in \Gamma$, with $\widehat{\gamma_1} = \widehat{\gamma_2}$. Then $\gamma_2 = \gamma_1 \delta$ for some $\delta \in \Gamma_K$. Thus $E_\delta \subseteq K$, and so $E_\delta K = K$. It follows that

$$E_{\gamma_2} K = E_{\gamma_1 \delta} K = E_{\gamma_1} E_\delta K = E_{\gamma_1} K.$$

Therefore, it makes sense to set $E_{\widehat{\gamma}} = E_\gamma K$ for every $\gamma \in \Gamma$. For any $\gamma, \gamma_1, \gamma_2 \in \Gamma$ we have

$$E_{\widehat{\gamma}} = K \Longleftrightarrow E_\gamma K = K \Longleftrightarrow E_\gamma \subseteq K \Longleftrightarrow \gamma \in \Gamma_K \Longleftrightarrow \widehat{\gamma} = \widehat{e},$$

$$E_{\widehat{\gamma_1}} E_{\widehat{\gamma_2}} = (E_{\gamma_1} K)(E_{\gamma_2} K) = E_{\gamma_1} E_{\gamma_2} K = E_{\gamma_1 \gamma_2} K = E_{\widehat{\gamma_1} \widehat{\gamma_2}}.$$

Let S be a set of representatives of the quotient group Γ/Γ_K. Then $\{ \sigma\Gamma_K \mid \sigma \in S \}$ is a partition of Γ, hence

$$E = \sum_{\gamma \in \Gamma} E_\gamma = \left(\sum_{\gamma \in \bigcup_{\sigma \in S} \sigma\Gamma_K} E_\gamma \right) K = \sum_{\sigma \in S} \left(\sum_{\delta \in \sigma\Gamma_K} (E_\delta K) \right)$$

$$= \sum_{\sigma \in S} \left(\sum_{\lambda \in \Gamma_K} (E_{\sigma\lambda} K) \right) = \sum_{\sigma \in S} \left(\sum_{\lambda \in \Gamma_K} (E_\sigma E_\lambda K) \right)$$

$$= \sum_{\sigma \in S} (E_\sigma K) = \sum_{\widehat{\gamma} \in \Gamma/\Gamma_K} E_{\widehat{\gamma}}.$$

This proves that E/K is a Γ/Γ_K-Clifford extension.

(2) For any $\Delta \leqslant \Gamma$ we have $(\Phi \circ \Psi)(\Delta) = \Gamma_{E_\Delta} = \Delta$. Indeed, the inclusion $\Delta \subseteq \Gamma_{E_\Delta}$ is obvious. For the opposite inclusion, let $\gamma \in \Gamma_{E_\Delta}$. Then $E_\gamma \subseteq E_\Delta = \sum_{\delta \in \Delta} E_\delta$, hence necessarily $\gamma \in \Delta$ since the assumption that E/F is a strongly Γ-graded extension means that $\sum_{\lambda \in \Gamma} E_\lambda = \bigoplus_{\lambda \in \Gamma} E_\lambda$. $\qquad\square$

PROPOSITION 10.1.8. *The following assertions are equivalent for a Γ-Clifford extension E/F with Γ a torsion group.*

(1) *Every subextension K/F of E/F is a strongly Γ_K-graded extension.*

(2) *E/F is a strongly Γ-graded extension, and the map*

$$\Phi : \underline{\text{Intermediate}}(E/F) \longrightarrow \underline{\text{Subgroups}}(\Gamma), \quad \Phi(K) = \Gamma_K,$$

is a lattice isomorphism.

(3) *E/F is a strongly Γ-graded extension, and the maps*

$$\Phi : \underline{\text{Intermediate}}(E/F) \longrightarrow \underline{\text{Subgroups}}(\Gamma), \quad \Phi(K) = \Gamma_K,$$
$$\Psi : \underline{\text{Subgroups}}(\Gamma) \longrightarrow \underline{\text{Intermediate}}(E/F), \quad \Psi(\Delta) = E_\Delta$$

are isomorphisms of lattices, inverse to one another.

(4) *E/F is a strongly Γ-graded extension, and every intermediate field K of E/F has a vector space basis over F consisting of homogeneous elements.*

(5) *E/F is a strongly Γ-graded extension, and every intermediate field K of E/F is obtained by adjoining to F a set of homogeneous elements of E.*

(6) *E/F is a strongly Γ-graded extension, and every nonzero element x ∈ E has its homogeneous components in F(x).*

(7) *E/K is a strongly Γ/Γ_K-graded extension for every intermediate field K of E/F.*

PROOF. (1) $\Longrightarrow$ (2): Taking $K = E$, we obtain from hypothesis that E/F is a strongly Γ_E-graded extension, hence a strongly Γ-graded extension, since obviously $\Gamma_E = \Gamma$. From Lemma 10.1.7 (2) we deduce that Φ is surjective.

We claim that $K = \bigoplus_{\gamma \in \Gamma_K} E_\gamma$ for every intermediate field K of E/F. By hypothesis, K/F is a strongly Γ_K-graded extension, hence $K = \bigoplus_{\gamma \in \Gamma_K} K_\gamma$. Since $E = \bigoplus_{\gamma \in \Gamma} E_\gamma$, we clearly have $K_\gamma \subseteq E_\gamma$ for every $\gamma \in \Gamma_K$. Conversely, for any $\gamma \in \Gamma_K$ we have $E_\gamma \subseteq K$, hence $E_\gamma \subseteq K_\gamma$. This proves our claim.

We are going to show that Φ is injective. Let K_1, K_2 be two intermediate fields of E/F such that $\Phi(K_1) = \Phi(K_2)$, i.e., $\Gamma_{K_1} = \Gamma_{K_2}$. Then

$$K_1 = \bigoplus_{\gamma \in \Gamma_{K_1}} E_\gamma = \bigoplus_{\gamma \in \Gamma_{K_2}} E_\gamma = K_2.$$

(2) $\Longleftrightarrow$ (3) follows immediately from Lemma 10.1.7 (2).

(3) $\Longrightarrow$ (4): Let $K \in \underline{\text{Intermediate}}(E/F)$. Then $K = (\Psi \circ \Phi)(K) = E_{\Gamma_K} = \bigoplus_{\gamma \in \Gamma_K} E_\gamma$. This shows that K, viewed as an F-vector space has a basis consisting of homogeneous elements $\{ x_\gamma \mid \gamma \in \Gamma_K \}$, where x_γ is an arbitrary nonzero element of E_γ for every $\gamma \in \Gamma_K$.

(4) $\Longrightarrow$ (5) is obvious.

(5) $\Longrightarrow$ (6) By Lemma 10.1.3 (6), the extension E/F is algebraic. Let $x \in E^*$. By assumption, there exists a finite subset $\{\gamma_1, \dots, \gamma_n\}$

of Γ and $x_{\gamma_i} \in E^*_{\gamma_i}$, $1 \leqslant i \leqslant n$, such that $F(x) = F(x_{\gamma_1}, \ldots, x_{\gamma_n}) = F[x_{\gamma_1}, \ldots, x_{\gamma_n}]$. Thus, every element of $F(x)$ can be written as a finite sum of monomials $ax^{i_1}_{\gamma_1} \cdots x^{i_n}_{\gamma_n}$, with $a \in F$ and $i_1, \ldots, i_n \in \mathbb{N}$. This implies that $F(x) = E_\Delta$, where $\Delta = \langle \gamma_1, \ldots, \gamma_n \rangle$. But E/F is a strongly Γ-graded extension, hence every homogeneous component of x belongs to E_Δ, i.e., to $F(x)$.

(6) $\Longrightarrow$ (7): Let K/F be a subextension of E/F. By Lemma 10.1.3 (6), E/F is an algebraic extension, so $K = F[A]$ for some $\varnothing \neq A \subseteq K$. Let C be the set of all nonzero homogeneous components of all elements of A. By assumption, we have $K = F[C]$. Let $D = \{ \gamma \in \Gamma \, | \, \exists \, c \in C, \, c \in E_\gamma \}$ and set $\Delta = \langle D \rangle$. Then, as in the proof of implication (5) $\Longrightarrow$ (6), it follows that $K = E_\Delta = \sum_{\delta \in \Delta} E_\delta = \bigoplus_{\delta \in \Delta} E_\delta$, that is, K/F is a strongly Δ-graded extension.

From Lemma 10.1.7 (2) we deduce that $\Delta = (\Phi \circ \Psi)(\Delta) = \Gamma_{E_\Delta} = \Gamma_K$. According to Lemma 10.1.7 (1), E/K is a Γ/Γ_K-Clifford extension, with

$$E = \sum_{\widehat{\gamma} \in \Gamma/\Gamma_K} E_{\widehat{\gamma}} \quad \text{and} \quad E_{\widehat{\gamma}} = E_\gamma K, \ \forall \, \gamma \in \Gamma.$$

Let S be a set of representatives of the quotient group Γ/Γ_K. Since $\{ \Gamma_K \sigma \, | \, \sigma \in S \}$ is a partition of Γ, then as in the proof of Lemma 10.1.7 (1) we have

$$E = \bigoplus_{\gamma \in \Gamma} E_\gamma = \bigoplus_{\gamma \in \bigcup_{\sigma \in S} \Gamma_K \sigma} E_\gamma = \bigoplus_{\sigma \in S} \left(\bigoplus_{\gamma \in \Gamma_K \sigma} E_\gamma \right)$$

$$= \bigoplus_{\sigma \in S} \left(\bigoplus_{\lambda \in \Gamma_K} (E_{\lambda \sigma}) \right) = \bigoplus_{\sigma \in S} \left(\bigoplus_{\lambda \in \Gamma_K} (E_\lambda E_\sigma) \right)$$

$$= \bigoplus_{\sigma \in S} \left(\left(\sum_{\lambda \in \Delta} E_\lambda \right) E_\sigma \right) = \bigoplus_{\sigma \in S} (K E_\sigma) = \bigoplus_{\widehat{\gamma} \in \Gamma/\Gamma_K} E_{\widehat{\gamma}},$$

which shows exactly that E/K is a strongly Γ/Γ_K-graded extension.

(7) $\Longrightarrow$ (1) If $\gamma \in \Gamma_F$, then $E_\gamma \subseteq F$, hence $1 = \dim_F(E_\gamma) \leqslant \dim_F(F) = 1$. Thus $E_\gamma = F$, and then $\gamma = e$ by Definition 10.1.2 (3). This shows that $\Gamma_F = \{e\}$. Now, taking $K = F$ in (7), we deduce that E/F is a strongly Γ-graded extension, hence $E = \bigoplus_{\gamma \in \Gamma} E_\gamma$.

Let K be an intermediate field of the extension E/F. Then E/K is a strongly Γ/Γ_K-graded extension by hypothesis. Hence

$$E = \bigoplus_{\widehat{\gamma} \in \Gamma/\Gamma_K} E_{\widehat{\gamma}} = \bigoplus_{\sigma \in S} (E_\sigma K),$$

where S is a fixed set of representatives of the quotient group Γ/Γ_K. As in the proof of implication (6) $\Longrightarrow$ (7), we have

$$E = \bigoplus_{\gamma \in \Gamma} E_\gamma = \bigoplus_{\sigma \in S} \left(\left(E_\sigma \left(\sum_{\lambda \in \Gamma_K} E_\lambda \right) \right) \right) \leqslant \bigoplus_{\sigma \in S} (E_\sigma K) = E,$$

since clearly

$$E_{\Gamma_K} = \bigoplus_{\lambda \in \Gamma_K} E_\lambda \leqslant K.$$

Using the definition of the internal direct sum of submodules of a module, we deduce that

$$E_\sigma \left(\sum_{\lambda \in \Gamma_K} E_\lambda \right) - E_\sigma K$$

for every $\sigma \in S$. By Definition 10.1.2, this implies that, for a fixed $\sigma \in S$ we have

$$\bigoplus_{\lambda \in \Gamma_K} E_\lambda = \sum_{\lambda \in \Gamma_K} E_\lambda = F \left(\sum_{\lambda \in \Gamma_K} E_\lambda \right) = (E_{\sigma^{-1}} E_\sigma) \left(\sum_{\lambda \in \Gamma_K} E_\lambda \right)$$

$$= E_{\sigma^{-1}} \left(E_\sigma \left(\sum_{\lambda \in \Gamma_K} E_\lambda \right) \right) = E_{\sigma^{-1}} (E_\sigma K) = (E_{\sigma^{-1}} E_\sigma) K = FK = K,$$

which proves that the extension K/F is strongly Γ-graded. $\qquad\square$

DEFINITION 10.1.9. *A field extension E/F is said to be Γ-Clifford-Cogalois if it a finite separable Γ-Clifford extension which satisfies one of the equivalent conditions of Proposition 10.1.8.* $\qquad\square$

PROPOSITION 10.1.10. *The following statements hold for a finite field extension E/F.*

(1) *If the extension E/F is G-Cogalois for some G with $F^* \leqslant G \leqslant E^*$, then E/F is a G/F^*-Clifford-Cogalois extension.*

(2) *Conversely, if E/F is a Γ-Clifford-Cogalois extension for some finite group Γ, then there exists a uniquely determined group G such that $F^* \leqslant G \leqslant E^*$, $\Gamma \cong G/F^*$, E/F is G-Cogalois, and $U^h(E) = G$.*

PROOF. (1) By Proposition 10.1.6 (4), E/F is a strongly G/F^*-graded extension, and by Theorem 4.2.7, the maps

$$\varphi : \mathcal{E} \longrightarrow \mathcal{G}, \quad \varphi(K) = K \cap G,$$

$$\psi : \mathcal{G} \longrightarrow \mathcal{E}, \quad \psi(H) = F(H),$$

are isomorphisms of lattices, inverse to one another, where

$$\mathcal{G} = \{\, H \mid F^* \leqslant H \leqslant G \,\} \text{ and } \mathcal{E} = \mathrm{Intermediate}(E/F).$$

Clearly, the canonical map

$$\rho : \mathrm{Subgroups}(G/F^*) \longrightarrow \mathcal{G}, \ H/F^* \mapsto H,$$

is a lattice isomorphism, so the composed map

$$\psi \circ \rho : \mathrm{Subgroups}(G/F^*) \longrightarrow \mathrm{Intermediate}(E/F)$$

is a lattice isomorphism.

By the proof of Proposition 10.1.6 (1) we have

$$E_{H/F^*} = \sum_{\widehat{h} \in H/F^*} Fh = F[H] = F(H),$$

so

$$(\psi \circ \rho)(H/F^*) = F(H) = E_{H/F^*}$$

for every $H \in \mathcal{G}$. This shows that $\psi \circ \rho$ is precisely the canonical map

$$\Psi : \mathrm{Subgroups}(G/F^*) \longrightarrow \mathrm{Intermediate}(E/F), \ \Psi(H/F^*) = E_{H/F^*},$$

defined just above Lemma 10.1.7. Thus, the canonical map Φ in Proposition 10.1.8 is a lattice isomorphism by Lemma 10.1.7 (2). By Proposition 10.1.8, it follows that E/F is a G/F^*-Clifford-Cogalois extension.

(2) According to Proposition 10.1.6 (5), there exists a group G such that $F^* \leqslant G \leqslant E^*$, $\Gamma \cong G/F^*$, E/F is G-Kneser, and $U^h(E) = G$.

Preserve the notation from the proof of Proposition 10.1.6 (2). The group isomorphism $\omega : \Gamma \longrightarrow G/F^*$ induces the lattice isomorphism

$$\Upsilon : \mathcal{G} \longrightarrow \mathrm{Subgroups}(\Gamma), \ H \mapsto \{\, \gamma \in \Gamma \mid E_\gamma^* \subseteq H \,\}.$$

We claim that for any $\gamma \in \Gamma$,

$$E_\gamma^* \subseteq H \Longleftrightarrow E_\gamma \subseteq F(H).$$

One implication is clear. For the other one, if $E_\gamma \subseteq F(H)$, then we have $E_\gamma^* \subseteq F(H) \cap G = H$ by Proposition 2.1.11. Therefore, $\Upsilon(H) = \Gamma_{F(H)}$.

Consequently, if we compose the lattice isomorphism

$$\Upsilon : \mathcal{G} \longrightarrow \mathrm{Subgroups}(\Gamma), \ H \mapsto \{\, \gamma \in \Gamma \mid E_\gamma^* \subseteq H \,\} = \Gamma_{F(H)},$$

with the lattice isomorphism

$$\Psi : \underline{\text{Subgroups}}(\Gamma) \longrightarrow \underline{\text{Intermediate}}(E/F), \quad \Psi(\Delta) = E_\Delta,$$

given by Proposition 10.1.8, we obtain the lattice isomorphism

$$\Phi \circ \Upsilon : \mathcal{G} \longrightarrow \underline{\text{Intermediate}}(E/F), \quad H \mapsto E_{\Gamma_{F(H)}}.$$

But $E_{\Gamma_{F(H)}} = F(H)$ by Proposition 10.1.8, hence $\Phi \circ \Upsilon$ is exactly the map

$$\psi : \mathcal{G} \longrightarrow \mathcal{E}, \quad \psi(H) = F(H).$$

Now, apply Theorem 4.2.7 to conclude that E/F is G-Cogalois. The uniqueness of G follows from Corollary 4.4.2. $\square$

The next result is the graded version of the uniqueness of the Kneser group of a G-Cogalois extension (see Corollary 4.4.2).

COROLLARY 10.1.11. *Let E/F be a finite extension which is simultaneously Γ-Clifford-Cogalois and Δ-Clifford-Cogalois. Then the groups Γ and Δ are isomorphic.*

PROOF. Apply Proposition 10.1.10 and Corollary 4.4.2. $\square$

PROPOSITION 10.1.12. *Let E/F be a Γ-Clifford-Cogalois, let Λ be a finite nonempty subset of Γ, and let $\{ x_\lambda \mid \lambda \in \Lambda \} \subseteq U^h(E)$. Then $\sum_{\lambda \in \Lambda} x_\lambda$ is a primitive element of the extension E/F if and only if $\langle \Lambda \rangle = \Gamma$.*

PROOF. Denote $K = F(\sum_{\lambda \in \Lambda} x_\lambda)$. By Proposition 10.1.8, we have $x_\lambda \in K$ for every $\lambda \in \Lambda$, hence $F(\{ x_\lambda \mid \lambda \in \Lambda \}) \subseteq K$. The other inclusion $K \subseteq F(\{ x_\lambda \mid \lambda \in \Lambda \})$ is clear, hence $K = F(\{ x_\lambda \mid \lambda \in \Lambda \})$.

On the other hand, $E_{\langle \Lambda \rangle} = F(\{ x_\lambda \mid \lambda \in \Lambda \})$ by Lemma 10.1.5. Thus, by Proposition 10.1.8, we have

$$E = K \iff E_\Gamma = E_{\langle \Lambda \rangle} \iff \langle \Lambda \rangle = \Gamma.$$

$\square$

REMARK 10.1.13. The characterization of finite G-Kneser and G-Cogalois extensions in terms of graded algebras given in Proposition 10.1.6 (4)-(6) and Proposition 10.1.10, respectively, also hold for infinite extensions. This can be easily seen if in their proofs one simply invokes Proposition 11.1.2 and Theorem 11.2.6 instead of Corollary 2.1.10 and Theorem 4.2.7, respectively. See also Exercise 5 in Chapter 11 and Exercise 14 in Chapter 12. Note that Corollary 10.1.11 has also an infinite variant (see Exercise 15, Chapter 12). $\square$

10.2. Cogalois extensions and Hopf algebras

In this section we describe the property of a finite field extension being G-Kneser, in particular Cogalois, in terms of Hopf algebras. To do that, we present first the main concepts and facts on Hopf algebras which will be used in that description.

Throughout this section K will denote a fixed field, and H will denote a K-Hopf algebra. Tensor products are assumed to be over K, unless stated otherwise. We shall denote by K-Alg the category of all associative K-algebras with unit, and by K-Mod the category of all K-vector spaces. If $A \in K$-Alg, then Mod-A will denote the category of all unital right A-modules.

10.2.1. ALGEBRA AND COALGEBRA. We first present an alternative definition for the classical concept of an associative K-algebra with unit via maps and diagrams, so that we may dualize it.

A K-*algebra* is a triple (A, μ, u), where A is a K-vector space, $\mu : A \otimes A \longrightarrow A$ and $u : K \longrightarrow A$ are morphisms of K-vector spaces, called the *multiplication map* and the *unit map* of A, respectively, such that the following diagrams are commutative:

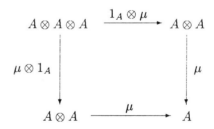

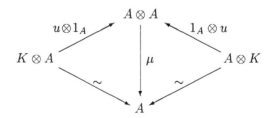

We denoted by 1_A the identity map of A, and the unnamed arrows from the second diagram are the canonical isomorphisms.

We now dualize the notion of algebra. A *K-coalgebra* is a triple (C, Δ, ε), where C is a K-vector space, $\Delta : C \longrightarrow C \otimes C$ and $\varepsilon : C \longrightarrow K$ are morphisms of K-vector spaces, called the *comultiplication map* and the *counit map* of C, respectively, such that the following diagrams are commutative:

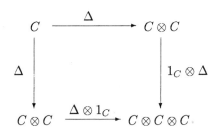

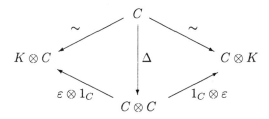

Let (C, Δ, ε) be a K-coalgebra and let $c \in C$. Then

$$\Delta(c) = \sum_{i=1}^{n} c_{i1} \otimes c_{i2}$$

for some $n \in \mathbb{N}^*$ and $c_{i1}, c_{i2} \in C$, $1 \leqslant i \leqslant n$. The *sigma notation*, introduced by Heyneman and Sweedler, suppresses the subscript i, so that,

$$\Delta(c) = \sum c_1 \otimes c_2.$$

If (A, μ_A, u_A) and (B, μ_B, u_B) are two K-algebras, then observe that a K-linear map $f : A \longrightarrow B$ is an *algebra morphism* if and only if the following diagrams are commutative:

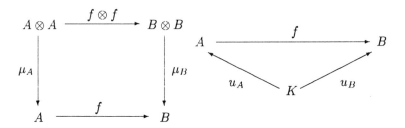

This observation allows us to define the dual concept of coalgebra morphism as follows. Let $(C, \Delta_C, \varepsilon_C)$ and $(D, \Delta_D, \varepsilon_D)$ be two K-coalgebras. Then, a K-linear map $g : C \longrightarrow D$ is a *coalgebra morphism* if the following diagrams are commutative:

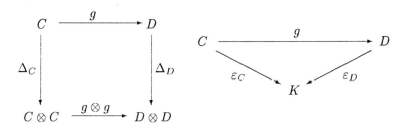

The commutativity of the first diagram may be written in the sigma notation as:

$$\Delta_D(g(c)) = \sum g(c)_1 \otimes g(c)_2 = \sum g(c_1) \otimes g(c_2).$$

10.2.2. COMODULE. In the same way as in the previous subsection, where we gave an alternative definition for an algebra using only morphisms and diagrams, we begin this subsection by presenting an alternative definition of the concept of left module over an algebra.

Let (A, μ, u) be a K-algebra. A left A-*module* is a pair (X, ν), where X is a K-vector space and $\nu : A \otimes X \longrightarrow X$ is a morphism of K-vector spaces, such that the following diagrams are commutative:

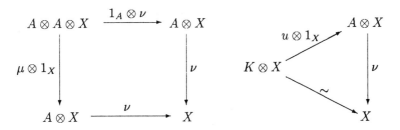

Similarly, one can define right modules over the algebra A, the only difference being that the structure map of the right A-module X has the form $\nu : X \otimes A \longrightarrow X$. If X is a left or right A-module, one also says that A *acts* on X.

By dualization, we obtain the concept of a comodule over a coalgebra. Let (C, Δ, ε) be a K-coalgebra. A right *C-comodule* is a pair (M, ρ), where M is a K-vector space and $\rho : M \longrightarrow M \otimes C$ is a morphism of K-vector spaces, such that the following diagrams are commutative:

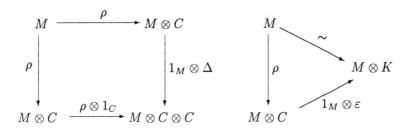

If M is a right C-comodule, one also says that C *coacts* on M.

We are now going to define the morphisms of comodules. To do that, we present first the definition of a module morphism using only commutative diagrams.

Let A be a K-algebra, and let $(X, \nu_X), (Y, \nu_Y)$ be two left A-modules. The K-linear map $f : X \longrightarrow Y$ is a morphism of A-modules if the following diagram is commutative:

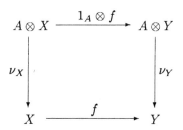

By dualization, we define the concept of a *comodule morphism*. Let C be a K-coalgebra, and let (M, ρ_M), (N, ρ_N) be two right C-comodules. The K-linear map $g : M \longrightarrow N$ is called a morphism of C-comodules if the following diagram is commutative:

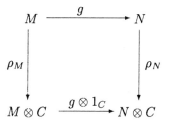

For any right C-comodule M with structure map $\rho : M \longrightarrow M \otimes C$ and any $m \in M$, we shall also use the *sigma notation*:

$$\rho(m) = \sum m_0 \otimes m_1.$$

Thus, the commutativity of the diagram above involving the definition of a comodule morphism, may be written in the sigma notation as:

$$\rho_N(g(m)) = \sum g(m_0) \otimes m_1, \ m \in M.$$

For a K-coalgebra C we shall denote by Comod-C the category of all right C-comodules.

10.2.3. BIALGEBRA AND HOPF ALGEBRA. A K-*bialgebra* is a K-vector space B endowed with an algebra structure (B, μ, u) and with a coalgebra structure (B, Δ, ε), such that Δ and ε are algebra morphisms, or equivalently, μ and u are coalgebra morphisms.

A map $f : B \longrightarrow B'$ of bialgebras is called a *bialgebra morphism* if it is both an algebra morphism and a coalgebra morphism.

A K-*Hopf algebra* is a bialgebra $(H, \mu, u, \Delta, \varepsilon)$ having a K-linear map $S : H \longrightarrow H$ satisfying the condition

$$\sum S(h_1)h_2 = \sum h_1 S(h_2) = \varepsilon(h)1$$

for every $h \in H$, where 1 is the unit element of H. The map S is called an *antipode* for H.

A map $f : H \longrightarrow H'$ between two K-Hopf algebras H and H' is called a *Hopf morphism* if it is a morphism of bialgebras. Any Hopf morphism preserves antipodes.

DEFINITION 10.2.4. *A left H-module algebra is a K-algebra A such that A is a left H-module with structure map*

$$H \otimes A \longrightarrow A, \ h \otimes a \mapsto h \cdot a, \ h \in H, \ a \in A,$$

and such that the multiplication $\mu : A \otimes A \longrightarrow A$, $\mu(a \otimes b) = ab$ is a morphism of H-modules, where $A \otimes A$ is a left H-module via

$$h \cdot (a \otimes b) = \sum (h_1 \cdot a) \otimes (h_2 \cdot b), \ h \in H, \ a, b \in A.$$

$\square$

DEFINITION 10.2.5. *A right H-comodule algebra is a K-algebra A such that A is a right H-comodule with structure map*

$$\rho : A \longrightarrow A \otimes H, \ \rho(a) = \sum a_0 \otimes a_1, \ a \in A,$$

and ρ is an algebra morphism. $\square$

DEFINITION 10.2.6. *A Galois H-object is a right H-comodule algebra A such that the linear map*

$$\gamma : A \otimes A \longrightarrow A \otimes H, \ \gamma(a \otimes b) = \sum ab_0 \otimes b_1, \ a, b \in A,$$

is bijective. $\square$

DEFINITION 10.2.7. *Let A be a right H-comodule algebra. A right (H, A)-Hopf module, or a Doi-Hopf module is a K-vector space M such that the following three conditions are satisfied.*

(1) M is a right H-comodule via the map

$$\nu : M \longrightarrow M \otimes H, \; \nu(m) = \sum m_0 \otimes m_1, \; m \in M.$$

(2) M is a right A-module via the map

$$M \otimes A \longrightarrow M, \; m \otimes a \mapsto m \cdot a, \; m \in M, \; a \in A.$$

(3) $\nu(m \cdot a) = \sum m_0 \cdot a_0 \otimes m_1 \cdot a_1, \forall \, m \in M, \; \forall \, a \in A.$ □

In the sequel, $\mathcal{M}_A^H$ will denote the category of all right (H, A)-Hopf modules, where the morphisms are the maps which are simultaneously morphisms of right A-modules and morphisms of right H-comodules.

EXAMPLES 10.2.8. (1) Clearly, $A \in \mathcal{M}_A^H$, with the usual structure of right A-module of A, and with the map $\rho : A \longrightarrow A \otimes H$ arising from the structure of right H-comodule algebra of A.

(2) If $X \in K$-Mod, then $A \otimes X \in \mathcal{M}_A^H$, with action

$$(a \otimes x) \cdot b = ab \otimes x,$$

and coaction

$$a \otimes x \mapsto \sum (a_0 \otimes x) \otimes a_1,$$

for $a, b \in A$, $x \in X$.

DEFINITION 10.2.9. Let $M \in$ Comod-H with comodule structure given by the map $\nu : M \longrightarrow M \otimes H$. The set

$$M^{\mathrm{co}H} = \{\, m \in M \mid \nu(m) = m \otimes 1 \,\}$$

is a K-vector subspace of M which is called the subspace of H-coinvariants of M. In particular, for any right H-comodule algebra A we can consider $A^{\mathrm{co}H}$, which is even a subalgebra of A.   □

PROPOSITION 10.2.10. Let A be an H-comodule algebra, and consider the functors:

$$F : K\text{-Mod} \longrightarrow \mathcal{M}_A^H, \quad F(X) = A \otimes X, \; X \in K\text{-Mod},$$
$$G : \mathcal{M}_A^H \longrightarrow K\text{-Mod}, \quad G(M) = M^{\mathrm{co}H}, \; M \in \mathcal{M}_A^H.$$

Then, G is a right adjoint of F.

PROOF. See Caenepeel [42, Proposition 8.1.3].   □

THEOREM 10.2.11. The following statements are equivalent for a Hopf algebra H with bijective antipode and for a right H-comodule algebra A.

(1) A is a Galois H-object.

(2) *The functors F and G in Proposition 10.2.10 are inverse equivalences, or equivalently, for all $X \in K$-Mod and all $M \in \mathcal{M}_A^H$, the natural maps*

$$X \longrightarrow G(F(X)) = (A \otimes X)^{\mathrm{co}H}, \ x \mapsto 1 \otimes x,$$
$$F(G(M)) = A \otimes M^{\mathrm{co}H} \longrightarrow M, \ a \otimes m \mapsto m \cdot a,$$

are isomorphisms.

Moreover, if (1) *or* (2) *holds, then* $A^{\mathrm{co}H} = K$.

PROOF. See Caenepeel [**42**, Definition 8.1.5 and Theorem 8.1.6]. $\qquad\square$

COROLLARY 10.2.12. *Let A be a Galois H-object. Then, for any $M \in \mathcal{M}_A^H$, the canonical map*

$$\mathrm{Subobjects}(M) \longrightarrow \mathrm{Subspaces}(M^{\mathrm{co}H}), \ N \mapsto N^{\mathrm{co}H},$$

is a lattice isomorphism.

PROOF. By Theorem 10.2.11, the functor

$$G : \mathcal{M}_A^H \longrightarrow K\text{-Mod}, \ G(M) = M^{\mathrm{co}H},$$

is an equivalence of categories. The corollary now follows from a general result saying that whenever we have an equivalence

$$G : \mathcal{C} \longrightarrow \mathcal{D}$$

between two arbitrary categories $\mathcal{C}$ and $\mathcal{D}$, then for every $X \in \mathcal{C}$, the functor G induces a bijective map between the subobjects of X and the subobjects of $G(X)$. $\qquad\square$

Next, we will illustrate the concepts of right H-comodule algebra and Galois H-object with a basic example. Let G be any multiplicative group with identity element e, and let $H = K[G]$ be the group algebra of G over the field K. It is well-known that $K[G]$ is a Hopf algebra over K via the comultiplication given by $\Delta(g) = g \otimes g$, the counit given by $\varepsilon(g) = 1$, and the antipode given by $S(g) = g^{-1}$, for all $g \in G$.

LEMMA 10.2.13. *Let G be a group, and let M be a K-vector space. Assume that there exists a family $(M_g)_{g \in G}$ of K-subspaces of M such that $M = \sum_{g \in G} M_g$. Then, $\rho : M \longrightarrow M \otimes_K K[G], \ x \mapsto x \otimes g$, for all $x \in M_g$ is a well-defined map if and only if $M = \bigoplus_{g \in G} M_g$, and in this case M is a right $K[G]$-comodule via the linear map ρ.*

Conversely, if the K-vector space M is a right $K[G]$-comodule, then M is a G-graded K-vector space, that is, $M = \bigoplus_{g \in G} N_g$ is a direct sum of K-subspaces N_g, $g \in G$.

PROOF. Assume that ρ is a well-defined map, and let $g \in G$. If $m \in M_g \cap \sum_{h \in G \setminus \{g\}} M_h$, then $m = \sum_{h \in G \setminus \{g\}} m_h$ with $m_h \in M_h$. Apply ρ to both sides of this equality, and then apply $1_M \otimes p_g$ to both sides to obtain $m \otimes 1 = 0$, where $p_g : K[G] \longrightarrow K$ denotes the linear map defined by $p_g(\sum_{h \in G} k_h h) = k_g$. This implies that $m = 0$, hence $M = \bigoplus_{g \in G} M_g$. Conversely, if $M = \bigoplus_{g \in G} M_g$, then clearly ρ is a well-defined linear map.

It can be easily checked that the map ρ endows M with a structure of right $K[G]$-comodule, and any right $K[G]$-comodule is a G-graded K-vector space (see e.g., Montgomery [**82**, Example 1.6.7]). $\square$

PROPOSITION 10.2.14. *The following assertions hold for a group G and a K-vector space A.*

(1) *A is a right $K[G]$-comodule algebra if and only if A is a G-graded K-algebra.*

(2) *A is a Galois $K[G]$-object if and only if A is a strongly G-graded K-algebra and $A_e = K$.*

PROOF. For (1) use Lemma 10.2.13, and for (2) see Caenepeel [**42**, Proposition 8.2.1] or Montgomery [**82**, Theorem 8.1.7]. $\square$

We are now in a position to present the connection between Cogalois field extensions and Hopf algebras.

THEOREM 10.2.15. *The following assertions are equivalent for a finite G-radical field extension E/F with G/F^* finite.*

(1) *E/F is a G-Kneser extension.*

(2) *E is a Galois $F[G/F^*]$-object via the comodule structure given by the linear map $E \longrightarrow E \otimes_F F[G/F^*]$, $x \mapsto x \otimes \widehat{g}$, for all $x \in Fg$ and $g \in G$.*

PROOF. (1) $\implies$ (2): By Proposition 10.1.6 (4), E/F is a strongly G/F^*-graded extension via the decomposition $E = \bigoplus_{\widehat{g} \in G/F^*} E_{\widehat{g}}$, with $E_{\widehat{g}} = Fg$ for every $g \in G$. Since $E_{\widehat{1}} = F1 = F$, we deduce by Proposition 10.2.14 (2) that E is a Galois $F[G/F^*]$-object.

(2) $\implies$ (1): By Lemma 10.2.13 and Proposition 10.2.14 (2), the extension E/F is strongly G/F^*-graded. Observe that $U^h(E) = G$. Now, apply Proposition 10.1.6 (6) to deduce that E/F is G-Kneser. $\square$

COROLLARY 10.2.16. *The following assertions are equivalent for a finite radical field extension E/F.*

(1) E/F *is a Cogalois extension.*

(2) E *is a Galois* $F[\operatorname{Cog}(E/F)]$*-object via the comodule structure given by the linear map* $E \longrightarrow E \otimes_F F[\operatorname{Cog}(E/F)]$, $x \mapsto x \otimes \widehat{g}$, *for all* $x \in Fg$ *and* $g \in T(E/F)$.

PROOF. The result follows at once from Theorem 10.2.15, since an extension E/F is Cogalois if and only if it is $T(E/F)$-Kneser, and $\operatorname{Cog}(E/F) = T(E/F)/F^*$. $\qquad\square$

Next, we will present the concept of H-Galois extension, as well as its connection with that of Galois H-object.

DEFINITION 10.2.17. *Let A be a right H-comodule algebra with structure map* $\rho : A \longrightarrow A \otimes H$. *We say that* $A^{\operatorname{co}H} \subset A$ *is an H-extension.*

The H-extension $A^{\operatorname{co}H} \subset A$ *is said to be an H-Galois extension if the map*

$$\beta : A \otimes_{A^{\operatorname{co}A}} H \longrightarrow A \otimes_K H, \ a \otimes b \mapsto ab_0 \otimes b_1$$

is bijective. $\qquad\square$

LEMMA 10.2.18. *Let H be a Hopf algebra over K, and suppose that the antipode of H is bijective. Then, the following statements hold for a right H-comodule algebra A.*

(1) A *is a Galois H-object.*

(2) $A^{\operatorname{co}H} = K$, *and the linear map* $A \otimes A \longrightarrow A \otimes H$, $a \otimes b \mapsto \sum ab_0 \otimes b_1$ *is bijective.*

PROOF. See Caenepeel [**42**, Theorem 8.1.6]. $\qquad\square$

The next result establishes the connection between Galois H-objects and H-Galois extensions.

PROPOSITION 10.2.19. *Let H be a Hopf algebra over K, and suppose that the antipode of H is bijective. Then, the following statements hold for a right H-comodule algebra A.*

(1) *If A is a Galois H-object, then $K = A^{\operatorname{co}H} \subset A$ is an H-Galois extension.*

(2) *If $A^{\operatorname{co}H} \subset A$ is an H-Galois extension and $A^{\operatorname{co}H} = K$, then A is a Galois H-object.*

PROOF. Apply Lemma 10.2.18. $\qquad\square$

We end this section by examining the connection between classical Galois field extensions and H-Galois extensions.

Recall that if H is a K-Hopf algebra and M is a left H-module, then the set

$$M^H := \{\, m \in M \mid hm = \varepsilon(h)m, \ \forall\, h \in H \,\}$$

is a K-vector subspace of M which is called the *subspace of H-invariants* of M.

If H is a finite dimensional K-Hopf algebra, then the K-vector space $H^* = \operatorname{Hom}_K(H, K)$ has a canonical structure of a K-Hopf algebra which is called the *dual Hopf algebra* of the K-Hopf algebra H.

LEMMA 10.2.20. *Let H be any finite dimensional K-Hopf algebra, and let A be any K-algebra. Then A is a right H-comodule algebra if and only if A is a left H^*-module algebra. Moreover, in this case we also have that $A^{H^*} = A^{coH}$.*

PROOF. See e.g., Dăscălescu, Năstăsescu, and Raianu [**54**, Proposition 6.2.4] or Montgomery [**82**, Lemma 1.7.2 and p. 41].   □

In particular, if E/F is any field extension, and $G \leqslant \operatorname{Gal}(E/F)$ is any finite group of F-automorphisms of E, then it is easily verified that E is a left $F[G]$-module algebra. Consequently, by Lemma 10.2.20, E is a right $F[G]^*$-comodule algebra and

$$E^{co F[G]^*} = E^{F[G]} = \operatorname{Fix}(G).$$

PROPOSITION 10.2.21. *The following assertions are equivalent for a field extension E/F and a finite group G with $G \leqslant \operatorname{Gal}(E/F)$.*

(1) *E/F is a Galois field extension with $G = \operatorname{Gal}(E/F)$.*

(2) *E is a Galois $F[G]^*$-object.*

(3) *$F \subset E$ is an $F[G]^*$-Galois extension and $F = \operatorname{Fix}(G)$.*

PROOF. Denote $K = \operatorname{Fix}(G)$ and $H = F[G]^*$. Then, as we observed above, we have $E^{coH} = K$.

(1) $\implies$ (2): If E/F is a Galois extension with $G = \operatorname{Gal}(E/F)$, then we have $F = K = E^{coH}$. Now continue as in Dăscălescu, Năstăsescu, and Raianu [**54**, Example 6.4.3 (1)] (see also Montgomery [**82**, 8.1.2]).

(2) $\implies$ (1): If E is a Galois $F[G]^*$-object, then by definition we have $F = K = E^{coH}$. By the Artin Theorem (see 1.2.9), E/K is a Galois field extension and $G = \operatorname{Gal}(E/K) = \operatorname{Gal}(E/F)$, as desired.

(2) $\iff$ (3) follows from Proposition 10.2.19.   □

10.3. Exercises to Chapter 10

1. Let K/F be a field extension, let $u \in K$ be an algebraic element over F of degree $n > 1$, and set $E = F(u)$ and $E_k = Fu^k$ for every $k \in \mathbb{Z}$.

   (a) Prove that the decomposition $E = \sum_{k \in \mathbb{Z}} E_k$ defines E/F as a $\mathbb{Z}$-Clifford extension if and only if $u^m \notin F$ for every $m \in \mathbb{N}^*$.

   (b) For every k, $0 \leqslant k < n$ set $E_{\widehat{k}} = Fu^k$. Prove that the decomposition $E = \bigoplus_{\widehat{k} \in \mathbb{Z}_n} E_{\widehat{k}}$ defines E/F as a strongly $\mathbb{Z}_n$-graded extension if and only if $u^n = a \in F$, or equivalently, $\mathrm{Min}(u, F) = X^n - a$.

2. (*Ştefan* [99]). Show that the following assertions hold for any Γ-Clifford extension E/F and any $\Delta_1, \Delta_2 \leqslant \Gamma$.

   (a) $E_{\Delta_1} E_{\Delta_2} = E_{\Delta_1 \Delta_2}$.

   (b) $E_{\Delta_1} \cap E_{\Delta_2} = E_{\Delta_1 \cap \Delta_2}$ if E/F is a strongly Γ-graded extension.

3. (*Ştefan* [99]). Let E/F be a Γ-Clifford extension, let K be an intermediate field of E/F, and set $\Gamma_K = \{ \gamma \in \Gamma \mid E_\gamma \subseteq K \}$. Prove that the following statements hold.

   (a) $\Gamma_K \leqslant \Gamma$.

   (b) $E_{\Gamma_K} \subseteq K$.

   (c) If $\Delta \leqslant \Gamma$ and $K = E_\Delta$, then $E_{\Gamma_K} = E_\Delta$.

4. (*Ştefan* [99]). Show that the following statements hold for a strongly Γ-graded extension E/F, with Γ a group of order 4 and $\mathrm{Char}(F) \neq 2$.

   (a) If Γ is a cyclic group, γ is a generator of Γ, $x \in E_\gamma^*$, and $z \in E^*$ is such that $z^2 \in F$, then either $z \in F \cup E_{\gamma^2}$, or there exists an $a \in F^*$ such that $x^4 = -a^2$ and $z = b(ax + x^3)$ for some $b \in F$.

   (b) If $\Gamma \cong \mathbb{Z}_2 \times \mathbb{Z}_2$ and $z \in E^*$ is such that $z^2 \in F$, then $z \in U^h(E)$.

5. (*Ştefan* [99]). Let E/F be a separable Γ-Clifford extension, with Γ a finite group of exponent n, and let p be a prime number dividing $|\Gamma|$. Assume that for every $q \in \mathcal{P}_n$ and for every cyclic subgroup Δ of order q of Γ one has $[E_\Delta : F] = q$. Prove the following statements.

   (a) $[E_{\Gamma_0} : F] = p^2$ for any subgroup Γ_0 of order p^2 of Γ.

(b) Whenever Δ_1 and Δ_2 are subgroups of Γ such that $|\Delta_1| = p$, $|\Delta_2| = p^r$ for some $r \in \mathbb{N}^*$, $\Delta_1 \subseteq \Delta_2$ and $E_{\Delta_1} = E_{\Delta_2}$, then necessarily $\Delta_1 = \Delta_2$.

6. (*Ştefan* [**99**]). Prove that the following assertions are equivalent for a separable Γ-Clifford extension E/F with Γ a finite group of exponent n.
   (a) E/F is a strongly Γ-graded extension.
   (b) For every $p \in \mathcal{P}_n$ and for every cyclic subgroup Δ of order p of Γ one has $[E_\Delta : F] = p$.
   (c) For every odd prime p, $\zeta_p \in U^h(E) \Longrightarrow \zeta_p \in F$, and $1 + \zeta_4 \in U^h(E) \Longrightarrow \zeta_4 \in F$.
   (d) $\mu_p(E) \cap U^h(E) \subseteq F$ and $(1 + \mu_4(E)) \cap U^h(E) \subseteq F$.

7. (*Ştefan* [**99**]). Prove that the following assertions are equivalent for a separable Γ-Clifford extension E/F with Γ a finite group of exponent n.
   (a) E/F is a Γ-Clifford-Cogalois extension.
   (b) $\mu_p(E) = \mu_p(F)$ for all $p \in \mathcal{P}_n$.

8. (*Ştefan* [**99**]). Let E/F be a finite strongly Γ-graded extension, such that Γ is the internal direct sum of two subgroups Γ_1 and Γ_2. Prove the following assertions.
   (a) E and $E_{\Gamma_1} \otimes_F E_{\Gamma_2}$ are isomorphic F-algebras.
   (b) If E/F is additionally an Abelian extension, then

$$\mathrm{Gal}(E/F) \cong \mathrm{Gal}(E_{\Gamma_1}/F) \times \mathrm{Gal}(E_{\Gamma_2}/F).$$

9. (*Ştefan* [**99**]). Let E/F be a Γ-Clifford-Cogalois extension, with Γ a finite group of exponent n. Assume that $\zeta_n \in E$, and let $\Delta \leqslant \Gamma$ be such that $E_\Delta = F(\zeta_n)$. Prove that E/F is a Galois extension and that there exists an exact sequence of groups:

$$1 \longrightarrow \Gamma/\Delta \longrightarrow \mathrm{Gal}(E/F) \longrightarrow \Delta \longrightarrow 1.$$

10. (*Ştefan* [**99**]). Show that if E/F is any Galois Γ-Clifford-Cogalois extension, with Γ a finite group of exponent a prime number, then $\mathrm{Gal}(E/F) \cong \Gamma$. (*Hint:* Use Exercise 9.)

11. (*Ştefan* [**99**]). Show that if E/F is any Galois Γ-Clifford-Cogalois extension, with Γ a finite group of exponent n, and $\mu_n(E) = \mu_n(F)$, then $\mathrm{Gal}(E/F) \cong \Gamma$. (*Hint:* Use Exercise 9.)

10.4. Bibliographical comments to Chapter 10

Section 10.1. The concepts of a *Clifford system* and a *Clifford extension* were invented in 1970 by Dade [50], [51]. Dade also introduced ten years later the concept of *strongly Γ-graded ring* [52]. A good account of graded rings and modules can be found in Năstăsescu and Van Oystaeyen [84]. The interpretation of finite *G*-radical, *G*-Kneser, and *G*-Cogalois extensions via Clifford and strongly graded extensions is due to Ştefan [99].

The results of this section are taken from Albu [14], where most of Ştefan's results were generalized from finite to infinite field extensions.

Section 10.2. We followed Sweedler [102], Montgomery [82], Caenepeel [42], Dăscălescu, Năstăsescu, and Raianu [54] for the standard concepts and facts of Hopf algebras presented in this section.

The connection between Cogalois extensions and Hopf algebras is mentioned in passing in Greither and Harrison [63]. The explicit connections provided by Theorem 10.2.15 and Corollary 10.2.16 is due to Albu [14].

The concept of *Galois H-coobject* is defined and investigated in the theory of Hopf algebras and is the formal dual of the Galois *H*-object (see, e.g., Caenepeel [42, Section 8.7]). It is not clear how this concept is related to that of Cogalois field extension.

Part 2

INFINITE COGALOIS
THEORY

CHAPTER 11

INFINITE KNESER EXTENSIONS

The aim of this chapter is to introduce and investigate one of the basic concepts of the Infinite Cogalois Theory, namely that of infinite Kneser extension.

Infinite G-Kneser extensions are introduced and characterized in Section 11.1. One says that an extension E/F, which is not necessarily finite, is G-*Kneser* if G is a subgroup of E^* containing F^* such that $E = F(G)$ and the factor group G/F^* is a torsion group having a set of representatives which is linearly independent over F. We show that the Kneser Criterion, which characterizes finite separable G-Kneser extensions, can be generalized to infinite extensions.

Section 11.2 introduces and investigates infinite strongly G-Kneser extensions. The key result of this section is Theorem 11.2.4 which, roughly speaking, states that whenever two of the extensions K/F, E/K, E/F in a tower of fields $F \subseteq K \subseteq E$ are Kneser, then so is the third one.

11.1. Infinite G-Kneser extensions

The aim of this section is to introduce and characterize the notion of infinite G-*Kneser extension*. We also present the infinite variant of the Kneser Criterion.

By Corollary 2.1.10, a finite G-radical extension E/F is G-Kneser if and only if there exists a set of representatives of G/F^* which is linearly independent over F. This property suggests we generalize the concept of G-Kneser extension to *arbitrary* field extensions as follows:

DEFINITION 11.1.1. *An extension E/F, which is not necessarily finite, is said to be G-Kneser if it is a G-radical extension such that there exists* *a set of representatives of the factor group G/F^* which is linearly independent over F. The extension E/F is called* Kneser *if it is G-Kneser for some group G.* □

The next result is a restatement of Proposition 2.1.8 in terms of infinite and finite Kneser extensions.

PROPOSITION 11.1.2. *The following assertions are equivalent for an arbitrary G-radical extension E/F.*

(1) *E/F is a G-Kneser extension.*

(2) *Every set of representatives of G/F^* is linearly independent over F.*

(3) *Every set of representatives of G/F^* is a vector space basis of E over F.*

(4) *There exists a set of representatives of G/F^* which is a vector space basis of E over F.*

(5) *Every subset of G consisted of elements having distinct cosets in the group G/F^* is linearly independent over F.*

(6) *Every finite subset $\{x_1, \ldots, x_n\} \subseteq G$ such that $\widehat{x_i} \neq \widehat{x_j}$ for every $i, j \in \{1, \ldots, n\}$, $i \neq j$, is linearly independent over F.*

(7) *For every subgroup H of G containing F^*, the extension $F(H)/F$ is H-Kneser.*

(8) *For every subgroup H of G such that $F^* \leqslant H$ and H/F^* is a finite group, the finite extension $F(H)/F$ is H-Kneser.* □

COROLLARY 11.1.3. *If E/F is a G-Kneser extension, then*

$$|G/F^*| = [E : F].$$

PROOF. Since E/F is a G-Kneser extension, there exists a set of representatives of G/F^* which is linearly independent over F. Thus $|G/F^*| \leqslant [E : F]$. By Lemma 2.1.6 we also have the opposite inequality $|G/F^*| \geqslant [E : F]$, and consequently $|G/F^*| = [E : F]$. □

The converse of Corollary 11.1.3 is not true, as the following example shows.

EXAMPLE 11.1.4. Let $p_1, \ldots, p_n, \ldots$ be the sequence of all positive prime numbers, and set $E = \mathbb{Q}(\zeta_3, \sqrt{p_1}, \ldots, \sqrt{p_n}, \ldots)$. We have

$$\left| \mathbb{Q}\langle \zeta_3, \sqrt{p_1}, \ldots, \sqrt{p_n}, \ldots \rangle / \mathbb{Q}^* \right| = \left[\mathbb{Q}(\zeta_3, \sqrt{p_1}, \ldots, \sqrt{p_n}, \ldots) : \mathbb{Q} \right] = \aleph_0.$$

However, the extension $E/\mathbb{Q}$ is not $\mathbb{Q}^* \langle \zeta_3, \sqrt{p_1}, \ldots, \sqrt{p_n}, \ldots \rangle$-Kneser. Indeed, set $G = \mathbb{Q}^* \langle \zeta_3, \sqrt{p_1}, \ldots, \sqrt{p_n}, \ldots \rangle$ and $H = \mathbb{Q}^* \langle \zeta_3 \rangle$. If the extension $E/\mathbb{Q}$ were G-Kneser, then it would follow that the finite extension

$\mathbb{Q}(H)/\mathbb{Q}$ would be H-Kneser by Proposition 11.1.2, which contradicts Example 2.1.15 (1) since $\mathbb{Q}(H) = \mathbb{Q}(\zeta_3)$. $\square$

Next, we state and prove the infinite variant of the Kneser Criterion (Theorem 2.2.1).

THEOREM 11.1.5 (THE INFINITE KNESER CRITERION). *Let E/F be an arbitrary separable G-radical extension E/F. Then, the following assertions are equivalent.*

(1) *E/F is a G-Kneser extension.*
(2) *For every odd prime p, $\zeta_p \in G \implies \zeta_p \in F$, and $1 \pm \zeta_4 \in G \implies \zeta_4 \in F$.*
(3) *$\mu_p(G) = \mu_p(F)$ for every odd prime p, and $1 \pm \zeta_4 \in G \implies \zeta_4 \in F$.*

PROOF. (1) $\implies$ (2): Suppose that the extension E/F is G-Kneser. If $\zeta_p \in G$, where p is an odd prime, then set $H = F^* \langle \zeta_p \rangle$, and observe that $\zeta_p \in H$, $F^* \leqslant H \leqslant G$, and H/F^* is a finite group. By Proposition 11.1.2, the finite extension $F(H)/F$ is H-Kneser. Now apply the (finite) Kneser Criterion (Theorem 2.2.1) to deduce that $\zeta_p \in F$.

If $1 \pm \zeta_4 \in G$, then use a similar argument, by considering as H the subgroup $F^* \langle 1 + \zeta_4 \rangle$ of G.

(2) $\implies$ (1): Suppose that (2) is true. We will show that condition (8) in Proposition 11.1.2 is verified, which will prove that the extension E/F is G-Kneser.

Let $H \leqslant G$ with $F^* \leqslant G$ and H/F^* finite. Then $F(H)/F$ is a finite extension. In order to show that this extension is H-Kneser, we will use again the (finite) Kneser Criterion. If $\zeta_p \in H$, where p is an odd prime, then $\zeta_p \in G$, and by assumption it follows that $\zeta_p \in F$. If $1 \pm \zeta_4 \in H$, then $1 \pm \zeta_4 \in G$, and by (2) it follows that $\zeta_4 \in F$. So, the finite extension $F(H)/F$ is H-Kneser, and by Proposition 11.1.2 the extension E/F is G-Kneser.

(2) $\Longleftrightarrow$ (3) as in the proof of Theorem 2.2.1. $\square$

EXAMPLE 11.1.6. Any G-radical extension E/F, with E a subfield of the field $\mathbb{R}$ of real numbers is a G-Kneser extension since it verifies condition (2) of Theorem 11.1.5. In particular, if $A \subseteq \mathbb{R}$ is any set of real roots of positive rational numbers, then the extension $\mathbb{Q}(A)/\mathbb{Q}$ is $\mathbb{Q}^* \langle A \rangle$-Kneser. $\square$

11.2. Infinite strongly Kneser extensions

The aim of this section is to introduce the infinite variant of the concept of strongly Kneser extension discussed in Section 4.2, and to investigate its main properties.

The key result of this section is Theorem 11.2.4 which, roughly speaking, states that if in a tower of fields $F \subseteq K \subseteq E$ two extensions are Kneser, then so is the third one.

Throughout this section, E/F will denote a fixed extension and G a group such that $F^* \leqslant G \leqslant E^*$. We shall use in the sequel the following notation:

$$\mathcal{G} = \{\, H \mid F^* \leqslant H \leqslant G \,\},$$

$$\mathcal{E} := \underline{\text{Intermediate}}\,(E/F) = \{\, K \mid F \subseteq K,\ K \text{ subfield of } E \,\}.$$

LEMMA 11.2.1. *Let E/F be a G-Kneser extension. Then, the extension $F(H)/F$ is H-Kneser and $F(H) \cap G = H$ for every $H \in \mathcal{G}$.*

PROOF. The extension $F(H)/F$ is H-Kneser by Proposition 11.1.2. The proof of the equality $F(H) \cap G = H$ is literally the same as that in the case of finite extensions (see Proposition 2.1.11), and therefore is omitted. $\square$

The next result, which is fundamental in generalizing the Finite Cogalois Theory to infinite extensions, is the infinite variant of Proposition 4.2.1. Its proof is different from and more complicated than the one in the finite case.

PROPOSITION 11.2.2. *The following assertions are equivalent for a G-Kneser extension E/F and an intermediate field K of E/F.*

(1) *K/F is H-Kneser for some $H \in \mathcal{G}$.*
(2) *K/F is $K^* \cap G$-Kneser.*
(3) *E/K is K^*G-Kneser.*

PROOF. $(1) \implies (3)$: Suppose that K/F is H-Kneser for some $H \in \mathcal{G}$. Then $K = F(H)$ and $F(H) \cap G = H$ by Lemma 11.2.1. In order to show that E/K is K^*G-Kneser it is sufficient, by definition, to prove that there exists a set of representatives of $(K^*G)/K^*$ which is linearly independent over K.

Observe that any set of representatives of G/H will be also a set of representatives of $(K^*G)/K^*$, because of the group isomorphism

$$G/H \xrightarrow{\sim} (K^*G)/K^*$$

induced by the composition of the canonical maps

$$G \hookrightarrow K^*G \longrightarrow (K^*G)/K^*.$$

Let $\{\, g_i \,|\, i \in I \,\}$ be a fixed set of representatives of G/H, and assume that $\{\, g_i \,|\, i \in I \,\}$ is linearly dependent over K. Then, there exists a finite subset $\{g_{i_1}, \ldots, g_{i_m}\}$ of $\{\, g_i \,|\, i \in I \,\}$ and $\mu_1, \ldots, \mu_m \in K^*$ such that

$$\mu_1 g_{i_1} + \cdots + \mu_m g_{i_m} = 0.$$

For simplicity, set $g_{i_1} = g_1, \ldots, g_{i_m} = g_m$. By hypothesis, $g_i H \neq g_j H$ for every $i \neq j$ in I.

Since $E = F(H) = F[H]$, for each $i = 1, \ldots, m$ we can write

$$\mu_i = \sum_{j=1}^{n_i} \lambda_{ij} h_{ij},$$

with $n_i \in \mathbb{N}^*$, $\lambda_{ij} \in F^*$, $h_{ij} \in H$, and $h_{ij} F^* \neq h_{ik} F^*$ for all $1 \leqslant j \neq k \leqslant n_i$. Thus,

$$\sum_{i=1}^m \mu_i g_i = \sum_{i=1}^m \sum_{j=1}^{n_i} \lambda_{ij} h_{ij} g_i = 0.$$

Now, $h_{ij} g_i F^* \neq h_{ik} g_i F^*$ for $j \neq k$, for otherwise it would follow that $h_{ij} F^* = h_{ik} F^*$, which is a contradiction. On the other hand, we have

$$h_{ij} g_i F^* \neq h_{ks} g_k F^*$$

for any $i \neq k$, for otherwise it would follow that $g_i g_k^{-1} \in H$, i.e., $g_i H = g_k H$ for $i \neq k$, which is a contradiction.

Consequently, $\{\, h_{ij} g_i \,|\, 1 \leqslant i \leqslant m, \, 1 \leqslant j \leqslant n_i \,\}$ is a subset of G having distinct cosets modulo F^*. Since E/F is G-Kneser, these elements are linearly independent over F by Proposition 11.1.2. Thus, $\lambda_{ij} = 0$ for every $1 \leqslant i \leqslant m$, $1 \leqslant j \leqslant n_i$, which is a contradiction. So, $\{\, g_i \,|\, i \in I \,\}$ is linearly independent over K, i.e., E/K is K^*G-Kneser.

$(3) \implies (2)$: Suppose that E/K is K^*G-Kneser, and set $K^* \cap G = H$. Then

$$(K^*G)/K^* \cong G/(K^* \cap G) = G/H.$$

We know that $F(H)/F$ is H-Kneser by Proposition 11.1.2, so it will be sufficient to show that $K = F(H)$.

Consider the tower of fields

$$F \subseteq F(H) \subseteq K \subseteq E = F(G),$$

and set $L = F(H)$. Since E/K is K^*G-Kneser by hypothesis, any set of representatives of

$$(K^*G)/K^* \cong G/(K^* \cap G) = G/H$$

is a basis of E over K. But L/F is H-Kneser, hence using the implication (1) $\Longrightarrow$ (3) we already proved, we deduce that E/L is L^*G-Kneser; so, any set of representatives of G/H is a basis of E over L because

$$(L^*G)/L^* \cong G/(L^* \cap G) = G/(F(H) \cap G) = G/H.$$

Now, consider the tower of fields

$$L \subseteq K \subseteq E,$$

which has the property that there exists a basis of E over K which is also a basis of E over L. We claim that $K = L$, which will finish the proof.

Indeed, assume that $K \neq L$, and take a basis $\{\, x_i \,|\, i \in I \,\}$ of E over K which is also a basis of E over L. Let $\{\, y_j \,|\, j \in J \,\}$ be a basis of K over L with $y_{j_1} = 1$ and $y_{j_2} = y$, where j_1 and j_2 are two distinct fixed elements in J and y is a fixed element in $K \setminus L$. Then, by the Tower Law,

$$B := \{\, x_i y_j \,|\, (i,j) \in I \times J \,\}$$

is a basis of E over L, in particular, the subset

$$C := \{\, x_i \,|\, i \in I \,\} \cup \{\, x_i y \,|\, i \in I \,\}$$

of B is linearly independent over L.

But $\{\, x_i \,|\, i \in I \,\}$ is a basis of E over L, hence, for a fixed $i_0 \in I$, $x_{i_0} y$ is an L-linear combination of $\{\, x_i \,|\, i \in I \,\}$, which contradicts the linear independence of C over L. This proves our claim.

(2) $\Longrightarrow$ (1) is obvious. $\square$

PROPOSITION 11.2.3. *Let E/F be an extension, and let G be a group such that $F^* \leqslant G \leqslant E^*$. If K is an intermediate field of E/F such that K/F is $K^* \cap G$-Kneser and E/K is K^*G-Kneser, then E/F is G-Kneser.*

PROOF. First of all, observe that by the proof of Proposition 4.2.2, E/F is a G-radical extension.

Set $H = K^* \cap G$. By Proposition 11.1.2, any set of representatives of H/F^* is a basis of K over F since K/F is a $K^* \cap G$-Kneser extension, and any set of representatives of $(K^*G)/K^* \cong G/(K^* \cap G) = G/H$ is a basis of

E over K since E/K is a K^*G-Kneser extension. So, if $\{\, h_i \,|\, i \in I \,\}$ is a set of representatives of H/F^* and $\{\, g_j \,|\, j \in J \,\}$ is a set of representatives of G/H, then, by the Tower Law, $\{\, h_i g_j \,|\, (i,j) \in I \times J \,\}$ is a basis of E over F.

On the other hand, we claim that $\{\, h_i g_j \,|\, (i,j) \in I \times J \,\}$ is a set of representatives of G/F^*, which will finish the proof. Indeed, let $g \in G$; then $gH = g_j H$ for some $j \in J$, i.e., $gg_j^{-1} \in H$. Then $gg_j^{-1} F^* = h_i F^*$ for some $i \in I$, i.e., $gF^* = h_i g_j F^*$. Now,

$$h_i g_j F^* = h_s g_t F^* \implies g_j H = g_t H \implies j = t \implies h_i F^* = h_s F^* \implies i = s.$$

This finishes the proof. □

Propositions 11.2.2 and 11.2.3 can be reformulated together as follows:

THEOREM 11.2.4. *Let $F \subseteq K \subseteq E$ be a tower of fields, and let G be a group such that $F^* \leqslant G \leqslant E^*$. Consider the following statements:*

(1) K/F *is* $K^* \cap G$-*Kneser.*
(2) E/K *is* K^*G-*Kneser.*
(3) E/F *is* G-*Kneser.*

Then, any two of the assertions (1)-(3) imply the remaining one. □

Now, we are in a position to introduce a basic concept of the Infinite Cogalois Theory.

DEFINITION 11.2.5. *An extension E/F is said to be* strongly G-Kneser *if it is a G-radical extension such that for any intermediate field K of E/F, the extension E/K is K^*G-Kneser, or equivalently, by Proposition 11.2.2, the extension K/F is $K^* \cap G$-Kneser.*

The extension E/F is called strongly Kneser *if it is strongly G-Kneser for some group G.* □

Clearly, any strongly G-Kneser extension is G-Kneser, but not conversely, as Example 4.2.4 shows.

Now consider an arbitrary G-radical extension E/F. Recall that we introduced in Section 4.1 the following notation:

$$\mathcal{G} = \{\, H \mid F^* \leqslant H \leqslant G \,\},$$
$$\mathcal{E} = \{\, E \mid F \subseteq E,\ E \text{ subfield of } E \,\},$$
$$\varphi : \mathcal{E} \longrightarrow \mathcal{G}, \quad \varphi(E) = E \cap G,$$
$$\psi : \mathcal{G} \longrightarrow \mathcal{E}, \quad \psi(H) = F(H),$$

We have seen that the maps φ and ψ define a Cogalois connection

$$\mathcal{E} \underset{\psi}{\overset{\varphi}{\rightleftarrows}} \mathcal{G}$$

which we have called the standard Cogalois connection associated with E/F.

The next two results are the infinite variants of Theorem 4.2.7 and Proposition 4.2.8, respectively. Their proofs are literally the same as in the case of finite extensions, and therefore will be omitted.

THEOREM 11.2.6. *The following assertions are equivalent for an arbitrary G-radical extension E/F.*

(1) E/F *is strongly G-Kneser.*

(2) E/F *is G-Kneser, and the map* $\psi : \mathcal{G} \longrightarrow \mathcal{E}$, $\psi(H) = F(H)$ *is surjective.*

(3) E/F *is G-Kneser, and every element of $\mathcal{E}$ is a closed element in the standard Cogalois connection associated with E/F.*

(4) E/F *is G-Kneser, and the map* $\varphi : \mathcal{E} \longrightarrow \mathcal{G}$, $\varphi(K) = K \cap G$ *is injective.*

(5) E/F *is a G-Kneser extension with G/F^*-Cogalois correspondence.*

(6) E/F *is G-Kneser, and the maps* $- \cap G : \mathcal{E} \longrightarrow \mathcal{G}$, $F(-) : \mathcal{G} \longrightarrow \mathcal{E}$ *are isomorphisms of lattices, inverse to one another.* □

PROPOSITION 11.2.7. *The following assertions hold for any strongly G-Kneser extension E/F and for any intermediate field K of E/F.*

(1) K/F *is strongly $K^* \cap G$-Kneser.*

(2) E/K *is strongly K^*G-Kneser.* □

11.3. Exercises to Chapter 11

1. Let E/F be an extension, which is simultaneously G-Kneser and H-Kneser. Prove that if $H \leqslant G$, then $H = G$.

2. Let $F \subseteq K \subseteq E$ be a tower of fields, and let G be a group such that $F^* \leqslant G \leqslant E^*$. Assume that K/F is a strongly $K^* \cap G$-Kneser extension and E/K is a strongly K^*G-Kneser extension. Is it true that E/F is a strongly G-Kneser extension?

3. Let $E_r = \mathbb{Q}(\cos(\pi/2^{r+1}))$, $r \in \mathbb{N}^*$, and $E_\infty = \bigcup_{r \geqslant 1} E_r$. Prove that $E_\infty/\mathbb{Q}$ is a Galois extension of infinite degree.

4. Let $r \in \mathbb{N}$, $r \geqslant 2$, $\nu_r = \underbrace{\sqrt{1 + \sqrt{1 + \sqrt{1 + \ldots + \sqrt{2}}}}}_{r \text{ radicals}}$, $F_r = \mathbb{Q}(\nu_r)$,

   and $F_\infty = \bigcup_{r \geqslant 1} F_r$. Prove that $F_\infty/\mathbb{Q}$ is a non Galois extension of infinite degree.

5. Prove that the following assertions hold for a field extension E/F.
   (a) Let G be a group with $F^* \leqslant G \leqslant E^*$. If E/F is an arbitrary G-Kneser extension, then E/F is a strongly G/F^*-graded extension.
   (b) Conversely, if E/F is a strongly Γ-graded extension for some torsion group Γ, then there exists a group G such that $F^* \leqslant G \leqslant E^*$, $\Gamma \cong G/F^*$, E/F is G-Kneser, and $U^h(E) = G$.
   (c) If E/F is a strongly G/F^*-graded extension, where G is a group such that $F^* \leqslant G \leqslant T(E/F)$ and $U^h(E) \subseteq G$, then E/F is a G-Kneser extension.

6. Prove that the following assertions are equivalent for a G-radical field extension E/F.
   (a) E/F is a G-Kneser extension.
   (b) E is a Galois $F[G/F^*]$-object via the comodule structure given by the linear map $E \longrightarrow E \otimes_F F[G/F^*]$, $x \mapsto x \otimes \hat{g}$, for all $x \in Fg$ and $g \in G$.

11.4. Bibliographical comments to Chapter 11

Section 11.1. The concept of an infinite Kneser extension is due to Albu and Ţena [**25**]. The contents of this section follows closely the presentation in Albu and Ţena [**25**].

Section 11.2. The contents of this section closely follows the presentation in Albu and Ţena [**25**] where the notion of an infinite strongly Kneser extension was introduced and investigated.

CHAPTER 12

INFINITE G-COGALOIS EXTENSIONS

In this chapter we generalize the concept of finite G-Cogalois extension to infinite extensions, and show that almost all of the results of Chapter 4 concerning finite G-Cogalois extensions also hold for infinite G-Cogalois extensions.

In Section 12.1 we introduce the concept of an arbitrary G-Cogalois extension. The main result of this section is the *General Purity Criterion* (Theorem 12.1.4) which characterizes G-Cogalois extensions in terms of $\mathcal{P}_G$-purity. This result is applied to show that a separable G-Kneser extension E/F is G-Cogalois if and only if the group G/F^* has a prescribed structure. We will use this criterion in the next chapter to provide large classes of infinite G-Cogalois extensions. Also, we particularize these results to bounded G-Cogalois extensions.

Section 12.2 deals with infinite Cogalois extensions. We show that most of the results on finite Cogalois extensions established in Chapter 3 also hold for infinite Cogalois extensions.

12.1. The General Purity Criterion and its applications

The most interesting G-radical extensions E/F with G/F^*-Cogalois correspondence are those which are also separable. The finite extensions of this type were called G-Cogalois extensions in Section 4.3. We will preserve this terminology also in the case of infinite extensions. We characterize these extensions in terms of $\mathcal{P}_G$-purity, and show that, as in the case of finite extensions, they can be described within the class of separable G-Kneser extensions by the structure of their uniquely determined Kneser groups.

DEFINITION 12.1.1. *An extension, which is not necessarily finite, is said to be G-Cogalois if it is separable and strongly G-Kneser.* □

The next result shows that the property of an extension being G-Cogalois behaves nicely with respect to subextensions and quotient extensions

PROPOSITION 12.1.2. *Let E/F be a G-Cogalois extension. Then, for every intermediate field K of E/F, the following assertions hold.*

(1) K/F *is* $K^* \cap G$*-Cogalois.*

(2) E/K *is* K^*G*-Cogalois.*

PROOF. The result follows at once from Proposition 11.2.7.   □

Recall that by $\mathbb{P}$ we have denoted the set of all strictly positive prime numbers, by $\mathcal{P}$ the set $(\mathbb{P} \setminus \{2\}) \cup \{4\}$, by $\mathbb{D}_n$ the set of all positive divisors of a given number $n \in \mathbb{N}$, and by $\mathcal{P}_n$ the set $\mathcal{P} \cap \mathbb{D}_n$.

DEFINITION 12.1.3. *Let $\mathcal{Q}$ be a nonempty subset of $\mathcal{P}$. An extension E/F is said to be $\mathcal{Q}$-pure if $\mu_p(E) \subseteq F$.*   □

Note that the concepts of pure and n-pure extension are particular cases of this more general notion: an extension is pure precisely when it is $\mathcal{P}$-pure, and n-pure when it is $\mathcal{P}_n$-pure.

Recall that for an arbitrary torsion group T we have denoted in Section 1.4 by $\mathcal{O}_T$ the set $\{\, \mathrm{ord}(x) \mid x \in T \,\}$.

For any G-radical extension E/F we shall use throughout this chapter the following notation:

$$\mathcal{P}_G := \mathcal{P} \cap \mathcal{O}_{G/F^*}.$$

Note that when the G-radical extension E/F is n-bounded (see Definition 5.1.1), then $\mathcal{O}_G = \mathbb{D}_n$, hence $\mathcal{P}_G = \mathcal{P} \cap \mathbb{D}_n = \mathcal{P}_n$.

Next, we provide a purity criterion for arbitrary G-Cogalois extensions, which generalizes to infinite extensions the n-Purity Criterion (see Theorem 4.3.2).

THEOREM 12.1.4 (THE GENERAL PURITY CRITERION). *Let E/F be an arbitrary G-radical separable extension. Then, the following assertions are equivalent.*

(1) E/F *is G-Cogalois.*

(2) E/F *is $\mathcal{P}_G$-pure.*

PROOF. (1) $\implies$ (2): Suppose that the extension E/F is G-Cogalois and let p be an odd prime in $\mathcal{P}_G$. By definition, there exists $g_p \in G$ with $\mathrm{ord}(\widehat{g_p}) = p$ in the quotient group G/F^*. Continue as in the proof of implication (2) $\implies$ (7) in Theorem 4.3.2 to deduce that $\zeta_p \in F$.

Now suppose that $4 \in \mathcal{P}_G$ and $\zeta_4 \in E \setminus F$. Then necessarily $\mathrm{Char}(F) \neq 2$, and there exists $g_4 \in G$ with $\mathrm{ord}(\widehat{g_4}) = 4$ in the quotient group G/F^*. As in the proof of implication $(2) \implies (7)$ in Theorem 4.3.2 one deduces that $\zeta_4 \in G$. Denote $K = F((1 + \zeta_4)g_4)$. Since the extension E/F is strongly G-Kneser, E/K is K^*G-Kneser. But $1 + \zeta_4 = (1 + \zeta_4)g_4 g_4^{-1} \in K^*G$, hence $\zeta_4 \in K$ by the Infinite Kneser Criterion (Theorem 11.1.5). Now continue as in the proof of implication $(2) \implies (7)$ in Theorem 4.3.2 to deduce that $\zeta_4 \in F^*$. But this contradicts our initial assumption that $\zeta_4 \notin F$. Consequently, E/F is $\mathcal{P}_G$-pure.

$(2) \implies (1)$: Assume that E/F is $\mathcal{P}_G$-pure, and let K be an arbitrary subfield of E containing F. We are going to show that K/F is $K^* \cap G$-Kneser. To do that, we will use the Infinite Kneser Criterion cited above.

Let p be an odd prime such that $\zeta_p \in K^* \cap G$. Since $\zeta_p^p = 1$, we deduce that either $\mathrm{ord}(\widehat{\zeta_p}) = 1$ or $\mathrm{ord}(\widehat{\zeta_p}) = p$. If $\mathrm{ord}(\widehat{\zeta_p}) = 1$ we have $\zeta_p \in F$, as desired. If $\mathrm{ord}(\widehat{\zeta_p}) = p$, it follows that $p \in \mathcal{P}_G$ since $\zeta_p \in G$, hence $\zeta_p \in F$ by $\mathcal{P}_G$-purity.

Now suppose that $1 + \zeta_4 \in K^* \cap G$, and set $g = 1 + \zeta_4$. Then necessarily $\mathrm{Char}(F) \neq 2$, for otherwise, it would follow that $\zeta_4 = 1$, hence $1 + \zeta_4 = 0 \in K^* \cap G$, which is a contradiction.

Set $m = \mathrm{ord}(\widehat{g})$. Since $g^4 = -4 \in F^*$ one has $m \in \{1, 2, 4\}$. If $m = 1$ then $1 + \zeta_4 \in F^*$, hence $\zeta_4 \in F$. If $m = 2$ then $g^2 = 2\zeta_4 \in F^*$, hence $\zeta_4 \in F$ since $\mathrm{Char}(F) \neq 2$. If $m = 4$ then $4 \in \mathcal{P}_G$. But $1 + \zeta_4 \in K^* \cap G \subseteq E$ implies that $\zeta_4 \in E$, hence $\zeta_4 \in F$ by $\mathcal{P}_G$-purity.

Thus E/F is strongly G-Kneser, and since it is also separable, we conclude that E/F is G-Cogalois. The proof is complete. $\qquad \square$

REMARK 12.1.5. The proof of the implication $(2) \implies (1)$ in Theorem 12.1.4 shows that any separable G-radical extension E/F with $\mathcal{P}_G = \varnothing$ is G-Cogalois. Notice that $\mathcal{P}_G = \varnothing$ if and only if either G/F^* is a group of exponent 2 or $G = F^*$. $\qquad \square$

Theorem 12.1.4, one of the key results in the Infinite Cogalois Theory, has many consequences. First, we will show that the property of an extension being G-Cogalois behaves nicely with respect to directed unions. This explains why most results from Finite Cogalois Theory also hold for infinite extensions.

PROPOSITION 12.1.6. *Let E/F be an extension. Assume that $E = \bigcup_{\alpha \in I} E_\alpha$ is a directed union of a family of intermediate fields E_α of E/F such that every extension E_α/F is G_α-radical and $G = \bigcup_{\alpha \in I} G_\alpha$ is a* *directed union of subgroups* G_α. *Then* E/F *is* G-*Cogalois if and only if* E_α/F *is* G_α-*Cogalois for every* $\alpha \in I$.

PROOF. Assume that E/F is G-Cogalois. Then E_α/F is $G \cap E_\alpha$-Cogalois for every $\alpha \in I$ by Proposition 12.1.2 (1). Observe that $G \cap E_\alpha = G \cap F(G_\alpha) = G_\alpha$ by Lemma 11.2.1.

Conversely, assume that every E_α/F is G_α-Cogalois. Note that E/F is a separable G-radical extension. Let $p \in \mathcal{P}_G$. Then, there exists $g_p \in G$ such that $\mathrm{ord}(\widehat{x_p}) = p$. Since $G = \bigcup_{\alpha \in I} G_\alpha$ we deduce that $g_p \in G_\beta$ for some $\beta \in I$. Assume that $\zeta_p \in E$. Then $\zeta_p \in E_\gamma$ for some $\gamma \in I$. Since $G = \bigcup_{\mu \in I} G_\mu$ is a directed union, there exists $\alpha \in I$ such that $E_\gamma \subseteq E_\alpha$ and $G_\beta \subseteq G_\alpha$. Then $\zeta_p \in E_\alpha$ and $p \in \mathcal{P}_{G_\alpha}$. But E_α/F is G_α-Cogalois, hence $\zeta_p \in F$ by the General Purity Criterion (Theorem 12.1.4). Thus, E/F is $\mathcal{P}_G$-pure, hence it is G-Cogalois again by the same criterion.   □

Another application of the General Purity Criterion is the determination of the structure of the group G/F^* for an arbitrary G-Cogalois extension which generalizes the results of Section 4.4 established for finite G-Cogalois extensions. To do that we need the next result which is the infinite variant of Lemma 6.1.6.

LEMMA 12.1.7. *Let* E/F *be a* G-*Cogalois extension, and let* $x \in E^*$ *be such that* $x^m \in F$ *for some* $m \in \mathbb{N}^*$. *Suppose that one of the following two conditions is satisfied.*

(1) $\mathcal{P}_m \subseteq \mathcal{P}_G$.
(2) $\mu_m(E) \subseteq F$ *(in particular, this holds if* $\zeta_m \in F$*).*
Then, we have $F(x) \subseteq E \Longleftrightarrow x \in G$.

PROOF. Suppose that $F(x) \subseteq E$. Set $K = F(x)$ and $H = F^*\langle x \rangle$. If $k = \mathrm{ord}(\widehat{x})$, then clearly $k \mid m$ and $k = \exp(H/F^*)$. Let $p \in \mathcal{P}_k$. Then $p \mid m$, hence $p \in \mathcal{P}_m \subseteq \mathcal{P}_G$ if condition (1) is satisfied. We deduce that $\mu_p(K) \subseteq \mu_p(E) \subseteq F$ since E/F is $\mathcal{P}_G$-pure by Theorem 12.1.4. If condition (2) is satisfied, then we have $\mu_p(K) \subseteq \mu_p(E) \subseteq \mu_m(E) \subseteq F$. Observe that K is a finite H-radical extension, hence, in both cases, K/F is a H-Cogalois extension by Theorem 4.3.2. But, the finite extension K/F is also $G \cap K^*$-Cogalois by Proposition 4.3.5 (1), so $H = G \cap K^*$ by Corollary 4.4.2. Thus, $x \in G$, as desired. Note that the result holds even if $\mathcal{P}_m = \varnothing$. The other implication is obvious.   □

Recall that if A is a multiplicative group with identity element e, then for every $p \in \mathbb{P}$ we have denoted by $t_p(A)$ the p-primary component of A.

Also, recall that for any extension E/F we have denoted by $\mathrm{Cog}(E/F)$ the torsion subgroup $t(E^*/F^*)$ of the quotient group E^*/F^*, and by $\mathrm{Cog}_2(E/F)$ the subgroup of $\mathrm{Cog}(E/F)$ consisting of all its elements of order $\leqslant 2$.

THEOREM 12.1.8. *The following statements hold for a separable G-Kneser extension E/F.*

(1) *Assume that $4 \in \mathcal{P}_G$. Then E/F is G-Cogalois if and only if*

$$G/F^* = \Big(\bigoplus_{p \in \mathcal{P}_G \setminus \{4\}} t_p(\mathrm{Cog}(E/F)) \Big) \bigoplus t_2(\mathrm{Cog}(E/F)).$$

(2) *Assume that $4 \notin \mathcal{P}_G$. Then E/F is G-Cogalois if and only if*

$$G/F^* = \Big(\bigoplus_{p \in \mathcal{P}_G} t_p(\mathrm{Cog}(E/F)) \Big) \bigoplus \mathrm{Cog}_2(E/F).$$

PROOF. (1) "$\Longrightarrow$": Clearly $t_p(G/F^*) = 0$ for all $p \in \mathbb{P} \setminus \mathcal{P}_G$, so

$$G/F^* = \Big(\bigoplus_{p \in \mathcal{P}_G \setminus \{4\}} t_p(G/F^*) \Big) \bigoplus t_2(G/F^*).$$

Using Lemma 12.1.7 we will show that

$$t_p(G/F^*) = t_p(\mathrm{Cog}(E/F))$$

for all $p \in (\mathcal{P}_G \setminus \{4\}) \cup \{2\}$, which will imply the desired equality

$$G/F^* = \Big(\bigoplus_{p \in \mathcal{P}_G \setminus \{4\}} t_p(\mathrm{Cog}(E/F)) \Big) \bigoplus t_2(\mathrm{Cog}(E/F)).$$

So, let $p \in (\mathcal{P}_G \setminus \{4\}) \cup \{2\}$, and let $x_p \in E^*$ with $\widehat{x_p} \in t_p(\mathrm{Cog}(E/F))$. Then $x_p^{p^s} \in F$ for some $s \in \mathbb{N}$. We can assume that $s \geqslant 2$. If p is odd then

$$\mathcal{P}_{p^s} = \mathcal{P} \cap \mathbb{D}_{p^s} = \{p\} \subseteq \mathcal{P}_G,$$

and if $p = 2$ then

$$\mathcal{P}_{2^s} = \mathcal{P} \cap \mathbb{D}_{2^s} = \{4\} \subseteq \mathcal{P}_G.$$

Now apply Lemma 12.1.7 to conclude that $x_p \in G$. Consequently

$$t_p(\mathrm{Cog}(E/F)) \subseteq t_p(G/F^*).$$

The other inclusion is obvious.

"$\Longleftarrow$": Let $p \in \mathcal{P}_G \setminus \{4\}$ and assume that $\zeta_p \in E$. Then $\zeta_p^p = 1$, hence $\widehat{\zeta_p}^{\,p} \in t_p(\mathrm{Cog}(E/F))$, and so $\widehat{\zeta_p} \in G/F^*$. Thus, $\zeta_p \in G$. But E/F

is G-Kneser by hypothesis, so by the Infinite Kneser Criterion (Theorem 11.1.5) we deduce that $\zeta_p \in F$.

By assumption, $4 \in \mathcal{P}_G$. If $\zeta_4 \in E$ and $\mathrm{Char}(F) \neq 2$, then $\widehat{1 + \zeta_4} \in t_2(\mathrm{Cog}(E/F)) \subseteq G/F^*$ since $(1 + \zeta_4)^4 = -4$, and so, $1 + \zeta_4 \in G$. Again by the Infinite Kneser Criterion, we deduce that $\zeta_4 \in F$. If $\mathrm{Char}(F) = 2$, then $\zeta_4 = 1 \in F$. This shows that E/F is $\mathcal{P}_G$-pure. Now apply Theorem 12.1.4 to conclude that E/F is G-Cogalois.

(2) "$\Longrightarrow$": As in case (1) one shows that the equality

$$t_p(G/F^*) = t_p(\mathrm{Cog}(E/F))$$

holds for any (odd) prime $p \in \mathcal{P}_G$.

The hypothesis $4 \notin \mathcal{P}_G$ implies that G/F^* contains no element of order 4, hence $t_2(G/F^*)$ is exactly the set of all elements of G/F^* of order $\leqslant 2$.

Let $x_2 \in E^*$ with $\widehat{x_2} \in \mathrm{Cog}_2(G/F^*)$. Then $x_2^2 \in F$, hence

$$\mathcal{P}_2 = \mathcal{P} \cap \mathbb{D}_2 = \varnothing \subseteq \mathcal{P}_G.$$

Apply Lemma 12.1.7 to conclude that $x \in G$. Therefore

$$t_2(G/F^*) = \mathrm{Cog}_2(G/F^*),$$

and consequently

$$G/F^* = \Big(\bigoplus_{p \in \mathcal{P}_G} t_p(\mathrm{Cog}(E/F)) \Big) \bigoplus \mathrm{Cog}_2(E/F).$$

"$\Longleftarrow$": Proceed as in case (1). $\square$

REMARK 12.1.9. The condition that E/F is a G-Kneser extension in Theorem 12.1.8 cannot be dropped since, otherwise, E/F may not be G-Cogalois as the example in Remarks 4.4.4 (1) shows. $\square$

The next result generalizes Corollary 4.4.2 to extensions which are not necessarily finite.

THEOREM 12.1.10. *Let E/F be an extension which is simultaneously G-Cogalois and H-Cogalois. Then $G = H$.*

PROOF. Let $x \in G$ be arbitrary, and set $K = F(x)$, $G' = F^*\langle x \rangle$. Since the extension E/F is H-Cogalois it follows by Theorem 11.2.6 that $K = F(H')$ for some H' with $F^* \leqslant H' \leqslant H$.

By Proposition 12.1.2 (1), the finite extension K/F is simultaneously $K^* \cap G$-Cogalois and also $K^* \cap H$-Cogalois. But $K^* \cap G = F(G') \cap G = G'$ and $K^* \cap H = F(H') \cap H = H'$ by Lemma 11.2.1. Consequently, the finite extension K/F is simultaneously G'-Cogalois and H'-Cogalois, and according to Corollary 4.4.2 it follows that $G' = H'$. Since $x \in G'$, we deduce that $x \in H'$, so $x \in H$. Hence $G \subseteq H$. The proof of the inverse inclusion $H \subseteq G$ is similar. $\square$

In view of Theorem 12.1.10, the group G of any G-Cogalois extension, finite or not, is uniquely determined. So, this leads to the following definition.

DEFINITION 12.1.11. *If E/F is a G-Cogalois extension, then the group G/F^* is called the* Kneser group *of the extension E/F and is denoted by* $\mathrm{Kne}(E/F)$. $\square$

Next, we particularize the General Purity Criterion (Theorem 12.1.4) we established for arbitrary G-Cogalois extensions to bounded G-Cogalois extensions. Recall that a G-radical extension E/F, which is not necessarily finite, is said to be bounded if G/F^* is a group of bounded order; in this case, if $\exp(G/F^*) = n$, one says that E/F is an n-bounded extension.

THEOREM 12.1.12 (THE INFINITE n-PURITY CRITERION). *The following assertions are equivalent for a separable n-bounded G-radical extension E/F.*

(1) E/F *is G-Cogalois.*
(2) E/F *is n-pure.*

PROOF. Since $\exp(G/F^*) = n$, we have $\mathcal{O}_{G/F^*} = \mathbb{D}_n$ by Remark 1.4.9, hence

$$\mathcal{P}_G = \mathcal{P} \cap \mathcal{O}_{G/F^*} = \mathcal{P} \cap \mathbb{D}_n = \mathcal{P}_n.$$

Now apply the General Purity Criterion. $\square$

EXAMPLE 12.1.13. Let $\{p_1, \ldots, p_k, \ldots\}$ be an infinite set of positive prime numbers, and set $E = \mathbb{Q}(\sqrt[n]{p_1}, \ldots, \sqrt[n]{p_k}, \ldots)$. We claim that the infinite extension $E/\mathbb{Q}$ is G-Cogalois, where $G = \mathbb{Q}^* \langle \sqrt[n]{p_1}, \ldots, \sqrt[n]{p_k}, \ldots \rangle$. Indeed, observe first that $E/\mathbb{Q}$ is clearly a G-radical extension, and the group $G/\mathbb{Q}^*$ has exponent n. Next, the extension $E/\mathbb{Q}$ is pure since $\mathbb{Q}(\sqrt[n]{p_1}, \ldots, \sqrt[n]{p_k}, \ldots) \subseteq \mathbb{R}$, and so, a fortiori, it is n-pure.

Thus, $E/\mathbb{Q}$ is an infinite G-Cogalois extension by Theorem 12.1.12. In particular, the lattice of all its subextensions, that is, the lattice of all subfields of E, can be easily understood if we know how the subgroups of its Kneser group $G/\mathbb{Q}^*$ look. If $n = 2$, the Abelian group $G/\mathbb{Q}^*$ has a canonical structure of vector space over the field $\mathbb{F}_2$ with 2 elements, and any of its subgroups is a vector subspace of $G/\mathbb{Q}^*$. Consequently, any subgroup of $G/\mathbb{Q}^*$ has a vector space basis over $\mathbb{F}_2$. This can help to describe effectively all the subfields of E. See also Exercise 7 for a more general case.   □

12.2. Infinite Cogalois extensions

As we have noticed in Examples 2.1.15 (2), a finite extension E/F is Cogalois precisely when it is $T(E/F)$-Kneser. Since we already have the concept of G-Kneser extension defined for extensions which are not necessarily finite, it is natural to use "Cogalois" to refer to an arbitrary extension E/F if it is $T(E/F)$-Kneser.

In this section we generalize, to infinite extensions, most of the results we established in Sections 3.1 and 3.2 for finite Cogalois extensions.

DEFINITION 12.2.1. *An extension E/F, which is not necessarily finite is said to be* Cogalois *if it is $T(E/F)$-Kneser.*

The next result is the infinite variant of the Greither-Harrison Criterion (Theorem 3.1.7).

THEOREM 12.2.2 (THE INFINITE GREITHER-HARRISON CRITERION). *The following assertions are equivalent for an arbitrary extension E/F.*

(1) *E/F is Cogalois.*
(2) *E/F is radical, separable, and pure.*

PROOF. (1) $\Longrightarrow$ (2): Suppose that E/F is a Cogalois extension, i.e., a $T(E/F)$-Kneser extension. In particular, E/F is a $T(E/F)$-radical extension, hence a radical extension.

Now we show that the extension E/F is separable. It is sufficient to prove that for any $x \in T(E/F)$, the extension $F(x)/F$ is separable. So, let $x \in T(E/F)$ be fixed, and set $H = F^*\langle x \rangle$ and $K = F(x) = F(H)$. Since the extension E/F is $T(E/F)$-Kneser and $H \leqslant T(E/F)$, we deduce that the finite extension $F(H)/F$ is H-Kneser by Proposition 11.1.2. Then, by Proposition 11.2.2, it follows that the extension K/F is $K^* \cap T(E/F) = T(K/F)$-Kneser, i.e., K/F is a Cogalois extension. Hence K/F is a finite Cogalois extension, so it is separable, according to the Greither-Harrison Criterion (Theorem 3.1.7) for finite extensions. Consequently, any element of K, in particular x, is separable over F, as desired.

In order to show that the extension E/F is pure, proceed exactly as in the proof of Theorem 3.1.7.

(2) $\implies$ (1): Assume that the extension E/F is radical, separable and pure. We are going to prove that the extension E/F is Cogalois, i.e., $T(E/F)$-Kneser.

Since E/F is a radical extension, it follows that E/F is also a $T(E/F)$-radical extension. For every odd prime p we have $\mu_p(T(E/F)) \subseteq \mu_p(E)$, and, by purity, $\mu_p(E) = \mu_p(F)$, so $\mu_p(T(E/F)) = \mu_p(F)$. If $1 \pm \zeta_4 \in T(E/F)$, then $1 \pm \zeta_4 \in E$, so $\zeta_4 \in E$, i.e., $\zeta_4 \in \mu_4(E)$. By purity, $\mu_4(E) \subseteq \mu_4(F)$, hence $\zeta_4 \in F$.

Since E/F is a separable extension, by the Infinite Kneser Criterion (Theorem 11.1.5) we deduce that the extension E/F is $T(E/F)$-Kneser, i.e., it is a Cogalois extension. $\qquad\Box$

The next result is the infinite variant of Theorem 7.5.1.

THEOREM 12.2.3. *Any Cogalois extension E/F is $T(E/F)$-Cogalois.*

PROOF. In view of the Infinite Greither-Harrison Criterion, the radical extension E/F is separable and pure, so, a fortiori, $\mathcal{P}_{T(E/F)}$-pure. Thus, E/F is $T(E/F)$-Cogalois by Theorem 12.1.4. $\qquad\Box$

An easy consequence of Theorems 12.2.2 and 12.2.3 is the following generalization of Proposition 3.2.2 and Theorem 3.2.3 from finite Cogalois extensions to arbitrary Cogalois extensions.

THEOREM 12.2.4. *The following statements hold for an arbitrary Cogalois extension E/F.*

(1) *The maps $- \cap T(E/F) : \mathcal{E} \longrightarrow \mathcal{C}$ and $F(-) : \mathcal{C} \longrightarrow \mathcal{E}$ are isomorphisms of lattices, inverse to one another, where $\mathcal{E} = \{ K \mid F \subseteq K, \ K \ subfield \ of \ E \}$ and $\mathcal{C} = \{ H \mid F^* \leqslant H \leqslant T(E/F) \}$.*

(2) *For every intermediate field $K \in \mathcal{E}$ one has $K = F(T(K/F))$.*

(3) *For every subgroup $H \in \mathcal{C}$ one has $\mathrm{Cog}(F(H)/F) = H/F^*$.*

(4) *For every intermediate field K of the extension E/F, E/K and K/F are both Cogalois extensions, and there exists a canonical group isomorphism*

$$\mathrm{Cog}(E/K) \cong \mathrm{Cog}(E/F)/\mathrm{Cog}(K/F).$$

PROOF. (1) Since $\mathrm{Cog}(E/F) = T(E/F)/F^*$, $\mathcal{C}$ is canonically isomorphic to the lattice $\underline{\mathrm{Subgroups}}(\mathrm{Cog}(E/F))$. The result follows immediately from Theorem 12.2.3 and Theorem 11.2.6.

(2) and (3) are consequences of (1) and of the following simple fact: for any $H \in \mathcal{C}$ one has $H = F(H) \cap T(E/F) = T(F(H)/F)$.

(4) By Theorem 12.2.3, the extension E/F is $T(E/F)$-Cogalois hence it is strongly $T(E/F)$-Kneser. It follows that K/F is $T(E/F) \cap K$-Kneser. But $T(E/F) \cap K = T(K/F)$, hence K/F is $T(K/F)$-Kneser. This means precisely that K/F is a Cogalois extension. By Theorem 12.2.2, E/F is radical, separable and pure. Then clearly E/K is also radical, separable and pure. Again by Theorem 12.2.2, we conclude that E/K is a Cogalois extension.

It remains only to prove the isomorphism

$$\operatorname{Cog}(E/K) \cong \operatorname{Cog}(E/F)/\operatorname{Cog}(K/F).$$

Since E/K is Cogalois, it is $T(E/K)$-Cogalois by Theorem 12.2.3. On the other hand, since E/F is $T(E/F)$-Cogalois, it follows by Proposition 12.1.2 (2) that E/K is $T(E/F)K^*$-Cogalois. Using Theorem 12.1.10 we obtain the equality $T(E/K) = T(E/F)K^*$. Hence

$$
\begin{aligned}
\operatorname{Cog}(E/K) &= T(E/K)/K^* = (T(E/F)K^*)/K^* \\
&\cong T(E/F)/(T(E/F) \cap K^*) = T(E/F)/T(K/F) \\
&\cong (T(E/F)/F^*)/(T(K/F)/F^*) \\
&= \operatorname{Cog}(E/F)/\operatorname{Cog}(K/F),
\end{aligned}
$$

and we are done. $\square$

PROPOSITION 12.2.5. *Let E/F be a radical extension, and let K be an intermediate field of E/F. If the extensions E/K and K/F are both Cogalois, then E/F is also a Cogalois extension.*

PROOF. Since E/K and K/F are separable extensions, so is E/F. Since E/K and K/F are pure, so is also E/F by Proposition 3.1.5. But E/F is a radical extension by hypothesis, hence E/F is Cogalois by Theorem 12.2.2. $\square$

EXAMPLES 12.2.6. (1) Proposition 3.2.6 shows that the result in Proposition 12.2.5 may fail if the extension E/F is not supposed to be radical.

(2) Consider the extension E_r/F discussed in Remark 5.3.2, where $E_r = \mathbb{Q}(\theta_r)$, $F = \mathbb{Q}$, and

$$\theta_r = \underbrace{\sqrt{2 + \sqrt{2 + \sqrt{2 + \ldots + \sqrt{2}}}}}_{r \text{ radicals}}.$$

Let $E_\infty = \bigcup_{r \geqslant 1} E_r$ be the directed union of subfields E_r of $\mathbb{R}$. Then E_∞/F is a Galois extension of infinite degree by Exercise 3, Chapter 11. We claim that the extension E_∞/F is not radical. Indeed, if it would be radical, then, since it is pure, it would be Cogalois by Theorem 12.2.2. Consequently, by Theorem 12.2.4 (4), any of its subextensions, in particular E_2/F, is Cogalois, which contradicts Proposition 5.3.1 (e).

Observe that the field E_∞ is a subfield of the maximal 2-primary Abelian extension of $\mathbb{Q}$ contained in $\mathbb{C}$:

$$\mathbb{Q}^{2,ab} := \bigcup_{r \geqslant 1} \mathbb{Q}(\zeta_{2^r}).$$

Since the extension $\mathbb{Q}(\zeta_{2^r})/\mathbb{Q}$ is $\mathbb{Q}^* \langle \zeta_{2^r} \rangle$-Kneser for every $r \geqslant 1$, it follows that the infinite Abelian extension $\mathbb{Q}^{2,ab}/\mathbb{Q}$ is $\mathbb{Q}^* \langle \{ \zeta_{2^r} \, | \, r \geqslant 1 \} \rangle$-Kneser. However, its subextension $E_\infty/\mathbb{Q}$ is not Kneser. $\square$

12.3. Exercises to Chapter 12

1. Prove that every separable 2-bounded G-radical extension is G-Cogalois.

2. Prove that a G-radical extension E/F is G-Cogalois if and only if every finite subextension K/F of E/F is $G \cap K$-Cogalois.

3. Let E/F be a G-Cogalois extension, and let $x \in T(E/F)$ with $m = \operatorname{ord}(\widehat{x})$. Show that $x \in G$ if and only if $\mathcal{P}_m \subseteq \mathcal{P}_G$.

4. Prove that the following statements hold for a separable n-bounded G-Kneser extension E/F.

   (1) Assume that $n \not\equiv 2 \pmod 4$. Then E/F is G-Cogalois if and only if

   $$G/F^* = \bigoplus_{p \in \mathbb{P}_n} t_p(\operatorname{Cog}(E/F)).$$

   (2) Assume that $n \equiv 2 \pmod 4$. Then E/F is G-Cogalois if and only if

   $$G/F^* = \Big(\bigoplus_{p \in \mathbb{P}_n \setminus \{2\}} t_p(\operatorname{Cog}(E/F)) \Big) \bigoplus \operatorname{Cog}_2(E/F).$$

5. Let E/F be a G-radical pure separable extension. Prove that E/F is simultaneously Cogalois and G-Cogalois, and $\mathrm{Cog}(E/F) = G/F^*$. (*Hint*: Adapt the proof of Corollary 4.4.3.)

6. Let $\{p_1, \ldots, p_k, \ldots\}$ be an infinite set of positive prime numbers, let $n \in \mathbb{N}$, $n \geqslant 2$, and set $E = \mathbb{Q}(\sqrt[n]{p_1}, \ldots, \sqrt[n]{p_k}, \ldots)$. Prove that

$$\mathrm{Cog}(E/\mathbb{Q}) = \mathbb{Q}^* \langle \sqrt[n]{p_1}, \ldots, \sqrt[n]{p_k}, \ldots \rangle / \mathbb{Q}^*.$$

(*Hint*: Use Exercise 5.)

7. Describe effectively all the subfields of the field E considered in Exercise 6.

8. (*The Infinite Gay-Vélez Criterion*). Prove that the following statements are equivalent for an extension E/F which is not necessarily finite.

   (1) E/F is a Cogalois extension.
   (2) E/F is radical, separable, and $\zeta_{2p} \notin E \setminus F$ for every $p \in \mathbb{P}$.
   (*Hint*: Adapt the proof of Theorem 3.1.9, by using the Infinite Greither-Harrison Criterion and Lemma 3.1.2.)

9. Let E/F be a separable n-bounded G-radical extension satisfying the following condition:

   $(*)$ $\mu_p(E) \subseteq F$ for any odd $p \in \mathbb{P}_n$, and
   $\mu_4(E) \subseteq F$ whenever n is even.

   Prove that E/F is G-Cogalois, and $G/F^* = \bigoplus_{p \in \mathbb{P}_n} t_p(\mathrm{Cog}(E/F))$. (*Hint*: Observe that the condition $\mu_4(E) \subseteq F$ for n even implies that $\mathrm{Cog}_2(G/F^*) = t_2(\mathrm{Cog}(E/F))$. Then use Exercise 4.)

10. Let E_1/F and E_2/F be Cogalois extensions such that their compositum E_1E_2/F is pure. Prove that E_1E_2/F is a Cogalois extension and that the following assertions are equivalent.

    (a) $T(E_1/F) \cap T(E_2/F) = F^*$.
    (b) The canonical map

$$\alpha : \mathrm{Cog}(E_1/F) \times \mathrm{Cog}(E_2/F) \longrightarrow \mathrm{Cog}(E_1E_2/F)$$

    defined as $\alpha(x_1 F^*, x_2 F^*) = x_1 x_2 F^*$, $x_1 \in E_1$, $x_2 \in E_2$, is a monomorphism of groups.
    (c) The canonical map α in (b) is an isomorphism of groups.
    (d) The fields E_1 and E_2 are linearly disjoint over F.
    (e) $E_1 \cap E_2 = F$.

11. Let $E_r = \mathbb{Q}(\cos(\pi/2^{r+1}))$, $r \in \mathbb{N}^*$, and $E_\infty = \bigcup_{r \geqslant 1} E_r$. Show that $\mathrm{Cog}\,(E_\infty/\mathbb{Q}) = \{\,\widehat{1}, \widehat{\sqrt{2}}\,\}$.

12. Let $r \in \mathbb{N}$, $r \geqslant 2$, $\nu_r = \underbrace{\sqrt{1 + \sqrt{1 + \sqrt{1 + \ldots + \sqrt{2}}}}}_{r\ \text{radicals}}$, $F_r = \mathbb{Q}(\nu_r)$,

and $F_\infty = \bigcup_{r \geqslant 1} F_r$. Prove that the non Galois extension of infinite degree $F_\infty/\mathbb{Q}$ (see Exercise 4, Chapter 11) is neither radical, nor Kneser, nor Cogalois.

13. With notation of Exercise 12, calculate $\mathrm{Cog}\,(F_\infty/\mathbb{Q})$.

14. The concept of Γ-Clifford-Cogalois can be obviously defined as in Definition 10.1.8 also for field extensions which are not necessarily finite. Prove that the following statements hold for a field extension E/F.
    (1) If E/F is G-Cogalois for some group G with $F^* \leqslant G \leqslant E^*$, then E/F is a G/F^*-Clifford-Cogalois extension.
    (2) Conversely, if E/F is a Γ-Clifford-Cogalois extension for some group Γ, then there exists a uniquely determined group G such that $F^* \leqslant G \leqslant E^*$, $\Gamma \cong G/F^*$, E/F is G-Cogalois, and $U^h(E) = G$.

15. Let E/F be a field extension which is simultaneously Γ-Clifford-Cogalois and Δ-Clifford-Cogalois. Prove that the groups Γ and Δ are isomorphic.

16. Prove that the following assertions are equivalent for a radical field extension E/F.
    (a) E/F is a Cogalois extension.
    (b) E is a Galois $F[\,\mathrm{Cog}(E/F)\,]$-object via the comodule structure given by the F-linear map $E \longrightarrow E \otimes_F F[\,\mathrm{Cog}(E/F)\,]$, $x \mapsto x \otimes \widehat{g}$, for all $x \in Fg$ and $g \in T(E/F)$.

12.4. Bibliographical comments to Chapter 12

Section 12.1. The concept of infinite G-Cogalois extension was introduced Albu and Ţena [**25**]. The General Purity Criterion (Theorem 12.1.4) and Theorem 12.1.8 giving the structure of the group G/F^* of a G-Cogalois extension are due to Albu [**8**]. The uniqueness of the Kneser group of a

G-Cogalois extension was established by Albu and Ţena [**25**].

Section 12.2. The notion of infinite Cogalois extension, as well as the infinite variant of the Greither-Harrison are due to Albu and Ţena [**25**]. The properties and examples of infinite Cogalois extensions presented in this section are taken from Albu [**8**].

CHAPTER 13

INFINITE KUMMER THEORY

The prototype of an infinite G-Cogalois extension is, by *Kummer Theory*, any *infinite classical Kummer extension*. In this section we show that the essential part of the Infinite Kummer Theory can be immediately deduced from our Cogalois Theory using the Infinite n-Purity Criterion given in Section 12.1. Moreover, this criterion allows us to provide large classes of infinite G-Cogalois extensions which generalize or are closely related to classical infinite Kummer extensions: *infinite generalized Kummer extensions, infinite Kummer extensions with few roots of unity,* and *infinite quasi-Kummer extensions.* The prototype of an infinite Kummer extensions with few roots of unity is any subextension of $\mathbb{R}/\mathbb{Q}$ of the form $\mathbb{Q}(\{\sqrt[n]{a_i} \mid i \in I\})/\mathbb{Q}$, where $\{a_i \mid i \in I\}$ is an arbitrary nonempty set of strictly positive rational numbers. Notice that, in general, these extensions are not Galois if $n \geqslant 3$.

Placing classical infinite Kummer extensions in the framework of infinite G-Cogalois extensions allows us to derive, from the Infinite Cogalois Theory, results on infinite generalized Kummer extensions. In particular, we obtain results on infinite Kummer extensions with few roots of unity, which are very similar to the known ones for infinite classical Kummer extensions.

13.1. Infinite classical Kummer extensions

In this section we present as an immediate application of the Infinite n-purity Criterion the basic properties of infinite classical Kummer extensions.

Recall that Ω is a fixed algebraically closed field containing F as a subfield; any considered overfield of F is supposed to be a subfield of Ω.

As in Section 7.1, for any nonempty subset A of F^* and any $n \in \mathbb{N}^*$ we will denote by $\sqrt[n]{A}$ the subset of $T(\Omega/F)$ defined by

$$\sqrt[n]{A} := \{\, x \in \Omega \mid x^n \in A \,\}.$$

LEMMA 13.1.1. *Let E/F be a bounded separable G-radical extension, and let $n \in \mathbb{N}^*$ be such that $G^n \subseteq F^*$. If the extension E/F is n-pure, then E/F is G-Cogalois.*

PROOF. Proceed exactly as in the proof of Lemma 7.1.4, but apply Theorem 12.1.12 instead of Theorem 4.3.2.   □

The next result essentially shows that the main part of Kummer Theory can be very easily derived from the Infinite Cogalois Theory.

THEOREM 13.1.2. *Let E/F be a classical n-Kummer extension, with $E = F(\sqrt[n]{A})$, where $n \in \mathbb{N}^*$ and $\varnothing \neq A \subseteq F^*$. Then, the following assertions hold.*

(1) *E/F is an $F^*\langle \sqrt[n]{A} \rangle$-Cogalois extension.*

(2) *The maps $H \mapsto F(\sqrt[n]{H})$ and $K \mapsto K^n \cap (F^{*n}\langle A \rangle)$ establish isomorphisms of lattices, inverse to one another, between the lattice of all subgroups H of $F^{*n}\langle A \rangle$ containing F^{*n} and the lattice of all intermediate fields K of E/F. Moreover, any subextension K/F of E/F is a classical n-Kummer extension.*

(3) *If H is any subgroup of $F^{*n}\langle A \rangle$ containing F^{*n}, then any set of representatives of the group $\sqrt[n]{H}/F^*$ is a vector space basis of $F(\sqrt[n]{H})$ over F, and $[\,F(\sqrt[n]{H}) : F\,] = |H/F^{*n}|$.*

PROOF. Proceed as in the proof of Theorem 7.1.6 using the n-bounded infinite variants of the involved results (e.g., Lemma 13.1.1 instead of Lemma 7.1.4, Theorem 12.1.12 instead of Theorem 4.3.2, etc.).   □

Let F be any field, and let $n \in \mathbb{N}^*$ be such that $\gcd(n, e(F)) = 1$ and $\mu_n(\Omega) \subseteq F$; these conditions mean precisely that $\zeta_n \in F$ and $\mathrm{ord}(\zeta_n) = n$ in Ω^*. In view of Theorem 7.1.2, any classical n-Kummer extension E of F, with E a subfield of Ω, has the form $F(\sqrt[n]{A})$ for some $\varnothing \neq A \subseteq F^*$. Since obviously $F(\sqrt[n]{F^*})/F$ is a classical n-Kummer extension, it follows that the set $\mathcal{K}_{n,\Omega}$ of all classical n-Kummer extensions of F contained in Ω, ordered by inclusion, has a greatest element, namely $F(\sqrt[n]{F^*})$. The field $F(\sqrt[n]{F^*})$ is called the *maximal classical n-Kummer extension of F contained in Ω*; clearly, it is the greatest Abelian extension of F contained in Ω for which its Galois group is a group of exponent a divisor of n. It will be denoted in the sequel by $F_{n,\Omega}^{ab}$, or more simply, by F_n^{ab}.

Taking $A = F^*$ in Theorem 13.1.2 (2) we obtain the next result, which is the Kummer Theory for the maximal classical n-Kummer extension F_n^{ab} of F.

COROLLARY 13.1.3. *Let F be any field and let $n \in \mathbb{N}^*$ be such that* $\gcd(n, e(F)) = 1$ *and* $\mu_n(\Omega) \subseteq F$. *Then, the maps* $H \mapsto F(\sqrt[n]{H})$ *and* $E \mapsto E^n \cap F^*$ *establish isomorphisms of lattices, inverse to one another, between the lattice of all subgroups H of F^* containing F^{*n} and the lattice of all classical n-Kummer extensions E of F (contained in Ω).* $\square$

The remaining part of the Infinite Kummer Theory, namely that concerning the description of the Kneser group of an infinite classical Kummer extension as the group of continuous characters of its Galois group, will be examined in Section 15.3.

13.2. Infinite generalized Kummer extensions

In this section we introduce and investigate the infinite variant of the concept of finite generalized Kummer extension discussed in Section 7.2. Since such extensions are bounded G-Cogalois extensions, we can easily deduce that the main properties they enjoy are very similar to those of infinite classical Kummer extensions.

DEFINITION 13.2.1. *We say that an extension E/F is a generalized n-Kummer extension, where $n \in \mathbb{N}^*$, if $E = F(B)$ for some $\varnothing \neq B \subseteq E^*$, with $\gcd(n, e(F)) = 1$, $B^n \subseteq F$, and $\mu_n(E) \subseteq F$.* $\square$

Clearly, a generalized n-Kummer extension E/F is finite if and only if the set B in Definition 13.2.1 can be chosen to be finite.

Observe that any classical n-Kummer extension $F(\sqrt[n]{A})/F$ of exponent n is a generalized n-Kummer extension, since if we denote $B = \sqrt[n]{A}$, then $B^n = A \subseteq F^*$. Conversely, if $A^n = B$, then $A \subseteq \sqrt[n]{B}$, but the inclusion may be strict.

THEOREM 13.2.2. *Let E/F be a generalized n-Kummer extension, with $\varnothing \neq B \subseteq E^*$, $E = F(B)$, $B^n \subseteq F$, $\gcd(n, e(F)) = 1$, and $\mu_n(E) \subseteq F$. If we denote $G = F^*\langle B \rangle$, then the following statements hold.*

(1) *The extension E/F is G-Cogalois.*

(2) *The maps* $H \mapsto F(\sqrt[n]{H} \cap G)$ *and* $K \mapsto K^n \cap (F^{*n} \langle B^n \rangle)$ *establish isomorphisms of lattices, inverse to one another, between the lattice of all subgroups* H *of* $F^{*n} \langle B^n \rangle$ *containing* F^{*n} *and the lattice of all intermediate fields* K *of* E/F. *Moreover, any subextension* K/F *of* E/F *is a generalized* n-*Kummer extension.*

(3) *If* H *is any subgroup of* $F^{*n} \langle B^n \rangle$ *containing* F^{*n}, *then any set of representatives of the group* $(\sqrt[n]{H} \cap G)/F^*$ *is a vector space basis of* $F(\sqrt[n]{H} \cap G)$ *over* F, *and* $[F(\sqrt[n]{H} \cap G) : F] = |H/F^{*n}|$.

PROOF. Proceed as in the proof of Theorem 7.2.3.   □

13.3. Infinite Kummer extensions with few roots of unity

The infinite Kummer extensions with few roots of unity are very particular cases of infinite generalized Kummer extensions. We present in this section their basic properties.

DEFINITION 13.3.1. *We say that an extension* E/F *is an* n-Kummer *extensions with few roots of unity if* $E = F(B)$ *for some* $\varnothing \neq B \subseteq E^*$, *with* $\gcd(n, e(F)) = 1$, $B^n \subseteq F$ *and* $\mu_n(E) \subseteq \{-1, 1\}$.   □

An immediate consequence of Theorem 13.2.2 is the following result.

THEOREM 13.3.2. *Let* E/F *be an* n-*Kummer extension with few roots of unity, with* $\varnothing \neq B \subseteq E^*$, $E = F(B)$, $B^n \subseteq F$, $\gcd(n, e(F)) = 1$, *and* $\mu_n(E) \subseteq \{-1, 1\}$. *If we denote* $G = F^* \langle B \rangle$, *then the following statements hold.*

(1) *The extension* E/F *is* G-*Cogalois.*

(2) *The maps* $H \mapsto F(\sqrt[n]{H} \cap G)$ *and* $K \mapsto K^n \cap (F^{*n} \langle B^n \rangle)$ *establish isomorphisms of lattices, inverse to one another, between the lattice of all subgroups* H *of* $F^{*n} \langle B^n \rangle$ *containing* F^{*n} *and the lattice of all intermediate fields* K *of* E/F. *Moreover, any subextension* K/F *of* E/F *is an* n-*Kummer extension with few roots of unity.*

(3) *If* H *is any subgroup of* $F^{*n} \langle B^n \rangle$ *containing* F^{*n}, *then any set of representatives of the group* $(\sqrt[n]{H} \cap G)/F^*$ *is a vector space basis of* $F(\sqrt[n]{H} \cap G)$ *over* F, *and* $[F(\sqrt[n]{H} \cap G) : F] = |H/F^{*n}|$.   □

13.4. Infinite quasi-Kummer extensions

In this section we present the infinite variant of the concept of finite quasi-Kummer extension discussed in Section 7.4.

DEFINITION 13.4.1. *An extension E/F is said to be a* quasi-Kummer *extension if $E = F(B)$ for some $\varnothing \neq B \subseteq E^*$, and there exists $n \in \mathbb{N}^*$ with $B^n \subseteq F$, $\gcd(n, e(F)) = 1$, such that $\zeta_p \in F$ for every $p \in \mathcal{P}_n$.*

THEOREM 13.4.2. *Any quasi-Kummer extension $F(B)/F$ with B as in Definition 13.4.1 is an $F^*\langle B \rangle$-Cogalois extension.*

PROOF. Let $G = F^*\langle B \rangle$. Then clearly $G^n \subseteq F^*$, and $E = F(G)$. So, E/F is a bounded G-radical extension. If $m = \exp(G/F^*)$, then $m \mid n$. Since $\gcd(n, e(F)) = 1$ by hypothesis, it follows that $\gcd(m, e(F)) = 1$, and consequently, E/F is a separable extension by Corollary 5.1.3. For every $p \in \mathcal{P}_n$ we have $\mu_p(E) \subseteq \mu_p(\Omega) \subseteq F$, so E/F is n-pure. By Lemma 13.1.1, it follows that $F(B)/F$ is an $F^*\langle B \rangle$-Cogalois extension. □

PROPOSITION 13.4.3. *Any Galois n-bounded G-Cogalois extension is a quasi-Kummer extension.*

PROOF. Let E/F be a Galois n-bounded G-Cogalois extension. Then $E = F(G)$, with $F^* \leqslant G \leqslant E^*$ and $\exp(G/F^*) = n$. Since the extension E/F is Galois, it follows that $\gcd(n, e(F)) = 1$ and $\zeta_n \in E$ by Corollary 5.1.6, so $\mu_n(\Omega) \subseteq \mu_n(E)$. By Theorem 12.1.12, the G-Cogalois extension E/F is n-pure, hence $\mu_p(E) \subseteq F$ for every $p \in \mathcal{P}_n$. Then $\mu_p(\Omega) \subseteq \mu_p(E) \subseteq F$ for any such p. Consequently, the extension E/F is quasi-Kummer. □

PROPOSITION 13.4.4. *Let E/F be a Galois generalized m-Kummer extension, with $E = F(B)$, $\varnothing \neq B \subseteq E^*$, $\gcd(m, e(F)) = 1$, $B^m \subseteq F$, and $\mu_m(E) \subseteq F$. Then E/F is $F^*\langle B \rangle$-Cogalois and $\mu_n(\Omega) \subseteq F$, where $n = \exp(F^*\langle B \rangle/F^*)$, i.e., E/F is a classical n-Kummer extension.*

PROOF. Let $p \in \mathcal{P}_n$. Since $n \mid m$, we have $\mu_p(E) \subseteq \mu_m(E) \subseteq F$, hence E/F is n-pure. Observe that E/F is G-radical, where $G = F^*\langle B \rangle$. By Theorem 12.1.12, E/F is a G-Cogalois extension. Now, in view of Proposition 5.1.6, we have $\gcd(n, e(F)) = 1$ and $\zeta_n \in E$. But $n \mid m$, hence

$$\zeta_n \in E \cap \mu_n(\Omega) = \mu_n(E) \subseteq \mu_m(E) \subseteq F.$$

Thus E/F is a classical n-Kummer extension. □

The last result of this section establishes the structure of Abelian bounded G-Cogalois extensions.

PROPOSITION 13.4.5. *The following assertions are equivalent for an extension $E/\mathbb{Q}$ with $E \subseteq \mathbb{C}$.*

(1) $E/\mathbb{Q}$ *is an Abelian n-bounded G-Cogalois extension.*
(2) *There exists $\varnothing \neq B \subseteq E^*$ such that $B^2 \subseteq \mathbb{Q}$ and $E = \mathbb{Q}(B)$.*
(3) *There exists $\varnothing \neq A \subseteq \mathbb{Q}^*$ such that $E = \mathbb{Q}(\sqrt{A})$, where $\sqrt{A}$ denotes the set $\{ z \in \mathbb{C} \mid z^2 \in A \}$, in other words, E is obtained by adjoining to $\mathbb{Q}$ an arbitrary nonempty set of square roots of rational numbers.*
(4) $E/\mathbb{Q}$ *is a classical 2-Kummer extension.*

PROOF. (1) $\Longrightarrow$ (2): By Proposition 13.4.3, $E/\mathbb{Q}$ is a quasi-Kummer extension. Then $E = \mathbb{Q}(B)$ for some $\varnothing \neq B \subseteq E^*$, there exists $n \in \mathbb{N}^*$ with $B^n \subseteq \mathbb{Q}$, and $\zeta_p \in \mathbb{Q}$ for every $p \in \mathcal{P}_n$. The last condition implies that $n = 1$ or $n = 2$, so $E = \mathbb{Q}(B)$ with $B^2 \subseteq \mathbb{Q}$.

(2) $\Longrightarrow$ (1): Any extension $\mathbb{Q}(B)/\mathbb{Q}$ with $B^2 \subseteq \mathbb{Q}$ is clearly Abelian and $\mathbb{Q}^* \langle A \rangle$-Cogalois. It is 1-bounded if $\mathbb{Q} = \mathbb{Q}(B)$, and 2-bounded otherwise.

All other implications are obvious. $\qquad\qquad\qquad\qquad\qquad\qquad$ $\square$

REMARK 13.4.6. Recall that we have denoted by $\mathbb{P}$ the set of all strictly positive prime numbers. Clearly, $\mathbb{Q}(\sqrt{\mathbb{P}})/\mathbb{Q}$ is an Abelian 2-bounded G-Cogalois extension, where $G = \mathbb{Q}^* \langle \sqrt{\mathbb{P}} \rangle$. The Kneser group of this extension is $\mathbb{Q}^* \langle \sqrt{\mathbb{P}} \rangle / \mathbb{Q}^*$, which is isomorphic to a countably infinite direct sum of copies of the cyclic group $\mathbb{Z}_2$ of order 2. On the other hand, the Galois group of this extension is isomorphic to a countably infinite direct product of copies of $\mathbb{Z}_2$.

Consequently, the Kneser group and the Galois group of the Abelian G-Cogalois extension $\mathbb{Q}(\sqrt{\mathbb{P}})/\mathbb{Q}$ are not isomorphic. This shows that Theorem 5.2.2, saying that the Kneser group and the Galois group of any finite Abelian G-Cogalois extension are isomorphic, is no more valid in the infinite case. The connection between $\mathrm{Gal}(E/F)$ and $\mathrm{Kne}(E/F)$ for an infinite Abelian G-Cogalois extension E/F, which involves also the Pontryagin duality, will be investigated in Section 15.3. $\qquad\qquad\qquad\qquad$ $\square$

13.5. Exercises to Chapter 13

1. Let F be a field, let $n \in \mathbb{N}^*$ be such that $\mu_n(\Omega) \subseteq F$, and let $F_{n,\Omega}^{ab}$ be the maximal classical n-Kummer extension of F contained in a fixed algebraically closed overfield Ω of F. Prove that

$$\operatorname{Hom}_c(\operatorname{Gal}(F_{n,\Omega}^{ab}/F), \mu_n(F)) \cong F^*/F^{*n}.$$

2. With the notation and hypotheses of Exercise 1, prove that

$$\operatorname{Gal}(F_{n,\Omega}^{ab}/F) \cong \operatorname{Hom}(F^*, \mu_n(F)).$$

3. Show that $\mathbb{Q}_{2,\mathbb{C}}^{ab} = \mathbb{Q}(i, \sqrt{2}, \sqrt{3}, \sqrt{5}, \dots)$.

4. Prove that the following assertions are equivalent for an extension E/F with $F \subseteq \mathbb{R}$ and $E \subseteq \mathbb{C}$.
   (a) E/F is a Galois n-bounded G-Cogalois extension.
   (b) E/F is an Abelian n-bounded G-Cogalois extension.
   (c) There exists $\varnothing \neq B \subseteq E^*$ such that $B^2 \subseteq F$ and $E = F(B)$.
   (d) There exists $\varnothing \neq A \subseteq F^*$ such that $E = F(\sqrt{A})$, where $\sqrt{A}$ denotes the set $\{z \in \mathbb{C} \mid z^2 \in A\}$, in other words, E is obtained by adjoining to F an arbitrary nonempty set of square roots of numbers from F.
   (e) E/F is a classical 2-Kummer extension.

13.6. Bibliographical comments to Chapter 13

Section 13.1. The concept of classical Kummer extension is rather basic and well-known, so it is difficult to give many precise attributions. A good account of the theory of such extensions can be found e.g., in Bourbaki [**40**], Karpilovsky [**76**], and Lang [**80**]. The idea to investigate classical Kummer extensions via Finite or/and Infinite Cogalois Theory is due to Albu and Nicolae [**19**] and Albu and Ţena [**25**]. Classical but longer proofs of Theorem 13.1.2 are provided e.g., in Bourbaki [**40**, Chap. V, §11, Théorème 4, p. 85], Karpilovsky [**76**, Chap. 7, Theorem 4.4, p. 412], or Lang [**80**, Chap. 8, Theorem 14].

Section 13.2. The results of this section are taken from Albu and Ţena [**25**], where the notion of infinite generalized Kummer extension was introduced and investigated.

Section 13.3. The concept of Kummer extension with few roots of unity was introduced and investigated by Albu [**3**] for finite extensions, and by Albu and Țena [**25**] for infinite extensions. The contents of this section follow the presentation in Albu and Țena [**25**].

Section 13.4. The notion of (finite) *neat presentation* is due to Greither and Harrison [**63**]. Albu and Nicolae [**19**] introduced the more general concept of (finite) *generalized neat presentation*. The infinite generalized neat presentations were introduced and investigated by Albu and Țena [**25**]. The results of this section on *quasi-Kummer extensions* are taken from Albu and Țena [**25**], where they were called generalized neat presentations.

CHAPTER 14

INFINITE GALOIS THEORY AND PONTRYAGIN DUALITY

In this chapter we present without proofs some basic facts on Infinite Galois Theory and Pontryagin Duality which will be used in the next chapter.

The reader is assumed to be familiar with the concept of projective limit and with the basic concepts and facts of General Topology.

14.1. Profinite groups and Infinite Galois Theory

The main aim of this section is to present without proofs the Fundamental Theorem of the Infinite Galois Theory and its consequences. The Galois group of any infinite Galois extension E/F turns out to be a profinite group, that is, a topological group which is the projective limit of a projective system of finite groups, each endowed with the discrete topology. Therefore we shall quote some standard properties of profinite groups which will be applied to Infinite Galois Theory. A good account of profinite groups can be found in Cassels and Fröhlich [46, Chapter 5] or Karpilovsky [76, Chapter 6, Section 2].

DEFINITION 14.1.1. *A topological group G is said to be a* profinite group *if G is a projective limit of a projective system $(G_i, \varphi_{ji})_{i,j \in I}$ of finite groups, each endowed with the discrete topology.* □

Recall briefly some definitions from General Topology. Let S be a topological space. A *fundamental system of neighborhoods* of a point x of S is a set $\mathcal{U}$ of neighborhoods of x such that for every neighborhood V of x there exists $U \in \mathcal{U}$ with $U \subseteq V$.

The space S is said to be *totally disconnected* if the connected component of each point of S consists of the point alone. The topological space S is said to be *compact* if for every family $(A_i)_{i \in I}$ of open subsets

A_i of S with $S = \bigcup_{i \in I} A_i$ there exists a finite subset J of I such that $S = \bigcup_{i \in J} A_i$.

For other basic terminology from General Topology, the reader is referred to Bourbaki [38].

PROPOSITION 14.1.2. *The following assertions are equivalent for a topological group G.*

(1) *G is a profinite group.*

(2) *G is a Hausdorff, compact group which has a fundamental system of open neighborhoods of the identity element e of G consisting of normal subgroups of G.*

(3) *G is a Hausdorff, compact, totally disconnected group.*

PROOF. See Karpilovsky [**76**, Proposition 2.8, p. 311]. □

The prototype of a profinite group is the Galois group Γ of an infinite Galois extension E/F. The group Γ can be endowed with the *Krull topology* as follows. Denote by $\mathcal{F}$ the set of all intermediate fields K of E/F such that K/F is a finite Galois extension, and for any $K \in \mathcal{F}$ set $\Gamma_K = \mathrm{Gal}(E/K)$. Notice that for any $K, K_1, K_2 \in \mathcal{F}$, one has $\Gamma_K \triangleleft \Gamma$, $K_1 K_2 \in \mathcal{F}$, and $\Gamma_{K_1 K_2} \subseteq \Gamma_{K_1} \cap \Gamma_{K_2}$ (see 1.2.9).

LEMMA 14.1.3. *Let G be a group, and let $\mathcal{N}$ be a nonempty set of normal subgroups of G such that for every $N_1, N_2 \in \mathcal{N}$ there exists an $N_0 \in \mathcal{N}$ with $N_0 \subseteq N_1 \cap N_2$. Then G becomes a topological group if one takes $\mathcal{N}$ as a fundamental system of open neighborhoods of the identity element e of G. A fundamental system of open neighborhoods in this topology of any $x \in G$ is the set $\{ xN \mid N \in \mathcal{N} \}$. Moreover, this group is Hausdorff if and only if $\bigcap_{N \in \mathcal{N}} N = \{e\}$.*

PROOF. See Bourbaki [**38**, Chap. III, §1, no.2, Example]. □

PROPOSITION 14.1.4. *Let E/F be a Galois extension. With the notation above, $\mathrm{Gal}(E/F)$ is a Hausdorff topological group if one takes the set $\{ \Gamma_K \mid K \in \mathcal{F} \}$ as a fundamental system of open neighborhoods of 1_E.*

PROOF. We have seen that the set $\{ \Gamma_K \mid K \in \mathcal{F} \}$ satisfies the conditions of Lemma 14.1.3. We claim that $\bigcap_{K \in \mathcal{F}} \Gamma_K = \{1_E\}$. Indeed, let $\sigma \in \bigcap_{K \in \mathcal{F}} \Gamma_K$ and $x \in E$. Set $L = F(x)$, and consider the normal closure of $\widetilde{L}/F$ of the finite extension L/F. Then $\widetilde{L} \in \mathcal{F}$, hence $\sigma \in \mathrm{Gal}(E/\widetilde{L})$, in particular we have $\sigma(x) = x$. This proves our claim. Now apply Lemma 14.1.3. □

The topology on $\mathrm{Gal}(E/F)$ defined in Proposition 14.1.4 is called the *Krull topology*; note that it is the discrete topology if and only if the Galois extension E/F is finite.

We are going to show that the topological group $\mathrm{Gal}(E/F)$ is a profinite group. To do that we will show first that the Krull topology on $\mathrm{Gal}(E/F)$ is nothing else than the finite topology. Recall that if X and Y are two nonempty sets, then the *finite topology* on the set Y^X of all maps from X to Y, identified with the Cartesian product $\prod_{x \in X} Y_x$, with $Y_x = Y$ for all $x \in X$, is the product topology on Y^X, where Y is endowed with the discrete topology. Taking into account the form of open sets in the product topology, we deduce at once that for an arbitrary $f \in Y^X$ a fundamental system of open neighborhoods of f consists of the sets

$$V_{\{x_1,\dots,x_n\}}(f) = \{\, g \in Y^X \mid g(x_i) = f(x_i), \ \forall\, i, \ 1 \leqslant i \leqslant n \,\},$$

where $\{x_1,\dots,x_n\}$ ranges over the finite subsets of X.

For any $Z \subseteq Y^X$, by the *finite topology* of Z we will understand the topology on Z induced by the finite topology on Y^X. If E/F is a Galois extension, then clearly $\mathrm{Gal}(E/F) \subseteq E^E$, hence it makes sense to consider the finite topology on $\mathrm{Gal}(E/F)$.

PROPOSITION 14.1.5. *The following statements hold for a Galois extension E/F with Galois group Γ.*

(1) *The Krull topology and the finite topology on Γ coincide.*

(2) *Γ is a Hausdorff, compact, and totally disconnected group.*

(3) *Γ is a profinite group.*

PROOF. (1) Let $\sigma \in \Gamma$, and let U be a neighborhood of σ in the finite topology. Then, with the notation above, there exists a finite subset $\{x_1,\dots,x_n\}$ of E such that $V_{\{x_1,\dots,x_n\}}(\sigma) \subseteq U$. Set $K = F(x_1,\dots,x_n)$. Since K/F is a finite extension, so also is its normal closure $\widetilde{K}/F$. But

$$
\begin{aligned}
V_{\{x_1,\dots,x_n\}}(\sigma) &= \{\, \tau \in \Gamma \mid \tau(x_i) = \sigma(x_i), \ \forall\, i, \ 1 \leqslant i \leqslant n \,\} \\
&= \{\, \tau \in \Gamma \mid \tau(x) = \sigma(x), \ \forall\, x \in K \,\} \\
&= \{\, \tau \in \Gamma \mid (\sigma^{-1} \circ \tau)(x) = 1_E(x), \ \forall\, x \in K \,\} \\
&= \sigma \mathrm{Gal}(E/K).
\end{aligned}
$$

Thus,

$$\sigma \mathrm{Gal}(E/\widetilde{K}) \subseteq \sigma \mathrm{Gal}(E/K) = V_{\{x_1,\dots,x_n\}}(\sigma) \subseteq U,$$

which shows that for every $\sigma \in \Gamma$, every neighborhood of σ in the finite topology includes a neighborhood of σ in the Krull topology.

Conversely, let $\sigma \in \Gamma$, and let W be a neighborhood of σ in the Krull topology. Then, there exists a finite Galois subextension L/F of E/F such that $\sigma \mathrm{Gal}(E/L) \subseteq W$. Let $\{y_1, \ldots, y_m\}$ be a basis of the finite dimensional vector space $_F L$. Then $L = F(y_1, \ldots, y_m)$, and so,

$$V_{\{y_1, \ldots, y_m\}}(\sigma) = \sigma \mathrm{Gal}(E/L),$$

as we have already shown. Thus, for every $\sigma \in \Gamma$, every neighborhood of σ in the Krull topology includes a neighborhood of σ in the finite topology. This proves (1).

(2) Γ is a Hausdorff topological group by Proposition 14.1.4. For every $x \in E$ set $E_x = E$, and let $\pi_x : \prod_{y \in E} E_y \longrightarrow E_x$ denote the canonical x-th projection. For any $x \in E$, the set $\pi_x(\Gamma) = \{\sigma(x) \mid \sigma \in \Gamma\}$ is the set of all conjugates of x over F, hence it is finite since x is algebraic over F. Thus, every $\pi_x(\Gamma)$ is a compact space, hence so also is $\prod_{x \in E} \pi_x(\Gamma)$.

We claim that Γ is a closed subset of $E^E = \prod_{x \in E} E_x$. To do that, let $f \in E^E$ be an element in the topological closure of Γ in E^E. By (1), the Krull topology on Γ is nothing else than the finite topology, hence, for every fixed $x, y \in E$ we have $\Gamma \cap V_{\{x,y,x+y,xy\}}(f) \neq \varnothing$, i.e., there exists $\sigma \in \Gamma$ such that

$$f(x) = \sigma(x), \ f(y) = \sigma(y), \ f(x+y) = \sigma(x+y), \ f(xy) = \sigma(xy).$$

It follows that $f(x+y) = f(x) + f(y)$ and $f(xy) = f(x)f(y)$. A similar argument shows that $f(a) = a$ for every $a \in F$. Thus $f : E \longrightarrow E$ is an F-morphism of the field E, which is an F-automorphism of E, i.e., $f \in \Gamma$ since E/F is an algebraic extension. This proves that Γ is a closed subset of E^E. Then Γ is a closed subset in the compact space $\prod_{x \in E} \pi_x(\Gamma)$, hence it is also compact.

Every $\sigma \in \Gamma$ has as a fundamental system of open neighborhoods the set $\{\sigma \mathrm{Gal}(E/K) \mid K \in \mathcal{F}\}$. Since every open subgroup of a topological group is closed, it follows that every member of this set is also closed. This implies that the connected component of any σ is contained in $\bigcap_{K \in \mathcal{F}} \sigma \mathrm{Gal}(E/K) = \{\sigma\}$, so it is reduced to $\{\sigma\}$; in other words, Γ is totally disconnected.

(3) follows from (2) and Proposition 14.1.2. A more accurate description of the realization of the Galois group of an infinite Galois extension as a projective limit of finite groups is given in Exercise 3. $\square$

As Exercise 4 shows, the bijective Galois correspondence between the lattice Subgroups(Γ) of all subgroups of the Galois group Γ of any finite

Galois extension E/F and the lattice $\underline{\text{Intermediate}}(E/F)$ of all intermediate fields of E/F, provided by the Fundamental Theorem of Finite Galois Theory, no longer remains valid if the Galois extension E/F is infinite. However, if we equip the Galois group Γ with the Krull topology, then the bijective Galois correspondence is restored when we restrict ourselves only to closed subgroups. Next we will present without proofs the basic facts of the Infinite Galois Theory.

THEOREM 14.1.6. (THE FUNDAMENTAL THEOREM OF INFINITE GALOIS THEORY). *The following statements hold for an arbitrary Galois extension E/F with Galois group Γ.*

(1) *The maps*
$$\alpha : \underline{\text{Intermediate}}\,(E/F) \longrightarrow \underline{\text{Closed Subgroups}}\,(\Gamma),$$
$$\alpha(K) = \text{Gal}(E/K)$$

and

$$\beta : \underline{\text{Closed Subgroups}}\,(\Gamma) \longrightarrow \underline{\text{Intermediate}}\,(E/F),$$
$$\beta(\Delta) = \text{Fix}(\Delta)$$

establish anti-isomorphisms of lattices, inverse to one another, between the lattice $\underline{\text{Intermediate}}\,(E/F)$ of all intermediate fields of the extension E/F and the lattice $\underline{\text{Closed Subgroups}}\,(\Gamma)$ of all closed subgroups of the group Γ endowed with the Krull topology.

(2) *If $K \in \underline{\text{Intermediate}}\,(E/F)$, then K/F is a finite extension if and only if $\text{Gal}(E/K)$ is an open subgroup of Γ, and in this case one has $(\Gamma : \text{Gal}(E/K)) = [K : F]$.*

(3) *If $K \in \underline{\text{Intermediate}}(E/F)$, then K/F is a Galois extension if and only if $\text{Gal}(E/K) \lhd \text{Gal}(E/F)$, and in this case, the restriction map $\text{Gal}(E/F) \longrightarrow \text{Gal}(K/F)$ yields an isomorphism of topological groups*
$$\text{Gal}(E/F)/\text{Gal}(E/K) \cong \text{Gal}(K/F).$$

PROOF. See Bourbaki [40, Théorème 4, A V.64] or Karpilovsky [76, Theorem 6.2, p. 343]. □

COROLLARY 14.1.7. *Let E/F and K/F be subextensions of an extension L/F. If $E/(E \cap K)$ is a Galois extension, then EK/K is also a Galois extension, and the restriction map:*
$$\text{Gal}(EK/K) \longrightarrow \text{Gal}(E/(E \cap K)), \ \sigma \mapsto \sigma|_E,$$
is an isomorphism of topological groups.

PROOF. See Karpilovsky [**76**, Proposition 6.5, p. 345].   □

COROLLARY 14.1.8. *Let E_1/F and E_2/F be Galois extensions, and assume that E_1, E_2 are subfields of some other field. Then $E_1 E_2/F$ is a Galois extension, and the map*

$$\mathrm{Gal}(E_1 E_2/F) \longrightarrow \mathrm{Gal}(E_1/F) \times \mathrm{Gal}(E_2/F), \quad \sigma \mapsto (\sigma|_{E_1}, \sigma|_{E_2}),$$

is a monomorphism of topological groups, which is an isomorphism when $E_1 \cap E_2 = F$.

PROOF. See Karpilovsky [**76**, Proposition 6.7, p. 346].   □

14.2. Character group and Pontryagin Duality

In this section we present some basic facts on characters of locally compact Abelian groups and Pontryagin Duality.

If G and H are topological groups, then by $\mathrm{Hom}_c(G, H)$ we denote the set of all *continuous* group morphisms from G to H, while $\mathrm{Hom}(G, H)$ denotes the set of all group morphisms from G to H. By $\mathbb{T}$ we denote the unit circle of the complex plane, that is, $\mathbb{T} = \{ z \mid z \in \mathbb{C}, |z| = 1 \}$. Observe that $\mathbb{T}$ is a subgroup of the multiplicative group $\mathbb{C}^*$. If we consider on $\mathbb{T}$ the induced topology of the usual topology of $\mathbb{C}$, then it becomes a topological group.

For any topological group G we denote by $\underline{\mathrm{Closed\ subgroups}}(G)$ the lattice of all closed subgroups of G, and by $\mathrm{Ch}\,(G)$ or by $\widehat{G}$ the *character group* of G, that is, the group of all continuous morphisms of G into the unit circle $\mathbb{T}$:

$$\mathrm{Ch}\,(G) = \widehat{G} = \mathrm{Hom}_c(G, \mathbb{T}).$$

By $\mathbf{1}$ we shall denote the 1-character of G, that is, $\mathbf{1}(x) = 1$ for all $x \in G$.

LEMMA 14.2.1. *The following assertions hold for a topological group G.*

(1) *If G is a torsion group, then*
$$\mathrm{Ch}(G) = \mathrm{Hom}_c(G, \mu(\mathbb{C})).$$

(2) *If G is an n-bounded group, then*
$$\mathrm{Ch}(G) = \mathrm{Hom}_c(G, \mu_n(\mathbb{C})).$$

(3) *If G is a profinite group, then $\mathrm{Ch}(G)$ is a torsion Abelian group, and*

$$\mathrm{Ch}(G) = \{ \chi \in \mathrm{Hom}(G, \mu(\mathbb{C})) \mid \mathrm{Ker}(\chi) \text{ is open } \}$$
$$= \mathrm{Hom}_c(G, \mu(\mathbb{C})).$$

(4) *If G is a finite discrete Abelian group, then $\mathrm{Ch}(G) \cong G$.*

(5) *If G is a locally compact Abelian group, then so is $\mathrm{Ch}(G)$.*

PROOF. (1) If G is a torsion group, then for every $x \in G$ there exists an $n_x \in \mathbb{N}^*$ such that $x^{n_x} = e$, where e is the identity element of G. For any $\chi \in \widehat{G}$ we have

$$1 = \chi(e) = \chi(x^{n_x}) = (\chi(x))^{n_x},$$

hence $\chi(x) \in \mu(\mathbb{C})$.

(2) If G is an n-bounded group, then $x^n = e$ for every $x \in G$, so, as in the proof of (1), we have $(\chi(x))^n = 1$ for every $\chi \in \widehat{G}$ and every $x \in G$, i.e., $\chi(x) \in \mu_n(\mathbb{C})$.

(3) See Karpilovsky [**76**, Corollary 5.3, p. 418].

(4) See Hewitt and Ross [**72**, 23.27 (d)].

(5) For any nonempty compact subset C of G and any real number $\varepsilon > 0$, let $U(C, \varepsilon)$ be the subset of $\widehat{G}$ defined by

$$U(C, \varepsilon) = \{ \chi \mid \chi \in \widehat{G}, \, |\chi(x) - 1| < \varepsilon, \, \forall x \in C \}.$$

If one takes the collection of all such $U(C, \varepsilon)$ as a fundamental system of neighborhoods of the identity element $1 \in \widehat{G}$, then $\widehat{G}$ becomes a locally compact Abelian group by Hewitt and Ross [**72**, Theorem 23.15]. $\square$

Throughout this section A will denote a fixed multiplicative Abelian group with identity element e. By Lemma 14.2.1 (5), for any locally compact Abelian group A, the character group $\mathrm{Ch}(A) = \widehat{A}$ of A is again a locally compact Abelian group, so it makes sense to consider the character group $\mathrm{Ch}(\mathrm{Ch}(A)) = \widehat{\widehat{A}}$ of $\widehat{A}$, called the *second character group of A*, and the canonical map

$$\omega_A : A \longrightarrow \widehat{\widehat{A}}, \quad \omega_A(a)(\chi) = \chi(a), \quad \chi \in \widehat{A}, \, a \in A.$$

The well-known *Duality Theorem*, due to Pontryagin, establishes the following fundamental result.

THEOREM 14.2.2. (THE PONTRYAGIN DUALITY). *The following assertions hold for any locally compact Abelian group A.*

(1) *The canonical map $\omega_A : A \longrightarrow \widehat{\widehat{A}}$ is a topological isomorphism.*

(2) *The assignment $A \mapsto \widehat{A}$ establishes an auto-duality of the category* **LCA** *of all locally compact Abelian groups.*

(3) *A is a compact group if and only if $\widehat{A}$ is a discrete group.*

(4) *A is a discrete group if and only if $\widehat{A}$ is a compact group.*

(5) *If A is a profinite group then $\widehat{A}$ is a discrete torsion group.*

PROOF. See Hewitt and Ross [**72**, Theorem 24.8] for (1), and Hewitt and Ross [**72**, Theorem 23.17] for (3) and (4). Since the assignment $A \mapsto \widehat{A}$ is functorial, (2) follows from (1). Finally, (5) follows from Proposition 14.1.2, Lemma 14.2.1, and (3). □

For any $\varnothing \neq X \subseteq A$ and $\varnothing \neq Y \subseteq \widehat{A}$ we shall use the following notation:

$$
\begin{aligned}
X^{\perp} &= \{ \chi \in \widehat{A} \mid \chi(a) = 1, \, \forall a \in X \}, \\
Y^{\perp} &= \{ a \in A \mid \chi(a) = 1, \, \forall \chi \in Y \}, \\
X^{\perp\perp} &= (X^{\perp})^{\perp}, \\
Y^{\perp\perp} &= (Y^{\perp})^{\perp}.
\end{aligned}
$$

LEMMA 14.2.3. *The following statements hold for any locally compact Abelian group A, any closed subgroup X of A, and any closed subgroup Y of $\widehat{A}$.*

(1) *$X^{\perp}$ is a closed subgroup of $\widehat{A}$.*

(2) *$Y^{\perp}$ is a closed subgroup of A.*

(3) *If $X \neq A$ and $x \in A \setminus X$, then there exists $\psi \in X^{\perp}$ such that $\psi(x) \neq 1$.*

(4) *$X = X^{\perp\perp}$.*

(5) *$Y = Y^{\perp\perp}$.*

PROOF. For (1) see Hewitt and Ross [**72**, Remarks 23.24 (c)]. Similarly, (2) is easily verified. For (3) see Hewitt and Ross [**72**, Corollary 23.26], and for (4) see Hewitt and Ross [**72**, Theorem 24.10].

We are going to prove (5). Assume that $Y \neq Y^{\perp\perp}$. This means that there exists $\chi \in Y^{\perp\perp} \setminus Y$. By (3), applied to the subgroup Y of $\widehat{A}$ and to the character χ, we deduce that there exists $a^* \in \widehat{\widehat{A}}$ such that $a^*(\psi) = 1$ for all $\psi \in Y$ and $a^*(\chi) \neq 1$. By Theorem 14.2.2 (1), there exists $a \in A$

such that $a^* = \omega_A(a)$, hence $a^*(\psi) = \omega_A(a)(\psi) = \psi(a)$ for all $\psi \in \widehat{A}$. In particular, we have $a^*(\chi) = \chi(a) \neq 1$.

On the other hand, $1 = a^*(\psi) = \psi(a)$ for every $\psi \in Y$, hence $a \in Y^\perp$. Since $\chi \in Y^{\perp\perp}$ we deduce that $\chi(b) = 1$ for all $b \in Y^\perp$, hence $\chi(a) = 1$, which is a contradiction. This proves (5). $\square$

PROPOSITION 14.2.4. *Let A be a locally compact Abelian group. Then, the maps:*

$$\text{Closed Subgroups}(A) \longrightarrow \text{Closed Subgroups}(\widehat{A}), \quad X \mapsto X^\perp,$$

$$\text{Closed Subgroups}(\widehat{A}) \longrightarrow \text{Closed Subgroups}(A), \quad Y \mapsto Y^\perp,$$

are anti-isomorphisms of lattices, inverse to one another.

PROOF. For simplicity, set

$$\mathcal{A} = \text{Closed Subgroups}(A) \quad \text{and} \quad \widehat{\mathcal{A}} = \text{Closed Subgroups}(\widehat{A}).$$

For every Z in $\mathcal{A}$ or $\widehat{\mathcal{A}}$ one has $Z = Z^{\perp\perp}$ by Lemma 14.2.3 (4) and (5), in other words, the order-reversing maps

$$\perp : \mathcal{A} \longrightarrow \widehat{\mathcal{A}} \quad \text{and} \quad \perp : \widehat{\mathcal{A}} \longrightarrow \mathcal{A}$$

are poset anti-isomorphisms, and consequently, lattice anti-isomorphisms, inverse to one another. $\square$

COROLLARY 14.2.5. *The following statements hold for a topological Abelian group A.*

(1) *If A is a compact group, then the map*

$$\text{Closed Subgroups}(A) \longrightarrow \text{Subgroups}(\widehat{A}), \quad X \mapsto X^\perp,$$

is an anti-isomorphism of lattices.

(2) *If A is a discrete group, then the map*

$$\text{Subgroups}(A) \longrightarrow \text{Closed Subgroups}(\widehat{A}), \quad X \mapsto X^\perp,$$

is an anti-isomorphism of lattices.

PROOF. By Theorem 14.2.2 (3) and (4), if A is a discrete (resp. compact) group, then $\widehat{A}$ is a compact (resp. discrete) group. Since

$$\text{Closed Subgroups}(D) = \text{Subgroups}(D)$$

for any discrete topological group D, the results follow at once from Proposition 14.2.4. $\square$

14.3. Exercises to Chapter 14

1. Show that the following statements hold for a topological group G.
   (a) Every open subgroup of G is closed.
   (b) Every closed subgroup of G of finite index is open.
   (c) If G is compact, then a subgroup of G is open if and only if it is closed and of finite index.

2. Let G be a compact topological group, and let $(N_i)_{i \in I}$ be a family of closed normal subgroups of finite index satisfying the following two conditions.
   (a) $\bigcap_{i \in I} N_i = \{e\}$.
   (b) For every $i_1, i_2 \in I$ there exists $i_0 \in I$ such that $N_{i_0} \subseteq N_{i_1} \cap N_{i_2}$.
   If we set $i \leqslant j$ for $N_i \supseteq N_j$ and $\varphi_{ij} : G/N_j \longrightarrow G/N_i$ for the canonical projection, then prove that $(G/N_i, \varphi_{ij})_{i,j \in I}$ is a projective system of topological groups, and there exists an isomorphism of topological groups
   $$G \cong \varprojlim_{i \in I} G/N_i.$$

3. Let E/F be a Galois extension, and denote by $\mathcal{F}$ the poset (ordered by inclusion) of all intermediate fields K of E/F such that K/F is a finite Galois extensions.
   (a) For any $K \subseteq L$ in $\mathcal{F}$ denote by
   $$\varphi_{KL} : \mathrm{Gal}(L/F) \longrightarrow \mathrm{Gal}(K/F)$$
   the restriction morphism. Show that $(\mathrm{Gal}(K/F), \varphi_{KL})_{K,L \in \mathcal{F}}$ is a projective system of discrete topological groups.
   (b) Prove that there exists an isomorphism of topological groups
   $$\mathrm{Gal}(E/F) \cong \varprojlim_{K \in \mathcal{F}} \mathrm{Gal}(K/F).$$

4. Let E be an algebraic closure of the field $F = \mathbb{F}_p$ of p elements, p prime, let $\Gamma = \mathrm{Gal}(E/F)$, let $\varphi \in \Gamma$ be the Frobenius automorphism of E, $\varphi(x) = x^p$, $x \in E$, and let $\Delta = \langle \varphi \rangle$. Show that there exists no $K \in \underline{\mathrm{Intermediate}}(E/F)$ such that $\Delta = \mathrm{Gal}(E/K)$. (*Hint*: See Karpilovsky [**76**, Example 5.5, p. 338].)

5. Let E/F be a Galois extension with Galois group Γ, let Δ be any subgroup of Γ, and let $\overline{\Delta}$ be the topological closure of Δ in Γ. Show that $\mathrm{Gal}(E/\mathrm{Fix}(\Delta)) = \overline{\Delta}$.

6. Let $\{1\} \neq S$ be a multiplicative subset of $\mathbb{N}^*$ (this means that $1 \in S$ and $mn \in S$ for any $m, n \in S$). Then S becomes a directed set if one orders S by the relation of divisibility. For every numbers $m \mid n$ in S let $\varphi_{mn} : \mathbb{Z}_n \longrightarrow \mathbb{Z}_m$ denote the canonical projection. Prove the following statements.

   (a) $(\mathbb{Z}_n, \varphi_{mn})_{m,n \in S}$ is a projective system of rings. Denote by $P(S)$ the corresponding projective limit, which is a compact topological ring if one endows each ring $\mathbb{Z}_n$ with the discrete topology. If $S = \mathbb{N}^*$ then $P(S)$ is denoted by $\widetilde{\mathbb{Z}}$ and is called the *Prüfer ring*. If $S = \{p^n \mid n \in \mathbb{N}\}$, where $p > 0$ is a prime number, then $P(S)$ is denoted by $\widetilde{\mathbb{Z}}_p$ and is called the *ring of p-adic integers*.

   (b) The underlying additive group of $\widetilde{\mathbb{Z}}$ (resp. $\widetilde{\mathbb{Z}}_p$) is an Abelian profinite group, having as open subgroups precisely the cyclic subgroups $n\widetilde{\mathbb{Z}}$, $n \in \mathbb{N}^*$ (resp. $p^m\widetilde{\mathbb{Z}}_p$, $m \in \mathbb{N}$).

   (c) There exist canonical group isomorphisms:
   $$\widetilde{\mathbb{Z}}/n\widetilde{\mathbb{Z}} \cong \mathbb{Z}/n\mathbb{Z} \quad \text{and} \quad \widetilde{\mathbb{Z}}_p/p^m\widetilde{\mathbb{Z}}_p \cong \mathbb{Z}/p^m\mathbb{Z}$$
   for any $n \in \mathbb{N}^*$ and any $m \in \mathbb{N}$.

   (d) There exists a canonical isomorphism of topological rings:
   $$\widetilde{\mathbb{Z}} \cong \prod_{p \in \mathbb{P}} \widetilde{\mathbb{Z}}_p.$$

7. Prove that for any profinite group G and for any given $g \in G$ there exists a unique continuous group morphism $\alpha_g : \widetilde{\mathbb{Z}} \longrightarrow G$ such that $\alpha_g(1) = g$.

8. (*Serre* [**97**, p. 198]). A field F is said to be *quasi-finite* if it is perfect and $\mathrm{Gal}(\overline{F}/F) \cong \widetilde{\mathbb{Z}}$. Show that any finite field and the field of formal power series $K((T))$ in the indeterminate T over any algebraically closed field K of characteristic 0 are quasi-finite fields.

9. Show that if E/F is an infinite Galois extension with Galois group Γ, then the cardinal numbers $[E : F]$ and $|\Gamma|$ are not necessarily equal.

10. Let $(E_i/F)_{1 \leqslant i \leqslant n}$ be a finite family of Galois extensions, and assume that all $E_1, \dots, E_n$ are subfields of some other field. If
    $$E_{i+1} \cap (E_1 \cdots E_i) = F$$

for every i, $1 \leqslant i \leqslant n - 1$, then show that there exists an isomorphism of topological groups

$$\mathrm{Gal}(E_1 \cdots E_n/F) \cong \mathrm{Gal}(E_1/F) \times \ldots \times \mathrm{Gal}(E_n/F).$$

11. Let E_1, E_2 be intermediate fields of a Galois extension E/F, and let $\Gamma = \mathrm{Gal}(E/F)$, $\Gamma_1 = \mathrm{Gal}(E_1/F)$, $\Gamma_2 = \mathrm{Gal}(E_2/F)$. Prove that the following assertions are equivalent.
    (a) The group Γ is the internal direct products of its subgroups Γ_1 and Γ_2.
    (b) E_1/F and E_2/F are both Galois extensions, $E_1 \cap E_2 = F$, and $E = E_1 E_2$.

12. (*Waterhouse* [**111**]). Prove that every profinite group is isomorphic to the Galois group of some field extension.

13. Prove that $\mathrm{Ch}(\widetilde{\mathbb{Z}}) \cong \mathbb{Q}/\mathbb{Z}$ and $\mathrm{Ch}(\widetilde{\mathbb{Z}}_p) \cong t_p(\mathbb{Q}/\mathbb{Z})$ for any $p \in \mathbb{P}$.

14. Show that the following statements hold for a locally compact Abelian group A.
    (a) The *evaluation map* on $\widehat{A}$

    $$\langle -, - \rangle : \widehat{A} \times A \longrightarrow \mathbb{T}, \quad \langle \chi, a \rangle = \chi(a)$$

    is bimultiplicative.
    (b) For any $\varnothing \neq X \subseteq A$ and any $\varnothing \neq Y \subseteq \widehat{A}$ one has

    $$\begin{aligned} X^{\perp} &= \{\, \chi \in \widehat{A} \,|\, \langle \chi, X \rangle = 1 \,\}, \\ Y^{\perp} &= \{\, a \in A \,|\, \langle Y, a \rangle = 1 \,\}. \end{aligned}$$

    (c) The evaluation map on $\widehat{A}$ is *nonsingular*, that is, $A^{\perp} = \{\mathbf{1}\}$ and $\widehat{A}^{\perp} = \{e\}$.

15. Give an example of a compact Abelian group A which is not profinite, but for which $\widehat{A}$ is a torsion group. This shows that the converse of Theorem 14.2.2 (5) is not true in general. (*Hint*: See Hewitt and Ross [**72**, 7.18 (d) and Theorem 24.26].)

14.4. Bibliographical comments to Chapter 14

Section 14.1. This section contains some well-known classical facts on Infinite Galois Theory and profinite groups. Our main sources of inspiration for this section were Bourbaki [**40**] and Karpilovsky [**76**]. An *abstract* Cogalois Theory, applicable to an arbitrary profinite group Γ acting continuously on a discrete subgroup of the additive group $\mathbb{Q}/\mathbb{Z}$, is developed in Albu and Basarab [**16**]. This theory is somewhat dual to the abstract Galois Theory which plays a major role in the general Class Field Theory as presented in Neukirch [**85**].

Section 14.2. A good account of Pontryagin Duality, only sketched in this section, can be found in Hewitt and Ross [**72**]. We also used the monograph of Karpilovsky [**76**] for our presentation. Proposition 14.2.4 is certainly known, but we could not locate it explicitly in the literature.

CHAPTER 15

INFINITE GALOIS G-COGALOIS EXTENSIONS

The aim of this chapter is two-fold: first, to generalize the main results of Chapter 6 from finite to infinite Galois extensions, and, second, to investigate infinite Abelian G-Cogalois extensions.

In Section 15.1 we study infinite Galois extensions E/F by means of continuous crossed homomorphisms of $\mathrm{Gal}(E/F)$ with coefficients in the group $\mu(E)$ of all roots of unity of E. We show that a considerable part of the results of Chapter 6 can be naturally generalized to infinite extensions.

Next, we are interested in finding a relationship between the Abelian groups $\mathrm{Kne}(E/F)$ and $\mathrm{Gal}(E/F)$ associated with an arbitrary Abelian G-Cogalois extension E/F. When the extension E/F is finite, then, by Theorem 5.2.2, these two groups are isomorphic, but not in a canonical way.

We show that in the infinite case, $\mathrm{Kne}(E/F)$ is isomorphic to the group $\mathrm{Ch}(\mathrm{Gal}(E/F))$ of characters of the profinite group $\mathrm{Gal}(E/F)$. To do that we cannot proceed, as in the finite case, by induction on the degree of the extension E/F. As we sketched in Remark 5.2.4, for any Abelian G-Cogalois extension E/F, finite or not, there exists a canonical lattice isomorphism between the lattices $\underline{\mathrm{Subgroups}}\,(\mathrm{Kne}(E/F))$ and $\underline{\mathrm{Subgroups}}\,(\mathrm{Ch}(\mathrm{Gal}(E/F)))$. So, it is natural to ask whether or not such a lattice isomorphism yields a group isomorphism between $\mathrm{Kne}(E/F)$ and $\mathrm{Ch}(\mathrm{Gal}(E/F))$.

Thus, we arrive at the following question: given two groups A and B, when does a lattice isomorphism $\varphi : \underline{\mathrm{Subgroups}}\,(A) \longrightarrow \underline{\mathrm{Subgroups}}(B)$ give rise a group isomorphism $f : A \longrightarrow B$? The groups A and B are called *lattice-isomorphic* if there exists a lattice isomorphism between their subgroup lattices $\underline{\mathrm{Subgroups}}\,(A)$ and $\underline{\mathrm{Subgroups}}(B)$. With this terminology, the problem we just stated can be reformulated briefly as follows: *when are two lattice-isomorphic groups isomorphic?* In general, lattice-isomorphic groups are not isomorphic, as a classical example discovered in 1928 by A. Rottlaender shows (see Baer [**27**]). However, if some restrictive conditions on one or both groups A and B are imposed, then any lattice-isomorphism between A and B produces a group isomorphism between them. By chance, such conditions are satisfied in our case, and so, the canonical lattice isomorphism between $\underline{\text{Subgroups}}$ (Kne (E/F)) and $\underline{\text{Subgroups}}$ (Ch (Gal (E/F))) yields a group isomorphism between the groups Kne (E/F) and Ch (Gal (E/F)).

In section 15.2 we present the basic notions and facts concerning lattice-isomorphic groups. Since many of these facts are quite technical, we shall avoid most of their proofs. Then, we apply in Section 15.3 these general results on lattice-isomorphic groups to our specific Galois-Cogalois case. Observe that this approach of the infinite case provides an alternative proof of Theorem 5.2.2 via lattice-isomorphisms of groups, which is more natural than that given in Section 5.2. Indeed, for any finite extension E/F, the finite group $\text{Gal}(E/F)$ is isomorphic, but not in a canonical way, to its character group Ch (Gal(E/F)).

15.1. The infinite Kneser group via crossed homomorphisms

The main purpose of this section is to generalize to infinite Galois extensions a series of results from Chapter 6 concerning the investigation of finite Galois extensions via crossed homomorphisms. As we know from Section 14.1, the Galois group of an infinite Galois extension is a totally disconnected compact topological group. In order to obtain these generalizations, one replaces *mutatis-mutandis* the (discrete) crossed homomorphisms by the continuous ones. We also characterize infinite radical extensions, Kneser extensions, and G-Cogalois extensions via crossed homomorphisms, generalizing the corresponding results established in Section 6.1 for finite extensions.

In Section 6.1 we reviewed some basic facts on finite *Galois Cohomology*. In particular we stated the (finite) *Hilbert's Theorem 90*. When E/F is an infinite field extension, then an *infinite* Hilbert's Theorem 90 still holds. As we have seen in Section 14.1, the Galois group Γ of E/F is a profinite group, or equivalently, a Hausdorff, compact, and totally disconnected topological group. A fundamental system of open neighborhoods of the identity element 1_E of Γ consists of normal subgroups of Γ of finite index, that is, of subgroups of the form $\mathrm{Gal}\,(E/N)$, with N/F is a finite normal subextension of E/F.

As in the finite case, let $M \leqslant E^*$ be such that $\sigma(M) \subseteq M$ for every $\sigma \in \Gamma$. A *continuous crossed homomorphism* or a *continuous 1-cocycle* of Γ with coefficients in M is a continuous function $f \in Z^1(\Gamma, M)$, where M is endowed with the discrete topology. The set of all continuous crossed homomorphisms of Γ with coefficients in M is a subgroup of $Z^1(\Gamma, M)$, and will be denoted in the sequel by $Z_c^1(\Gamma, M)$.

We claim that for every $\alpha \in M$, the 1-coboundary $f_\alpha : \Gamma \to M$ defined as in Section 6.1 by $f_\alpha(\sigma) = \sigma(\alpha){\cdot}\alpha^{-1}$, $\sigma \in \Gamma$, is a continuous map. To see this, observe that a map $h : P \to D$, where P is a profinite group and D is a discrete set is continuous if and only if h is locally constant, that is, there exists a normal subgroup H in P of finite index such that h factorizes through the canonical surjection map $P \to P/H$. Let $\alpha \in M$, and denote by N/F the normal closure of $F(\alpha)/F$, with $N \subseteq E$. Then N/F is a finite Galois extension, and for all $\sigma \in \mathrm{Gal}\,(E/F)$ and $\nu \in \mathrm{Gal}\,(E/N)$ one has and

$$f_\alpha(\sigma\nu) = (\sigma\nu)(\alpha) \cdot \alpha^{-1} = \sigma(\nu(\alpha)) \cdot \alpha^{-1} = \sigma(\alpha) \cdot \alpha^{-1} = f_\alpha(\sigma),$$

hence f_α factorizes through the canonical map

$$\mathrm{Gal}\,(E/F) \longrightarrow \mathrm{Gal}\,(E/F)/\mathrm{Gal}\,(E/N).$$

This proves our claim.

Consequently, the set $B^1(\Gamma, M) = \{\, f_\alpha \mid \alpha \in M \,\}$ of all 1-coboundaries of Γ with coefficients in M coincides with the set $B_c^1(\Gamma, M)$ of all *continuous* 1-coboundaries of Γ with coefficients in M. The quotient group $Z_c^1(\Gamma, M)/B_c^1(\Gamma, M)$ is called the first *continuous cohomology group* of Γ with coefficients in M, and is denoted by $H_c^1(\Gamma, M)$.

Note that when E/F is a finite Galois extension, then $Z_c^1(\Gamma, M) = Z^1(\Gamma, M)$, hence $H^1(\Gamma, M) = H_c^1(\Gamma, M)$. The *continuous* (or *infinite*) *Hilbert's Theorem 90* asserts that if E/F is an arbitrary Galois extension, finite or infinite, then $H_c^1(\Gamma, E^*) = \mathbf{1}$ (see e.g., Cassels and Fröhlich [**46**, Proposition 2.2, Chapter V] or Serre [**96**, Proposition 1, Chapitre 2]).

For an arbitrary extension E/F we have considered in Section 6.1 the map

$$f : \mathrm{Gal}\,(E/F) \times \mathrm{Cog}\,(E/F) \longrightarrow \mu(E),$$

$$f(\sigma, \widehat{\alpha}) = f_\alpha(\sigma) = \sigma(\alpha){\cdot}\alpha^{-1}, \ \sigma \in \mathrm{Gal}(E/F), \ \alpha \in T(E/F),$$

and for every integer $n \geqslant 1$, its restriction to $\mathrm{Cog}_n(E/F)$

$$f_n : \mathrm{Gal}\,(E/F) \times \mathrm{Cog}_n(E/F) \longrightarrow \mu_n(E).$$

For every fixed $\sigma \in \mathrm{Gal}\,(E/F)$, the partial map $f(\sigma, -)$ is clearly multiplicative on $\mathrm{Cog}\,(E/F)$, and for every fixed $\widehat{\alpha} \in \mathrm{Cog}(E/F)$, the partial map $f(-, \widehat{\alpha})$ is precisely the 1-coboundary $f_\alpha \in Z^1(\mathrm{Gal}\,(E/F), \mu(E))$, so f and f_n induce morphisms of groups

$$\psi : \mathrm{Cog}\,(E/F) \longrightarrow Z^1(\mathrm{Gal}\,(E/F), \mu(E)), \quad \psi(\widehat{\alpha})(\sigma) = f(\sigma, \widehat{\alpha}),$$

and

$$\psi_n : \mathrm{Cog}_n(E/F) \longrightarrow Z^1(\mathrm{Gal}\,(E/F), \mu_n(E)), \quad \psi_n(\widehat{\alpha})(\sigma) = f_n(\sigma, \widehat{\alpha}).$$

The Galois group $\mathrm{Gal}\,(E/F) = \Gamma$ is a profinite group, in particular it is a topological group with respect to its Krull topology. The morphisms ψ and ψ_n have images in the corresponding groups of continuous crossed homomorphisms since $\psi(\widehat{\alpha}) = f_\alpha \in B^1(\Gamma, \mu(E)) \subseteq Z_c^1(\Gamma, \mu(E))$. Consequently, we can assume that the canonical maps ψ and ψ_n are defined as follows:

$$\psi : \mathrm{Cog}\,(E/F) \longrightarrow Z_c^1(\mathrm{Gal}\,(E/F), \mu(E)), \quad \psi(\widehat{\alpha}) = f_\alpha,$$

and

$$\psi_n : \mathrm{Cog}_n(E/F) \longrightarrow Z_c^1(\mathrm{Gal}\,(E/F), \mu_n(E)), \quad \psi_n(\widehat{\alpha}) = f_\alpha.$$

LEMMA 15.1.1. *Let E/F be an arbitrary Galois extension, and let $n \in \mathbb{N}^*$. Then, the morphism*

$$\psi_n : \mathrm{Cog}_n(E/F) \longrightarrow Z_c^1(\mathrm{Gal}\,(E/F), \mu_n(E))$$

defined above is an isomorphism of groups.

PROOF. Denote by Γ the group $\mathrm{Gal}\,(E/F)$, and let $\alpha \in T_n(E/F)$. As in the proof of Lemma 6.1.1 one shows that ψ_n is a monomorphism.

We are going to prove that ψ_n is surjective. Let $h \in Z_c^1(\Gamma, \mu_n(E))$. Since clearly $h \in Z_c^1(\Gamma, E^*)$, by the continuous Hilbert's Theorem 90, there exists $\alpha \in E^*$ such that $h = f_\alpha$. Now continue as in the proof of Lemma 6.1.1 to deduce that $h = \psi_n(\widehat{\alpha})$, with $\widehat{\alpha} \in \mathrm{Cog}_n(E/F)$. $\qquad\square$

THEOREM 15.1.2. *For any Galois extension E/F, the map $\widehat{\alpha} \mapsto f_\alpha$ establishes a group isomorphism*

$$\mathrm{Cog}\,(E/F) \cong Z_c^1(\mathrm{Gal}\,(E/F), \mu(E)).$$

PROOF. For any $n \geqslant 1$, the restriction of the morphism

$$\psi : \mathrm{Cog}\,(E/F) \longrightarrow Z_c^1(\mathrm{Gal}\,(E/F), \mu(E))$$

to $\mathrm{Cog}_n(E/F)$ is ψ_n. Since $\Gamma = \mathrm{Gal}\,(E/F)$ is a profinite group, every $h \in Z_c^1(\Gamma, \mu(E))$ is locally constant, and so,

$$Z_c^1(\Gamma, \mu(E)) = \bigcup_{n \geqslant 1} Z_c^1(\Gamma, \mu_n(E)).$$

But ψ_n is bijective for all $n \in \mathbb{N}^*$ by Lemma 15.1.1, hence ψ is bijective too. $\qquad\square$

COROLLARY 15.1.3. *If E/F is a Galois extension with Galois group Γ, then the map*

$$\varphi : \{\, H \mid F^* \leqslant H \leqslant T(E/F)\,\} \longrightarrow \{\, U \mid U \leqslant Z_c^1(\Gamma, \mu(E))\,\},$$
$$\varphi(H) = \{\, f_\alpha \in Z_c^1(\Gamma, \mu(E)) \mid \alpha \in H \,\},$$

is a lattice isomorphism, which induces a canonical lattice isomorphism

$$\underline{\mathrm{Subgroups}}(\mathrm{Cog}\,(E/F)) \cong \underline{\mathrm{Subgroups}}\,(Z_c^1(\Gamma, \mu(E))).$$

For every cyclic subgroup C of $Z_c^1(\Gamma, \mu(E))$ there exists $\alpha \in T(E/F)$ such that $\varphi(F^*\langle\alpha\rangle) = \langle f_\alpha \rangle = C$. Moreover, $H/F^* \cong \varphi(H)$ for every H with $F^* \leqslant H \leqslant T(E/F)$. $\qquad\square$

For an arbitrary G-radical extension E/F we shall use the following notation

$$\mu_G(E) := \bigcup_{m \in \mathcal{O}_{G/F^*}} \mu_m(E).$$

Recall that

$$\mathcal{O}_{G/F^*} := \{\, \mathrm{ord}(\widehat{x}) \mid x \in G \,\}.$$

Since $\mathcal{O}_{G/F^*}$ is a directed set with respect to the divisibility relation by Lemma 1.4.3 (2), $\mu_G(E)$ is a subgroup of the group $\mu(E)$. Also, recall that by $\mathcal{P}_G$ we have denoted the set $\mathcal{O}_{G/F^*} \cap \mathcal{P}$.

Note that $\mu_G(E) = \mu_n(E)$ whenever E/F is n-bounded, that is, whenever G/F^* has finite exponent n, since then $\mathcal{O}_{G/F^*} = \mathbb{D}_n$. When the G-radical extension E/F is unbounded, the inclusion $\mu_G(E) \subseteq \mu(E)$ may be strict.

Let E/F be an arbitrary G-radical extension, and let $\alpha \in G$. If $m = \mathrm{ord}(\widehat{\alpha})$, then $m \in \mathcal{O}_{G/F^*}$ and

$$(f_\alpha(\sigma))^m = (\sigma(\alpha) \cdot \alpha^{-1})^m = \sigma(\alpha^m) \cdot \alpha^{-m} = \alpha^m \cdot \alpha^{-m} = 1,$$

hence $f_\alpha(\sigma)) \in \mu_m(E) \subseteq \mu_G(E)$ for all $\sigma \in \mathrm{Gal}\,(E/F)$, i.e., $\mathrm{Im}(f_\alpha) \in \mu_G(E)$.

We deduce that whenever E/F is a G-radical Galois extension, then the group isomorphism

$$\psi : \mathrm{Cog}\,(E/F) \longrightarrow Z_c^1(\mathrm{Gal}\,(E/F), \mu(E))$$

induces by restriction to G/F^* a monomorphism

$$\psi_G : G/F^* \longrightarrow Z_c^1(\mathrm{Gal}\,(E/F), \mu_G(E)).$$

The next result shows that if additionally the extension E/F is G-Cogalois, then the monomorphism ψ_G is also surjective, in other words, it is a group isomorphism.

THEOREM 15.1.4. *For any Galois G-Cogalois extension E/F, the map $\widehat{\alpha} \mapsto f_\alpha$ yields a group isomorphism*

$$\mathrm{Kne}\,(E/F) \cong Z_c^1(\mathrm{Gal}\,(E/F), \mu_G(E)).$$

PROOF. Denote by Γ the Galois group of E/F. Since Γ is a profinite group, every $f \in Z_c^1(\Gamma, \mu_G(E))$ is locally constant, and so, taking into account that $\mathrm{lcm}\,(s, t) \in \mathcal{O}_{G/F^*}$ for all $s, t \in \mathcal{O}_{G/F^*}$ by Lemma 1.4.3 (2), we deduce that

$$Z_c^1(\Gamma, \mu_G(E)) = \bigcup_{m \in \mathcal{O}_{G/F^*}} Z_c^1(\Gamma, \mu_m(E)).$$

Let $h \in Z_c^1(\Gamma, \mu_G(E))$. Then, there exists $m \in \mathcal{O}_{G/F^*}$ such that $h \in Z_c^1(\Gamma, \mu_m(E))$. But, the map $\psi_m : \mathrm{Cog}_m(E/F) \longrightarrow Z_c^1(\Gamma, \mu_m(E))$ is an isomorphism by Lemma 15.1.1, hence $h = \psi_m(\widehat{\alpha})$ for some $\alpha \in T_m(E/F)$. Since clearly $\mathcal{P}_m \subseteq \mathcal{P}_G$, Lemma 12.1.7 (1) implies that $\alpha \in G$. Thus $\widehat{\alpha} \in G/F^* = \mathrm{Kne}(E/F)$, and so $h = \psi_m(\widehat{\alpha}) = \psi_G(\widehat{\alpha})$, which proves that ψ_G is a surjective map, and we are done. $\square$

COROLLARY 15.1.5. *Let E/F be a Galois n-bounded G-Cogalois extension. Then*

$$\mathrm{Kne}\,(E/F) \cong Z_c^1(\mathrm{Gal}\,(E/F), \mu_n(E)).$$

PROOF. We have already noticed that $\mu_G(E) = \mu_n(E)$ whenever E/F is n-bounded. Now apply Theorem 15.1.4. $\square$

For any Galois G-Cogalois extension E/F, the map

$$f : \mathrm{Gal}\,(E/F) \times \mathrm{Cog}\,(E/F) \longrightarrow \mu(E),$$

considered at the beginning of this section yields by restriction the map

$$g : \mathrm{Gal}\,(E/F) \times \mathrm{Kne}\,(E/F) \longrightarrow \mu_G(E),$$

$$g(\sigma, \widehat{\alpha}) = f_\alpha(\sigma) = \sigma(\alpha) \cdot \alpha^{-1}.$$

For every $\Delta \leqslant \mathrm{Gal}\,(E/F)$ and $W \leqslant \mathrm{Kne}\,(E/F)$ let

$$\Delta^\top = \{\, c \in \mathrm{Kne}\,(E/F) \mid g(\sigma, c) = 1, \ \forall\, \sigma \in \Delta \,\},$$

$$W^\top = \{\, \sigma \in \mathrm{Gal}\,(E/F) \mid g(\sigma, c) = 1, \ \forall\, c \in W \,\}.$$

PROPOSITION 15.1.6. *For any Galois G-Cogalois extension E/F, the assignments $(-)^\top$ define mutually inverse anti-isomorphisms between the lattices* Closed subgroups $(\mathrm{Gal}\,(E/F))$ *and* Subgroups $(\mathrm{Kne}(E/F))$.

PROOF. The proof is literally the same as that of Proposition 6.1.8 if we replace the words "subgroup of $\mathrm{Gal}(E/F)$" with "closed subgroups of $\mathrm{Gal}(E/F)$", and therefore is omitted.   ⊔

Let E/F be a Galois extension with Galois group Γ. Then, by Theorem 15.1.2, there exists a canonical isomorphism $\mathrm{Cog}\,(E/F) \cong Z_c^1(\Gamma, \mu(E))$, hence the canonical map

$$f : \mathrm{Gal}\,(E/F) \times \mathrm{Cog}\,(E/F) \longrightarrow \mu(E),$$

$$f(\sigma, \widehat{\alpha}) = \sigma(\alpha) \cdot \alpha^{-1},$$

considered above produces, by replacing $\mathrm{Cog}\,(E/F)$ with its isomorphic copy $Z_c^1(\Gamma, \mu(E))$, precisely the *evaluation map*

$$\langle -, - \rangle : \Gamma \times Z_c^1(\Gamma, \mu(E)) \longrightarrow \mu(E), \ \ \langle \sigma, h \rangle = h(\sigma).$$

For any $\Delta \leqslant \Gamma$, $U \leqslant Z_c^1(\Gamma, \mu(E))$, and $\chi \in Z_c^1(\Gamma, \mu(E))$ we shall denote

$$\Delta^\perp = \{\, h \in Z_c^1(\Gamma, \mu(E)) \mid \langle \sigma, h \rangle = 1, \ \forall\, \sigma \in \Delta \,\},$$

$$U^\perp = \{\, \sigma \in \Gamma \mid \langle \sigma, h \rangle = 1, \ \forall\, h \in U \,\},$$

$$\chi^\perp = \{\, \sigma \in \Gamma \mid \langle \sigma, \chi \rangle = 1 \,\}.$$

One verifies easily that $\Delta^\perp \leqslant Z_c^1(\Gamma, \mu(E))$, $U^\perp \leqslant \Gamma$, and $\chi^\perp = \langle \chi \rangle^\perp$.

The next result, characterizing radical subextensions of a given Galois extension E/F by means of subgroups of $Z_c^1(\mathrm{Gal}\,(E/F), \mu(E))$, is the infinite variant of Theorem 6.2.1.

THEOREM 15.1.7. *Let E/F be a Galois extension with Galois group Γ, and let K be an intermediate field of the extension E/F. Then K/F is a radical extension (resp. a simple radical extension) if and only if there exists $U \leqslant Z_c^1(\Gamma, \mu(E))$ (resp. $\chi \leqslant Z_c^1(\Gamma, \mu(E))$) such that $\mathrm{Gal}(E/K) = U^\perp$ (resp. $\mathrm{Gal}(E/K) = \chi^\perp$).*

PROOF. The proof is the same as that of Theorem 6.2.1 except that we replace $Z^1(\Gamma, \mu(E))$) with $Z_c^1(\Gamma, \mu(E))$) and apply the infinite variant of the Fundamental Theorem of Galois Theory. $\square$

The next result is the infinite variant of Corollary 6.2.3.

COROLLARY 15.1.8. *Let E/F be a Galois extension with Galois group Γ, and let K/F be a G-radical subextension of E/F. For every H with $F^* \leqslant H \leqslant G$ set $U_H = \{\, f_\alpha \mid \alpha \in H \,\} \leqslant Z_c^1(\Gamma, \mu(E))$. Then, the following assertions hold.*

(1) *The extension K/F is G-Kneser if and only if $(\Gamma : U_H^\perp) = |U_H|$ for every H with $F^* \leqslant H \leqslant G$ and H/F^* finite.*

(2) *The extension K/F is G-Cogalois if and only if it is G-Kneser and the "perpendicular" map $V \mapsto V^\perp$ yields a bijection, or equivalently, an anti-isomorphism of lattices*

$$\{\, V \mid V \leqslant U_G \,\} \longrightarrow \{\, \Delta \mid U_G^\perp \leqslant \Delta \leqslant \Gamma, \ \Delta \ \text{closed subgroup of } \Gamma \,\}.$$

PROOF. (1) By Proposition 11.1.2, the extension K/F is G-Kneser if and only if for every H with $F^* \leqslant H \leqslant G$ and H/F^* finite, the finite extension $F(H)/F$ is H-Kneser, i.e., $[F(H) : F] = |H/F^*|$. If we set $\Delta_H = \mathrm{Gal}(E/F(H))$, then $[F(H) : F] = (\Gamma : \Delta_H)$ by Galois Theory, and $H/F^* \cong U_H$ by Corollary 15.1.3, so $|H/F^*| = |U_H|$. On the other hand, $\Delta_H = U_H^\perp$ according to Theorem 15.1.7. Summing up, we obtain

$$[F(H) : F] = |H/F^*| \iff (\Gamma : \Delta_H) = |U_H| \iff (\Gamma : U_H^\perp) = |U_H|.$$

(2) The proof can be achieved by replacing in the proof of Corollary 6.2.3 (2) the words "subgroup of $\mathrm{Gal}(E/F)$" with "closed subgroups of $\mathrm{Gal}(E/F)$", and by using the Fundamental Theorem of Infinite Galois Theory instead of the Fundamental Theorem of Finite Galois Theory. $\square$

Next, we generalize from finite to infinite extensions the results of Section 6.2 concerning the transfer under the base field change of the property of a Galois extension being radical, Kneser, or G-Cogalois.

For any extension E/F we have denoted in Section 6.2 by $\underline{\mathrm{Radical}}(E/F)$ the set of all subextensions K/F of E/F which are radical. If E/F is a

Galois extension, which is not necessarily finite, and L/F is any extension with $L \cap E = F$, and such that E and L are subfields of some other field, then according to Corollary 14.1.7, the map

$$\mathrm{Gal}\,(EL/L) \xrightarrow{\sim} \mathrm{Gal}\,(E/F), \ \sigma \mapsto \sigma_{|E},$$

is an isomorphism of topological groups.

Using this map, it follows as in Section 6.2 that there exists an injective map

$$\rho : \underline{\mathrm{Radical}}\,(E/F) \longrightarrow \underline{\mathrm{Radical}}\,(EL/L),$$

$$F(G)/F \mapsto F(G)L/L = L(GL^*)/L, \ F^* \leqslant G \leqslant T(E/F).$$

The next result is the infinite variant of Theorem 6.2.4.

THEOREM 15.1.9. *Let E/F be a Galois extension with Galois group Γ, and let L/F be an arbitrary field extension such that $E \cap L = F$. If E and L are subfields of some other field, and $\mu(EL) = \mu(E)$, then the following assertions hold.*

(1) $GL^* \cap E^* = G$ *for every* G *with* $F^* \leqslant G \leqslant T(E/F)$.
(2) $G_1 = (G_1 \cap E^*)L^*$ *for every* G_1 *with* $L^* \leqslant G_1 \leqslant T(EL/L)$.
(3) *The map*

$$\rho : \underline{\mathrm{Radical}}\,(E/F) \longrightarrow \underline{\mathrm{Radical}}\,(EL/L),$$

$$F(G)/F \mapsto L(GL^*)/L, \ F^* \leqslant G \leqslant T(E/F),$$

is bijective, and the map

$$\underline{\mathrm{Radical}}\,(EL/L) \longrightarrow \underline{\mathrm{Radical}}\,(E/F),$$

$$L(G_1)/L \mapsto F(G_1 \cap E^*)/F, \ L^* \leqslant G_1 \leqslant T(EL/L),$$

is its inverse.

PROOF. The proof is literally the same as that of Theorem 6.2.4, and therefore is left to the reader. $\square$

REMARK 15.1.10. The restriction of the bijection ρ in Theorem 15.1.9 to Kneser or to strongly Kneser subextensions of E/F and EL/L is still bijective by Exercise 4. $\square$

15.2. Lattice-isomorphic groups

The aim of this section is two-fold: first, to provide the terminology and notation concerning lattice-isomorphisms of groups, and second, to establish a basic result on lattice-isomorphic torsion Abelian groups, which will be needed in the next section.

The reader is assumed to be familiar with the main concepts and facts of Lattice Theory, as exposed e.g., in the monographs of Birkhoff [36], Crawley and Dilworth [49], or Grätzer [62].

The letters A, B, C, D, G, H, ... will be used to denote groups. The group morphisms will be denoted by f, g, h, ... For any group A, the lattice Subgroups(A) of all subgroups of A will be denoted briefly by $\mathbb{L}(A)$. The *index* of a subgroup B of a group A, which is either a positive integer or ∞, will be denoted as usually by $(A : B)$.

We shall denote by **Gr** the category of all groups, by **Pos** the category of all posets, and by **Lat** the category of all lattices. The morphisms in the category **Pos** are the order-preserving (i.e., increasing) maps. Recall that if L and L' are lattices, then a morphism in **Lat** from L to L' is a map $\alpha : L \longrightarrow L'$ satisfying the following two conditions:

$$\alpha(x \vee y) = \alpha(x) \vee \alpha(y) \quad \text{and} \quad \alpha(x \wedge y) = \alpha(x) \wedge \alpha(y), \ \forall x, y \in L.$$

It is easily seen that any lattice isomorphism between complete lattices commutes with arbitrary meets and joins. The morphisms in **Lat** will be denoted by small Greek letters α, β, γ, φ, ψ, ...

Clearly, any group morphism $f : A \longrightarrow B$ yields a morphism in **Pos**

$$\mathbb{L}(f) : \mathbb{L}(A) \longrightarrow \mathbb{L}(B), \quad H \mapsto f(H),$$

which is not necessarily a morphism in **Lat**.

A map $\alpha : L \longrightarrow L'$ is an isomorphism in **Lat** if and only if α is an order-preserving bijection such that its inverse α^{-1} is also an order-preserving map, in other words, α is an isomorphism in the category **Pos**. This implies that for any isomorphism $f : A \longrightarrow B$ in **Gr**, the map

$$\mathbb{L}(f) : \mathbb{L}(A) \longrightarrow \mathbb{L}(B), \quad H \mapsto f(H),$$

is an isomorphism in **Lat**. This fact can be described briefly by saying that we have a canonical map

$$\mathbb{L}_{A,B} : \operatorname{Isom}_{\mathbf{Gr}}(A, B) \longrightarrow \operatorname{Isom}_{\mathbf{Lat}}(\mathbb{L}(A), \mathbb{L}(B)), \quad f \mapsto \mathbb{L}(f),$$

where, if $\mathcal{C}$ is any category, then $\operatorname{Isom}_{\mathcal{C}}(X, Y)$ denotes the set, possibly empty, of all isomorphisms from the object X of $\mathcal{C}$ to the object Y of $\mathcal{C}$.

DEFINITION 15.2.1. *A* lattice-isomorphism *from a group A to a group B is any isomorphism* $\varphi \in \mathrm{Isom}_{\mathbf{Lat}}(\mathbb{L}(A), \mathbb{L}(B))$. *The lattice-isomorphism φ is said to be* induced by a group isomorphism *if there exists an isomorphism $f : A \longrightarrow B$ such that $\varphi = \mathbb{L}(f)$, and in that case, φ is said to be* induced by f. *The groups A and B are called* lattice-isomorphic *if* $\mathrm{Isom}_{\mathbf{Lat}}(\mathbb{L}(A), \mathbb{L}(B)) \neq \varnothing$, *and we denote this situation by $A \cong_{\mathrm{L}} B$.* □

Note that if A and B are isomorphic groups, then, as usually, we shall denote this situation by $A \cong B$. Clearly, if $A, B \in \mathbf{Gr}$ and $A \cong B$, then $A \cong_{\mathrm{L}} B$, but not conversely, as a classical example discovered in 1928 by A. Rottlaender shows (see Baer [**27**]). Therefore, the following natural question arises:

Given a class $\mathcal{X}$ of groups, what kind of conditions (C) on lattice-isomorphisms of groups should be imposed such that for every $A \in \mathcal{X}$ and for every $B \in \mathbf{Gr}$, every lattice-isomorphism $A \cong_{\mathrm{L}} B$ satisfying the conditions (C) implies that $A \cong B$?

The next definitions present three conditions for lattice-isomorphisms of groups, which will be involved in answering the question above for $\mathcal{X}$, the class of all Abelian torsion groups (see Theorem 15.2.7).

DEFINITIONS 15.2.2. *A lattice-isomorphism $\varphi \in \mathrm{Isom}_{\mathbf{Lat}}(\mathbb{L}(A), \mathbb{L}(B))$ between the groups A and B is said to be* index-preserving *(resp. strictly* index-preserving*) if*

$$(C : D) = (\varphi(C) : \varphi(D))$$

for every cyclic subgroup (resp. subgroup) C of A and for every $D \leqslant C$.

The lattice-isomorphism φ is said to be normal *if $\varphi(N) \lhd B$ for every $N \lhd A$.* □

The proof of the main result of this section is essentially based on the next four lemmas, whose proofs are too long and technical to be included here.

Recall that a group A is called *locally cyclic* if every finite subset of A generates a cyclic subgroup of A. In particular, any locally cyclic group is Abelian. In view of a classical result due to Ore (see e.g., Schmidt [**95**, Theorem 1.2.3]), a group A is locally cyclic if and only if the lattice $\mathbb{L}(A)$ is distributive. Therefore, such groups are also called *distributive groups*, or shortly, *D-groups*.

LEMMA 15.2.3. *Let A be a distributive group and let B be any group. A lattice-isomorphism $\varphi : \mathbb{L}(A) \xrightarrow{\sim} \mathbb{L}(B)$ is induced by a group isomorphism $A \xrightarrow{\sim} B$ if and only if φ is index-preserving.*

PROOF. See Baer [**27**, Corollary 4.4].   □

LEMMA 15.2.4. *For any normal lattice-isomorphism* $A \cong_\mathbb{L} B$, B *is Abelian whenever* A *is Abelian.*

PROOF. See Baer [**27**, Theorem 7.1]).   □

Recall that for any group A and for any prime number p we have denoted by $t_p(A)$ its *p-primary component*, that is,

$$t_p(A) = \{\, a \in A \,|\, \mathrm{ord}(a) = p^n \text{ for some } n \in \mathbb{N} \,\}.$$

It is well-known that any torsion Abelian group A can be decomposed into the internal direct sum of its p-primary components:

$$A = \bigoplus_{p \in \mathbb{P}} t_p(A).$$

See 1.1.2 for the concept of internal direct sum of a family of subgroups of a group.

LEMMA 15.2.5. *Let* A *be a torsion Abelian group, and let* B *be a group which is lattice isomorphic to* A *via* $\varphi : \mathbb{L}(A) \xrightarrow{\sim} \mathbb{L}(B)$. *Then* B *is a torsion group, and*

$$B = \bigoplus_{q \in \mathbb{P}} \varphi(t_q(A))$$

is an internal direct sum decomposition of the group B *into p-subgroups.*

PROOF. See Baer [**27**, 11.1, Theorem 2.4].   □

LEMMA 15.2.6. *Let* A *and* B *be two lattice-isomorphic Abelian groups. If* A *is an Abelian p-group which is not distributive, then* $A \cong B$.

PROOF. See Baer [**27**, Theorem 11.8 (a3) and (c)].   □

THEOREM 15.2.7. *Let* A *be a torsion Abelian group, and let* B *be a group which is lattice-isomorphic to* A *via* $\varphi : \mathbb{L}(A) \xrightarrow{\sim} \mathbb{L}(B)$. *If* φ *is index-preserving and normal, then* $A \cong B$.

PROOF. First of all, note that B is Abelian by Lemma 15.2.4. Since A is an Abelian torsion group, we can decompose A into the internal direct sum $A = \bigoplus_{p \in \mathbb{P}} t_p(A)$ of its p-primary components $t_p(A)$. The lattice-isomorphism $\varphi : \mathbb{L}(A) \xrightarrow{\sim} \mathbb{L}(B)$ obviously yields for every $p \in \mathbb{P}$ a lattice isomorphism $\varphi_p : \mathbb{L}(t_p(A)) \xrightarrow{\sim} \mathbb{L}(\varphi(t_p(A)))$, which is also index-preserving.

On the other hand, according to Lemma 15.2.5, $B = \bigoplus_{p \in \mathbb{P}} \varphi(t_p(A))$, so, if we could show that for every $p \in \mathbb{P}$ there exists an isomorphism $f_p : t_p(A) \xrightarrow{\sim} \varphi(t_p(A))$, then clearly $\bigoplus_{p \in \mathbb{P}} f_p : A \longrightarrow B$ will provide a desired isomorphism.

Thus, without loss of generality, we can suppose that A is an Abelian p-group and B is an Abelian group which is lattice-isomorphic to A.

If A is a distributive group, then $\varphi = \mathbb{L}(f)$ for some $f \in \text{Isom}_{\mathbf{Gr}}(A, B)$ by Lemma 15.2.3, hence $A \cong B$. If A is not a distributive group, then apply Lemma 15.2.6 to obtain an isomorphism $A \cong B$. $\qquad\square$

15.3. Infinite Abelian G-Cogalois extensions

In this section we investigate infinite Abelian G-Cogalois extensions. Mainly, we are interested to find out how the Abelian groups $\text{Kne}(E/F)$ and $\text{Gal}(E/F)$ associated with an arbitrary Abelian G-Cogalois extension E/F are related. In the case of finite extensions these two groups are isomorphic by Theorem 5.2.2, but not in a canonical way.

Before stating and proving the main result of this section, we shall reformulate Theorem 5.2.2 and Theorem 6.1.7 together, dealing with finite Abelian G-Cogalois extensions, in a form which is suitable to be generalized for infinite extensions.

PROPOSITION 15.3.1. *Let E/F be a finite Abelian G-Cogalois extension with $n = \exp(G/F^*)$. Then*

$$\text{Kne}(E/F) \cong \text{Ch}(\text{Gal}(E/F)) \cong Z^1(\text{Gal}(E/F), \mu_n(E)).$$

PROOF. By Theorem 6.1.7, there exists a canonical group isomorphism

$$\text{Kne}(E/F) \cong Z^1(\text{Gal}(E/F), \mu_n(E)),$$

and by Theorem 5.2.2, there exists a non-canonical group isomorphism

$$\text{Kne}(E/F) \cong \text{Gal}(E/F).$$

Further, the finite Abelian group $\text{Gal}(E/F)$ is isomorphic, but not in a canonical way, with its character group by Lemma 14.2.1 (4). $\qquad\square$

REMARKS 15.3.2. (1) Observe that, by Lemma 14.2.1 (2),

$$\text{Ch}(\text{Gal}(E/F)) = \text{Hom}(\text{Gal}(E/F), \mu_n(\mathbb{C}))$$

for any finite Abelian G-Cogalois extension E/F with $n = \exp(G/F^*)$. Since E/F is an n-bounded Galois extension, we have $\gcd(n, e(F)) = 1$ and

$\zeta_n \in E$ by Corollary 5.1.7, hence $\mu_n(E) = \mu_n(\Omega) \cong \mu_n(\mathbb{C})$. Consequently, there exists a *non-canonical* group isomorphism

$$Z^1(\text{Gal}(E/F), \mu_n(E)) \cong \text{Hom}(\text{Gal}(E/F), \mu_n(E)).$$

Note that both $Z^1(\text{Gal}(E/F), \mu_n(E))$ and $\text{Hom}(\text{Gal}(E/F), \mu_n(E))$ are subgroups of the group $\text{Maps}(\text{Gal}(E/F), \mu_n(E))$ of all maps from the group $\text{Gal}(E/F)$ to the group $\mu_n(E)$. Exercise 18 asks when these two isomorphic subgroups of $\text{Maps}(\text{Gal}(E/F), \mu_n(E))$ coincide.

(2) If additionally $\zeta_n \in F$, that is, if E/F is a finite classical Kummer extension of exponent n, then $\mu_n(E) = \mu_n(F)$, hence, every 1-cocycle f of $\text{Gal}(E/F)$ with coefficients in $\mu_n(E) = \mu_n(F)$ is actually a morphism of groups, since $f(\sigma\tau) = f(\sigma) \cdot \sigma(f(\tau)) = f(\sigma) \cdot f(\tau)$, for every $\sigma, \tau \in \text{Gal}(E/F)$. Thus,

$$Z^1(\text{Gal}(E/F), \mu_n(E)) = \text{Hom}(\text{Gal}(E/F), \mu_n(E)),$$

and then, we obtain a *canonical* group isomorphism

$$\text{Kne}(E/F) \cong \text{Hom}(\text{Gal}(E/F), \mu_n(E)).$$

Note that this yields a group isomorphism

$$\text{Kne}(E/F) \cong \text{Ch}(\text{Gal}(E/F)),$$

which is *not canonical* since it depends on the chosen isomorphism of cyclic groups $\mu_n(E) \overset{\sim}{\longrightarrow} \mu_n(\mathbb{C})$.   □

For the rest of this section we shall preserve the notation and terminology from Chapter 14. In particular, **LCA** denotes the category of all locally compact Abelian groups. For any group A, the lattice $\underline{\text{Subgroups}}(A)$ of all subgroups of A will be denoted shortly, as in the previous section, by $\mathbb{L}(A)$. Also, if A is a topological group, the lattice $\underline{\text{Closed Subgroups}}(A)$ of all closed subgroups of A will be denoted by $\overline{\mathbb{L}}(A)$.

As in Section 14.2, for any $A \in \textbf{LCA}$ and any subsets $\varnothing \neq X \subseteq A$ and $\varnothing \neq Y \subseteq \widehat{A}$ we shall use the following notation:

$$X^{\perp} = \{\, \chi \in \widehat{A} \mid \chi(a) = 1,\ \forall\, a \in X \,\},$$

$$Y^{\perp} = \{\, a \in A \mid \chi(a) = 1,\ \forall\, \chi \in Y \,\}.$$

PROPOSITION 15.3.3. *If E/F is any Abelian G-Cogalois extension, then, the groups $\text{Kne}(E/F)$ and $\text{Ch}(\text{Gal}(E/F))$ are lattice-isomorphic via the canonical lattice isomorphism*

$$\mathbb{L}(\text{Kne}(E/F)) \overset{\sim}{\longrightarrow} \mathbb{L}(\widehat{\text{Gal}(E/F)}),\ H/F^* \mapsto \text{Gal}(E/F(H))^{\perp}.$$

PROOF. Consider the following canonical maps

$\gamma_1 \; : \; \mathbb{L}(\mathrm{Kne}\,(E/F)) \longrightarrow \underline{\mathrm{Subextensions}}\,(E/F), \; H/F^* \mapsto F(H)/F,$

$\gamma_2 \; : \; \underline{\mathrm{Subextensions}}\,(E/F) \longrightarrow \overline{\mathbb{L}}(\mathrm{Gal}(E/F)), \; K/F \mapsto \mathrm{Gal}(E/K),$

$\gamma_3 \; : \; \overline{\mathbb{L}}(\mathrm{Gal}(E/F)) \longrightarrow \mathbb{L}(\widehat{\mathrm{Gal}(E/F)}), \; B \mapsto B^{\perp}.$

Since γ_1 is an isomorphism of lattices by Theorem 11.2.6, γ_2 is an anti-isomorphism of lattices by the Fundamental Theorem of Infinite Galois Theory (Theorem 14.1.6), and γ_3 is an anti-isomorphism of lattices by Proposition 14.1.5 and Lemma 14.2.5 (1), we deduce that their composition $\gamma = \gamma_3 \circ \gamma_2 \circ \gamma_1$ yields an isomorphism of lattices

$$\gamma : \mathbb{L}(\mathrm{Kne}\,(E/F)) \longrightarrow \mathbb{L}(\widehat{\mathrm{Gal}\,(E/F)}), \; H/F^* \mapsto \mathrm{Gal}(E/F(H))^{\perp},$$

and we are done.   □

Next, we are going to extend Proposition 15.3.1 to infinite extensions. To do that, we need some preparatory results.

LEMMA 15.3.4. *Let* $A \in \mathbf{LCA}$ *and* $B \in \overline{\mathbb{L}}(A)$. *Then*

$$0 \longrightarrow \widehat{A/B} \longrightarrow \widehat{A} \longrightarrow \widehat{B} \longrightarrow 0$$

is an exact sequence in the category $\mathbf{LCA}$ *of all locally compact Abelian groups. In particular, there exist canonical isomorphisms in* $\mathbf{LCA}$

$$\widehat{A/B} \cong B^{\perp} \quad and \quad \widehat{B} \cong \widehat{A}/B^{\perp}.$$

PROOF. See Bourbaki [**39**, Théorème 2, Chap. 2] or Hewitt and Ross [**72**, Theorem 23.25, Theorem 24.11, and Corollary 24.12].   □

LEMMA 15.3.5. *Let* E/F *be a* G-*Cogalois extension. Then*

$$|\,H_2/H_1\,| = [\,F(H_2) : F(H_1)\,]$$

for any subgroups H_1, H_2 *of* G, *with* $F^* \leqslant H_1 \leqslant H_2$.

PROOF. The extension $F(H_2)/F$ is H_2-Kneser by Proposition 11.1.2. Then, the extension $F(H_2)/F(H_1)$ is $F(H_1)^*H_2$-Kneser by Proposition 11.2.2, and $F(H_1)^* \cap H_2 = F(H_1) \cap H_2 = H_1$ by Lemma 11.2.1. Again by Proposition 11.1.2, every set of representatives of

$$(F(H_1)^*H_2)/F(H_1)^* \cong H_2/(F(H_1)^* \cap H_2) = H_2/H_1$$

is a vector space basis, possibly infinite, of $F(H_2)$ over $F(H_1)$. This implies the desired equality.   □

LEMMA 15.3.6. *Let $K_1 \subseteq K_2$ be two intermediate fields of an Abelian extension E/F. If K_2/K_1 is a finite extension, then*

$$(\text{Gal } (E/K_2)^{\perp} : \text{Gal } (E/K_1)^{\perp}) = [K_2 : K_1],$$

where "$\perp$" is taken in the group $\text{Ch}(\text{Gal}(E/F))$.

PROOF. For simplicity, set $A = \text{Gal}(E/F)$ and $B_i = \text{Gal}(E/K_i)$, $i = 1, 2$. By Lemma 15.3.4, the exact sequence of compact Abelian groups

$$0 \longrightarrow B_1/B_2 \longrightarrow A/B_2 \longrightarrow A/B_1 \longrightarrow 0$$

yields the exact sequence of discrete Abelian groups

$$0 \longrightarrow \widehat{A/B_1} \longrightarrow \widehat{A/B_2} \longrightarrow \widehat{B_1/B_2} \longrightarrow 0.$$

Also, we have $\widehat{A/B_i} \cong B_i^{\perp}$, $i = 1, 2$. Since the diagram below

$$
\begin{array}{ccc}
\widehat{A/B_1} & \longrightarrow & \widehat{A/B_2} \\
\downarrow{\imath} & & {\imath}\downarrow \\
B_1^{\perp} & \longrightarrow & B_2^{\perp}
\end{array}
$$

is clearly commutative, we deduce that

$$B_2^{\perp}/B_1^{\perp} \cong \widehat{B_1/B_2}.$$

Since E/F is an Abelian extension, we have

$$B_1/B_2 = \text{Gal}(E/K_1)/\text{Gal}(E/K_2) \cong \text{Gal}(K_2/K_1),$$

by Theorem 14.1.6 (3). But K_2/K_1 is a finite Abelian extension, hence its Galois group is a finite Abelian group, and so, by Lemma 14.2.1 (4), we have

$$\text{Gal}(K_2/K_1) \cong \widehat{\text{Gal}(K_2/K_1)}.$$

Thus,

$$(B_2^{\perp} : B_1^{\perp}) = |B_2^{\perp}/B_1^{\perp}| = |\widehat{B_1/B_2}| = |\widehat{\text{Gal}(K_2/K_1)}|$$
$$= |\text{Gal}(K_2/K_1)| = [K_2 : K_1].$$

$\square$

LEMMA 15.3.7. *If E/F is an Abelian G-Cogalois extension, then the isomorphism of lattices*

$$\gamma : \mathbb{L}(\text{Kne}(E/F)) \longrightarrow \mathbb{L}(\widehat{\text{Gal}(E/F)}), \ (H/F^*) \mapsto \text{Gal}(E/F(H))^{\perp}$$

given by Proposition 15.3.3 is strictly index-preserving.

PROOF. Let $C_1 \subseteq C_2$ be two subgroups of $\mathrm{Kne}(E/F) = G/F^*$ with the index $(C_2 : C_1)$ finite. Then, there exists $H_1, H_2 \in \mathbb{L}(G/F^*)$ such that

$$F^* \leqslant H_1 \leqslant H_2 \leqslant G \quad \text{and} \quad C_i = H_i/F^*, \; i = 1, 2.$$

We have to show that

$$(C_2 : C_1) = (\gamma(C_2) : \gamma(C_1)).$$

By Lemma 15.3.5 one has

$$(C_2 : C_1) = (H_2 : H_1) = [\, F(H_2) : F(H_1) \,],$$

so, it remains only to prove that

$$(\gamma(C_2) : \gamma(C_1)) = [\, F(H_2) : F(H_1) \,].$$

Set $K_i = F(H_i)$, $i = 1, 2$. Then $\gamma(C_i) = \mathrm{Gal}(E/K_i))^{\perp}$, hence

$$(\gamma(C_2) : \gamma(C_1)) = (\mathrm{Gal}(E/K_2)^{\perp} : \mathrm{Gal}(E/K_1)^{\perp})$$

$$= [\, K_2 : K_1 \,] = [\, F(H_2) : F(H_1) \,]$$

by Lemma 15.3.6, and we are done. $\qquad\square$

THEOREM 15.3.8. *If E/F is an Abelian G-Cogalois extension, then the discrete torsion Abelian groups $\mathrm{Kne}(E/F)$ and $\mathrm{Ch}(\mathrm{Gal}(E/F))$ are isomorphic.*

PROOF. By Proposition 15.3.3, the Abelian groups $\mathrm{Kne}(E/F)$ and $\mathrm{Ch}(\mathrm{Gal}(E/F))$ are lattice-isomorphic via the isomorphism of lattices

$$\gamma : \mathbb{L}(\mathrm{Kne}(E/F)) \longrightarrow \mathbb{L}(\widehat{\mathrm{Gal}(E/F)})$$

defined by

$$\gamma(H/F^*) = \mathrm{Gal}(E/F(H))^{\perp},$$

for every $H/F^* \leqslant \mathrm{Kne}(E/F) = G/F^*$, where $F^* \leqslant H \leqslant G$.

Since $\mathrm{Kne}(E/F) \leqslant \mathrm{Cog}(E/F)$, and $\mathrm{Cog}(E/F)$ is precisely the torsion subgroup of the Abelian group E^*/F^*, it follows that $\mathrm{Kne}(E/F)$ is a torsion Abelian group. Further, the lattice isomorphism γ is index-preserving by Lemma 15.3.7, and it is obviously normal, since $\mathrm{Ch}(\mathrm{Gal}(E/F))$ is an Abelian group. Now, apply Theorem 15.2.7 to obtain the desired result. $\qquad\square$

REMARKS 15.3.9. (1) When applied to finite Abelian G-Cogalois extensions, the proof of Theorem 15.3.8 provides an alternate proof of Theorem 5.2.2.

(2) A shorter approach to the result of Theorem 15.3.8, which avoids the technical facts involving lattice-isomorphic groups and which is based on Theorem 5.2.2, could be the one presented below.

Let E/F be an Abelian G-Cogalois extension. It is convenient to index in an obvious way the directed set

$$\{\, H \mid F^* \leqslant H \leqslant G, \ H/F^* \text{ is finite} \,\}$$

in order to obtain a direct system $(G_i, \varphi_{ij})_{i,j \in I}$ of groups. Thus, for every $i, j \in I$ we set $i \leqslant j$ if and only if $G_i \subseteq G_j$, and for any $i \leqslant j$ we define the morphism $\varphi_{ij} : G_i \longrightarrow G_j$ as being the canonical inclusion.

Clearly

$$\mathrm{Kne}\,(E/F) = G/F^* = \bigcup_{i \in I} (G_i/F^*) = \varinjlim_{i \in I} (G_i/F^*).$$

Further, for every $i \in I$ we have

$$\mathrm{Ch}\,(G_i/F^*) = \mathrm{Hom}_c(G_i/F^*, \mathbb{T}) \cong G_i/F^*$$

$$= \mathrm{Kne}\,(F(G_i)/F) \cong \mathrm{Gal}\,(F(G_i)/F)$$

by Theorem 5.2.2. Note that the two isomorphisms above are *not canonical*. Thus, we obtain the following isomorphisms of topological groups:

$$\mathrm{Ch}\,(\mathrm{Kne}\,(E/F)) = \mathrm{Ch}\,(G/F^*) = \mathrm{Hom}_c(G/F^*, \mathbb{T})$$

$$= \mathrm{Hom}_c(\varinjlim_{i \in I} (G_i/F^*), \mathbb{T}) \cong \varprojlim_{i \in I} \mathrm{Hom}_c(G_i/F^*, \mathbb{T})$$

$$\cong \varprojlim_{i \in I} \mathrm{Gal}\,(F(G_i)/F) \cong \mathrm{Gal}\,(E/F).$$

Note that the last isomorphism follows from Proposition 14.1.5 (see also Exercise 3, Chapter 14) since E/F is an Abelian extension, hence, for every $i \in I$ there exists a canonical group isomorphism

$$\mathrm{Gal}\,(F(G_i)/F) \cong \mathrm{Gal}\,(E/F)/\mathrm{Gal}\,(E/F(G_i)).$$

The approach above contains a serious gap which could affect its validity; namely, the two isomorphisms

$$\varprojlim_{i \in I} \mathrm{Hom}_c(G_i/F^*, \mathbb{T}) \cong \varprojlim_{i \in I} \mathrm{Gal}\,(F(G_i)/F) \cong \mathrm{Gal}\,(E/F)$$

considered are *not canonical*, hence it is not clear that we really obtain a projective system of groups which is coherent in such a way as to produce the Galois group $\mathrm{Gal}(E/F)$. $\square$

COROLLARY 15.3.10. *For any Abelian G-Cogalois extension E/F, the totally disconnected compact Abelian groups $\mathrm{Gal}(E/F)$ and $\mathrm{Ch}(\mathrm{Kne}(E/F))$ are topologically isomorphic.*

PROOF. By Theorem 15.3.8, there exists a group isomorphism

$$f : \mathrm{Kne}\,(E/F) \xrightarrow{\sim} \mathrm{Ch}\,(\mathrm{Gal}\,(E/F)),$$

which induces a topological isomorphism

$$\widehat{f} : \mathrm{Ch}\,(\mathrm{Kne}\,(E/F)) \xrightarrow{\sim} \mathrm{Ch}\,(\mathrm{Ch}\,(\mathrm{Gal}\,(E/F))).$$

By the Pontryagin Duality (see Theorem 14.2.2 (1)), the compact groups $\mathrm{Gal}\,(E/F)$ and $\mathrm{Ch}\,(\mathrm{Ch}\,(\mathrm{Gal}\,(E/F)))$ are canonically isomorphic. □

COROLLARY 15.3.11. *If E/F is an Abelian Cogalois extension, then the discrete torsion Abelian groups $\mathrm{Cog}\,(E/F)$ and $\mathrm{Ch}(\mathrm{Gal}\,(E/F))$ are isomorphic.*

PROOF. Any Cogalois extension E/F is $T(E/F)$-Cogalois by Theorem 12.2.3. Since $\mathrm{Kne}(E/F) = T(E/F)/F^* = \mathrm{Cog}(E/F)$, we can apply Theorem 15.3.8 to obtain the desired isomorphism. □

REMARK 15.3.12. Another approach to the proof of Theorem 15.3.8, which is sketched below, is due to the anonymous referee of this book.

LEMMA A. *Let E/F be an Abelian G-Cogalois extension. Assume that $G/F^* = \bigoplus_{i\in I}(G_i/F^*)$, and let $E_i = F(G_i)$ for every $i \in I$. If $\mathrm{Kne}(E_i/F) \cong \mathrm{Ch}(\mathrm{Gal}(E_i/F))$ for all $i \in I$, then $\mathrm{Kne}(E/F) \cong \mathrm{Ch}(\mathrm{Gal}(E/F))$.*

PROOF. We have $\mathrm{Kne}(E/F) = \bigoplus_{i\in I} \mathrm{Kne}(E_i/F)$ and $\mathrm{Gal}(E/F) \cong \prod_{i\in I} \mathrm{Gal}(E_i/F)$ with the product topology. Note that each extension E_i/F, $i \in I$, is an Abelian G_i-Cogalois extension. Since each continuous character of the product group kills almost all factors, we deduce that

$$\mathrm{Ch}(\mathrm{Gal}(E/F)) \cong \bigoplus_{i\in I} \mathrm{Ch}(\mathrm{Gal}(E_i/F)) \cong \bigoplus_{i\in I} \mathrm{Kne}(E_i/F) = \mathrm{Kne}(E/F).$$

Recall that for an Abelian group A and a prime number p we denote by $t_p(A)$ the p-primary component of A.

LEMMA B. *Let E/F be an Abelian G-Cogalois extension with Kneser group G/F^* a p-group, where p is a prime number. Assume that $t_p(\mu(F))$ is a finite group. Then, the following statements hold.*

(1) *If $t_p(\mu(E))$ is a finite group, then G/F^* is a group of bounded order.*

(2) *If $t_p(\mu(E))$ is an infinite group, then $G/F^* \cong t_p(\mathbb{Q}/\mathbb{Z}) \oplus B$, where B is a group of bounded order.*

PROOF. (1) Let $x \in G$ and $k = \mathrm{ord}(\widehat{x})$. Since $F(x)/F$ is $F^*\langle x \rangle$-Kneser, it follows that $\mathrm{Irr}(x, F) = X^k - x^k$, and so, all the conjugates of x over F are ζx, ζ running over $\mu_k(\overline{F})$, and they must lie in E. Therefore, $k \leqslant |t_p(\mu(E))|$, which shows that G/F^* is a group of bounded order.

(2) We denote by μ_{p^∞} the p-primary component $t_p(\mu(\overline{F}))$ of the group $\mu(\overline{F})$ of all roots of unity in an algebraic closure $\overline{F}$ of F. Since $t_p(\mu(E))$ is assumed to be an infinite group, then, by using Theorem 12.1.8, one shows that the group G/F^* contains a subgroup Q isomorphic to the quotient group $M := \mu_{p^\infty}/t_p(\mu(F))$. Now, observe that M is isomorphic to the quasi-cyclic group $\mathbb{Z}_{p^\infty} := t_p(\mathbb{Q}/\mathbb{Z})$ of type p^∞. Since M is a divisible Abelian group, we can write $G/F^* = Q \oplus (H/F^*)$ for some $F^* \leqslant H \leqslant G$. It remains to show that $B := H/F^*$ is a group of bounded order. But $K = F(H)$ cannot contain any p-power root of unity not already contained in F. Therefore, by the proof of (1), $\exp(B) \leqslant |t_p(\mu(F))|$, and we are done.

PROOF OF THEOREM 15.3.8. By decomposing the Kneser group of the given Abelian G-Cogalois extension E/F into its p-primary components, Lemma A shows that there is no loss of generality to assume that E/F is an Abelian G-Cogalois extension with Kneser group G/F^* a p-group. There are three cases:

Case 1: If $t_p(\mu(E))$ is a finite group, then G/F^* is a group of bounded order by Lemma B and therefore it is a direct sum of a family $(G_i/F^*)_{i \in I}$ of finite cyclic groups by Theorem 1.4.2. Hence E is the compositum of the family of fields $(E(G_i))_{i \in I}$. By Proposition 15.3.1, $\mathrm{Kne}(E_i/F) \cong \mathrm{Ch}(\mathrm{Gal}(E_i/F))$ for all $i \in I$. Lemma A now yields the result.

Case 2: If $t_p(\mu(F))$ is a finite group but $t_p(\mu(E))$ is an infinite group, then Lemma B allows us to write E as the compositum of $F_\infty := F(\mu_{p^\infty})$ and $K := F(H)$, where $F^* \leqslant H \leqslant G$ and H/F^* is a group of bounded order. By referring to the first case and using Lemma A we reduce the problem to $E = F_\infty$. In this case, $\mathrm{Kne}(E/F) \cong \mathbb{Z}_{p^\infty}$ and $\mathrm{Gal}(E/F)$ is isomorphic to the additive group of the ring $\widetilde{\mathbb{Z}}_p$ of p-adic integers. Note that on needs to use that $\zeta_{2p} \in F$ for both of these statements. Since $\mathrm{Ch}(\widetilde{\mathbb{Z}}_p) \cong \mathbb{Z}_{p^\infty}$ by Exercise 13, Chapter 14, we are done.

Case 3: If $t_p(\mu(F))$ is an infinite group, then E/F is a classical Kummer extension, and $\mathrm{Gal}(E/F)$ acts trivially on $t_p(\mu(E))$. Consequently,

$\mathrm{Ch}(\mathrm{Gal}(E/F)) = Z_c^1(\mathrm{Gal}(E/F), t_p(\mu(E)))$, and it suffices to apply Theorem 15.1.4. $\square$

15.4. Exercises to Chapter 15

1. Let E/F be an arbitrary Galois extension, and let K be any intermediate field of E/F. Show that the diagram below, where all the arrows are canonical morphisms,

$$
\begin{array}{ccc}
\mathrm{Cog}(E/F) & \longrightarrow & \mathrm{Cog}(E/K) \\
\downarrow{\wr} & & {\wr}\downarrow \\
Z_c^1(\mathrm{Gal}(E/F), \mu(E)) & \longrightarrow & Z_c^1(\mathrm{Gal}(E/K), \mu(E))
\end{array}
$$

   is commutative.

2. Let E/F be an arbitrary Galois Cogalois extension with Galois group Γ. Prove that the canonical morphism

$$Z_c^1(\Gamma, \mu(E)) \longrightarrow Z_c^1(\Delta, \mu(E))$$

   is surjective for every closed subgroup Δ of Γ. (*Hint*: Use Exercise 1 and Theorem 12.2.4 (4).)

3. Give an example of infinite G-radical extension E/F for which $\mu_G(E)$ is a strict subset of $\mu(E)$.

4. Let E/F be a Galois extension, and let L/F be an extension such that E and L are subfields of some other field, $E \cap L = F$, and $\mu(LE) = \mu(E)$. Let G be a group such that $F^* \leqslant G \leqslant T(E/F)$, and denote $G_1 = GL^*$. Prove the following statements.
   (a) $G/F^* \cong G_1/L^*$, and so, $\mathcal{O}_{G/F^*} = \mathcal{O}_{G_1/L^*}$.
   (b) The extension $F(G)/F$ is G-Kneser (resp. G-Cogalois) if and only if the extension $L(G_1)/L$ is G_1-Kneser (resp. G_1-Cogalois).

   (*Hint*: Apply the Infinite Kneser Criterion, Theorems 12.1.12, and Theorem 15.1.9.)

5. Let E/F be a Galois extension, and let L/F be an extension such that E and L are subfields of some other field and $E \cap L = F$. Let K be a subfield of E containing F such that KL/L is a G_1-radical extension and $\zeta_m \in E$ for all $m \in \mathcal{O}_{G_1/L^*}$, and set $G = G_1 \cap E^*$. Prove the following assertions.

(a) The extension K/F is G-radical and $G/F^* \cong G_1/L^*$.

(b) The extension K/F is G-Kneser (resp. G-Cogalois) if and only if the extension KL/L is G_1-Kneser (resp. G_1-Cogalois).

(*Hint*: Apply Corollary 15.1.3, Theorem 15.1.9, and Exercise 4.)

6. Let E/F be a Galois extension, and let L/F be an extension such that E and L are subfields of some other field and $E \cap L = F$. Let K be a subfield of E containing F such that KL/L is a G_1-radical extension with G_1/L^* a group of exponent n, and assume that $\zeta_n \in E$. If we set $G = G_1 \cap E^*$, then prove the following assertions.

(a) The extension K/F is G-radical and $G/F^* \cong G_1/L^*$.

(b) The extension K/F is G-Kneser (resp. G-Cogalois) if and only if the extension KL/L is G_1-Kneser (resp. G_1-Cogalois).

(*Hint*: Observe that $\mathcal{O}_{G_1/L^*} = \mathbb{D}_n$, and use Exercise 5.)

7. Assume that all fields below are subfields of some other field. Prove the following statements.

(a) If E_1/F and E_2/F are Abelian extensions, then so is E_1E_2/F.

(b) If E/F is an Abelian extension and K/F is any extension, then EK/K is an Abelian extension.

(c) Any subextension and any quotient extension of an Abelian extension are Abelian extensions.

8. Prove that a group G is finite if and only if the lattice $\mathbb{L}(G)$ is finite.

9. Let $\varphi, \psi \in \mathrm{Isom}_{\mathbf{Lat}}(\mathbb{L}(A), \mathbb{L}(B))$, where A and B are lattice-isomorphic groups. Show that $\varphi = \psi$ if and only if $\varphi(X) = \psi(X)$ for every cyclic subgroup X of A.

10. Prove that a group is distributive if and only if it is isomorphic to a subgroup of $\mathbb{Q}$ or of $\mathbb{Q}/\mathbb{Z}$.

11. Prove that a finite group is distributive if and only if it is cyclic.

12. (*Baer* [**27**]). Let A be a group. Prove that $A \cong_{\mathrm{L}} \mathbb{Z} \Longleftrightarrow A \cong \mathbb{Z}$.

13. (*Baer* [**27**]). Let $r, n_1, \dots, n_r \in \mathbb{N}^*$, and let $p_1, \dots, p_r \in \mathbb{P}$ be distinct primes. If A is a cyclic group of order $p_1^{n_1} \cdot \dots \cdot p_r^{n_r}$ and B is any group, then prove that $A \cong_{\mathrm{L}} B$ if and only if B is a cyclic group of order $q_1^{n_1} \cdot \dots \cdot q_r^{n_r}$ with distinct primes $q_1, \dots, q_r \in \mathbb{P}$.

14. Find classes $\mathcal{C}$ of groups for which the map

$$\mathbb{L}_{A,B} : \mathrm{Isom}_{\mathbf{Gr}}(A, B) \longrightarrow \mathrm{Isom}_{\mathbf{Lat}}(\mathbb{L}(A), \mathbb{L}(B)), \ f \mapsto \mathbb{L}(f),$$

is injective for all A, $B \in \mathcal{C}$.

15. We say that the groups A and B are *lattice-anti-isomorphic* if there exists a lattice anti-isomorphism between $\mathbb{L}(A)$ and $\mathbb{L}(B)$. Investigate the properties of lattice-anti-isomorphic groups.

16. (*Baer-Sadovskii Theorem*). Let A and B be two lattice-isomorphic groups, and let $\varphi : \mathbb{L}(A) \longrightarrow \mathbb{L}(B)$ be an isomorphism of lattices. Assume that there exists a nonempty set $\mathcal{H}$ of subgroups of A satisfying the following three conditions.
    (a) Every $H \in \mathcal{H}$ is finitely generated.
    (b) For every $a \in A$ there exists $H \in \mathcal{H}$ such that $a \in H$.
    (c) For any H_1, $H_2 \in \mathcal{H}$ there exists $H_3 \in \mathcal{H}$ with $H_1 \cup H_2 \subseteq H_3$.
    If for every $H \in \mathcal{H}$, $\varphi|_H$ is induced by an isomorphism of H onto $\varphi(H)$, then prove that φ is induced by an isomorphism $f : A \longrightarrow B$. (*Hint*: See Schmidt [**95**, Corollary 1.3.7].)

17. Let E/F be an Abelian G-Cogalois extension. By Theorem 15.3.8, there exists a non-canonical group isomorphism
$$f : \mathrm{Kne}\,(E/F) \overset{\sim}{\longrightarrow} \mathrm{Ch}\,(\mathrm{Gal}\,(E/F)),$$
and by Proposition 15.3.3, there exists a canonical lattice isomorphism
$$\gamma : \mathbb{L}(\mathrm{Kne}\,(E/F)) \overset{\sim}{\longrightarrow} \mathbb{L}(\mathrm{Ch}\,(\mathrm{Gal}\,(E/F))).$$
When is $\gamma = \mathbb{L}(f)$?

18. Characterize those n-bounded Abelian G-Cogalois extensions E/F for which the isomorphic groups $Z^1(\mathrm{Gal}\,(E/F), \mu_n(E))$ and $\mathrm{Hom}(\mathrm{Gal}\,(E/F), \mu_n(E))$ are equal.

15.5. Bibliographical comments to Chapter 15

Section 15.1. Except for Theorem 15.1.2, which is due to Barrera-Mora, Rzedowski-Calderón and Villa-Salvador [**30**], all of the results of this section are due to Albu [**9**].

Section 15.2. This section contains the basic terminology and notation for lattice-isomorphic groups, as well as the presentation without proofs, of some technical results on this material. We have mainly followed Baer [**27**] and Schmidt [**95**]. Theorem 12.1.7 is only mentioned en passant in the introduction of the fundamental paper of Baer [**27**], but it is neither explicitly stated nor proved in the rest of that paper. Surprisingly, we could not find it in other sources, though it provides an important case when lattice-isomorphic groups are isomorphic.

According to a result established by Whitman in 1946 (see e.g., Grätzer [62, Corollary 5, Chap. 4, §4]), every lattice is isomorphic to a sublattice of $\mathbb{L}(G)$ for some group G. The term *lattice-isomorphism* has various other names in the literature: *subgroup-isomorphism* in Baer [27], *projectivity* in Schmidt [95] and Suzuki [101], *L-isomorphism*, etc. The distributive groups are called *ideal-cyclic* groups in Baer [27, p. 8], *D*-groups in Suzuki [101], and *locally cyclic* groups in Kurosh [78].

Section 15.3. The results of this section are due to Albu [10], and Albu and Basarab [15]. Remark 15.3.12 reproduces the completely different approach to the proof of Theorem 15.3.8 which is due to the referee of this book and to whom the author is very thankful. The version in the framework of abstract Cogalois Theory of the group isomorphism given by Theorem 15.3.8 is discussed in Albu and Basarab [17].

Bibliography

[1]   M. ACOSTA DE OROZCO and W.Y. VÉLEZ, *The lattice of subfields of a radical extension*, J. Number Theory **15** (1982), 388-405.

[2]   M. ACOSTA DE OROZCO and W.Y. VÉLEZ, *The torsion group of a field defined by radicals*, J. Number Theory **19** (1984), 283-294.

[3]   T. ALBU, *Kummer extensions with few roots of unity*, J. Number Theory **41** (1992), 322-358.

[4]   T. ALBU, *From the irrationality of sums of radicals to Kneser and G-Cogalois field extensions (I)* (in Romanian), Gazeta Matematică **100** (1995), 421-430.

[5]   T. ALBU, *From the irrationality of sums of radicals to Kneser and G-Cogalois field extensions (II)* (in Romanian), Gazeta Matematică **100** (1995), 611-619.

[6]   T. ALBU, *Field extensions having a Cogalois correspondence - A survey*, in "Proceedings of the Annual Meeting of the Romanian Society of Mathematical Sciences, Bucharest, 1997", Tome 1, Eds. M. Becheanu, I. D. Ion, and A.Vernescu, 1998, pp. 13-22.

[7]   T. ALBU, *Some examples in Cogalois Theory with applications to elementary Field Arithmetic*, J. Algebra Appl. **1** (2002), 1-29.

[8]   T. ALBU, *Infinite field extensions with Cogalois correspondence*, Comm. Algebra **30** (2002), 2335-2353.

[9]   T. ALBU, *Infinite field extensions with Galois-Cogalois correspondence (I)*, Rev. Roumaine Math. Pures Appl. **47** (2002), to appear.

[10] T. ALBU, *Infinite field extensions with Galois-Cogalois correspondence (II)*, Rev. Roumaine Math. Pures Appl. **47** (2002), to appear.

[11] T. ALBU, *Field extensions with the unique subfield property, and G-Cogalois extensions*, Turkish J. Math. **27** (2003), to appear.

[12] T. ALBU, *Corrigendum and Addendum to my paper concerning Kummer extensions with few roots of unity*, J. Number Theory, to appear.

[13] T. ALBU, *On radical field extensions of prime exponent*, J. Algebra Appl. **2** (2003), to appear.

[14] T. ALBU, *Infinite Cogalois Theory, Clifford extensions, and Hopf algebras*, submitted.

[15] T. ALBU and Ş. BASARAB, *Lattice-isomorphic groups, and infinite Abelian G-Cogalois field extensions*, J. Algebra Appl. **1** (2002), to appear.

[16] T. ALBU and Ş. BASARAB, *Toward an abstract Cogalois Theory (I)*, submitted.

[17] T. ALBU and Ş. BASARAB, *Toward an abstract Cogalois Theory (II)*, in preparation.

[18] T. ALBU and C. NĂSTĂSESCU, "Relative Finiteness in Module Theory", Marcel Dekker, Inc., New York and Basel, 1984.

[19] T. ALBU and F. NICOLAE, Kneser field extensions with Cogalois correspondence, J. Number Theory 52 (1995), 299-318.

[20] T. ALBU and F. NICOLAE, G-Cogalois field extensions and primitive elements, in "Symposia Gaussiana", Conference A: Mathematics and Theoretical Physics, Eds. M. Behara, R. Fritsch, and R.G. Lintz, Walter de Gruyter & Co., Berlin New York, 1995, pp. 233-240.

[21] T. ALBU and F. NICOLAE, Heckesche Systeme idealer Zahlen und Knesersche Körpererweiterungen, Acta Arith. 73 (1995), 43-50.

[22] T. ALBU and F. NICOLAE, Finite radical field extensions and crossed homomorphisms, J. Number Theory 60 (1996), 291-309.

[23] T. ALBU, F. NICOLAE, and M. ŢENA, Some remarks on G-Cogalois field extensions, Rev. Roumaine Math. Pures Appl. 41 (1996), 145-153.

[24] T. ALBU and L. PANAITOPOL, Quartic field extensions with no proper intermediate field, Rev. Roumaine Math. Pures Appl. 47 (2002), to appear.

[25] T. ALBU and M. ŢENA, Infinite Cogalois Theory, Mathematical Reports 3 (53) (2001), 105-132.

[26] E. ARTIN, "Galoissche Theorie", B.G. Teubner Verlagsgesellschaft, Leipzig, 1959.

[27] R. BAER, The significance of the system of subgroups for the structure of the group, Amer. J. Math. 61 (1939), 1-44.

[28] A. BAKER and H.M. STARK, On a fundamental inequality in number theory, Ann. of Math. 94 (1971), 190-199.

[29] F. BARRERA-MORA, On subfields of radical extensions, Comm. Algebra 27 (1999), 4641-4649.

[30] F. BARRERA-MORA, M. RZEDOWSKI-CALDERÓN, and G. VILLA-SALVADOR, On Cogalois extensions, J. Pure Appl. Algebra 76 (1991), 1-11.

[31] F. BARRERA-MORA, M. RZEDOWSKI-CALDERÓN, and G. VILLA-SALVADOR, Allowable groups and Cogalois extensions, J. Pure Appl. Algebra 104 (1995), 1-11.

[32] F. BARRERA-MORA and W.Y. VÉLEZ, Some results on radical extensions, J. Algebra 162 (1993), 295-301.

[33] J.R. BASTIDA, "Field Extensions and Galois Theory", Addison-Wesley Publishing Company, Reading, Massachusetts, 1984.

[34] E. BECKER, R. GROBE, and M. NIERMANN, Radicals of binomial ideals, J. Pure Appl. Algebra 117 & 118 (1997), 41-79.

[35] A. BESICOVITCH, On the linear independence of fractional powers of integers, J. London Math. Soc. 15 (1940), 3-6.

[36] G. BIRKHOFF, "Lattice Theory", American Mathematical Society Colloquium Publications, Vol. XXV, 3rd Edition, Providence, 1984.

[37] Z.I. BOREVITCH and I.R. SHAFAREVITCH, "Number Theory", Academic Press, New York, 1966.

[38] N. BOURBAKI, "Topologie Générale", Hermann, Paris, 1966.

[39] N. BOURBAKI, "Théorie Spectrales", Hermann, Paris, 1967.

[40] N. BOURBAKI, "Algèbre", Chapitres 4 à 7, Masson, Paris, 1981.

[41] A. BRANDIS, *Über die multiplikative Struktur von Körpererweiterungen*, Math. Z. **87** (1965), 71-73.

[42] S. CAENEPEEL, *"Brauer Groups, Hopf Algebras, and Galois Theory"*, Kluver Academic Publishers, Dordrecht, 1998.

[43] A. CAPELLI, *Sulla riduttibilità delle equazioni algebriche I*, Rend. Accad. Sci. Fis. Mat. Napoli **3** (1897), 243-252.

[44] A. CAPELLI, *Sulla riduttibilità delle equazioni algebriche II*, Rend. Accad. Sci. Fis. Mat. Napoli **3** (1898), 84-90.

[45] A. CAPELLI, *Sulla riduttibilità della funzione $x^n - A$ in un campo qualunque di razionalità*, Math. Ann. **54** (1901), 602-603.

[46] J.W.S. CASSELS and A. FRÖHLICH, *"Algebraic Number Theory"*, Academic Press, London New York, 1967.

[47] S.U. CHASE and M. SWEEDLER, *"Hopf Algebras and Galois Theory"*, Lecture Notes in Mathematics, Vol. 97, Springer-Verlag, Berlin Heidelberg New York, 1969.

[48] L. CHILDS, *"A Concrete Introduction to Higher Algebra"*, Springer-Verlag, New York, Inc., 1979.

[49] P. CRAWLEY and R.P. DILWORTH, *"Algebraic Theory of Lattices"*, Prentice-Hall, Inc., Englewood Cliffs, New Jersey, 1973.

[50] E.C. DADE, *Compounding Clifford's Theory*, Ann. of Math. **91** (1970), 236-290.

[51] E.C. DADE, *Isomorphisms of Clifford extensions*, Ann. of Math. **92** (1970), 375-433.

[52] E.C. DADE, *Group-graded rings and modules*, Math. Z. **174** (1980), 241-262.

[53] G. DARBI, *Sulla riducibilità delle equazioni algebriche*, Ann. Mat. Pura Appl. (4), **4** (1926), 185-208.

[54] S. DĂSCĂLESCU, C. NĂSTĂSESCU, and Ş. RAIANU, *"Hopf Algebras: An Introduction"*, Marcel Dekker, Inc., New York and Basel, 2001.

[55] D.S. DUMMIT, *On the torsion in quotients of the multiplicative groups in Abelian extensions*, in: "Number Theory", Proceedings of the International Number Theory Conference held at Université Laval, 1987, Eds. J-M. De Koninck and C. Levesque, Walter de Gruyter, Berlin and New York, 1989.

[56] E.E. ENOCHS, J.R. GARCIA ROZAS, and L. OYONARTE, *Compact coGalois groups*, Math. Proc. Cambridge Philos. Soc. **128** (2000), 233-244.

[57] E.E. ENOCHS, J.R. GARCIA ROZAS, and L. OYONARTE, *Covering morphisms*, Comm. Algebra **28** (2000), 3823-3835.

[58] L. GAAL, *"Classical Galois Theory with Examples"*, Markham Publishing Co., Chicago, 1971.

[59] D.J.H. GARLING, *"A Course in Galois Theory"*, Cambridge University Press, Cambridge, 1986.

[60] D. GAY and W.Y. VÉLEZ, *On the degree of the splitting field of an irreducible binomial*, Pacific J. Math. **78** (1978), 117-120.

[61] D. GAY and W.Y. VÉLEZ, *The torsion group of a radical extension*, Pacific J. Math. **92** (1981), 317-327.

[62] G. GRÄTZER, *"General Lattice Theory"*, Birkhäuser Verlag, Basel and Stuttgart, 1978.

[63] C. GREITHER and D.K. HARRISON, *A Galois correspondence for radical extensions of fields*, J. Pure Appl. Algebra **43** (1986), 257-270.

[64] C. GREITHER and B. PAREIGIS, *Hopf Galois theory for separable field extensions*, J. Algebra **106** (1987), 239-258.

[65] F. HALTER-KOCH, *Über Radikalerweiterungen*, Acta Arith. **36** (1980), 43-58.

[66] F. HALTER-KOCH, *Körper, über denen alle algebraischen Erweiterungen der Kummerschen Theorie genügen*, J. Algebra **64** (1980), 391-398.

[67] H. HASSE, *"Zahlentheorie"*, Akademie-Verlag, Berlin, 1963.

[68] H. HASSE, *"Bericht über neuere Untersuchungen und Probleme aus der Theorie der algebraischen Zahlkörper, Teil II: Reziprozitätsgesetz"*, Physica-Verlag, Würzburg Wien, 1965.

[69] H. HASSE, *"Vorlesungen über Klassenkörpertheorie"*, Physica-Verlag, Würzburg Wien, 1967.

[70] E. HECKE, *Eine neue Art von Zetafunktionen und ihre Beziehungen zur Verteilung der Primzahlen (Zweite Mitteilung)*, Math. Z. **4** (1920), 11-51.

[71] E. HECKE, *"Vorlesungen über die Theorie der algebraischen Zahlen"*, Chelsea Publishing Company, New York, 1948.

[72] E. HEWITT and K.A. ROSS, *"Abstract Harmonic Analysis"*, Springer-Verlag, Berlin Göttingen Heidelberg, 1963.

[73] I. KAPLANSKY, *"Infinite Abelian Groups"*, The University of Michigan Press, Ann Arbor, 1971.

[74] I. KAPLANSKY, *"Fields and Rings"*, University of Chicago Press, Chicago, 1972.

[75] L.-C. KAPPE and B. WARREN, *An elementary test for the Galois group of a quartic polynomial*, Amer. Math. Monthly **96** (1989), 133-137.

[76] G. KARPILOVSKY, *"Topics in Field Theory"*, North-Holland, Amsterdam, New York, Oxford, and Tokyo, 1989.

[77] M. KNESER, *Lineare Abhängigkeit von Wurzeln*, Acta Arith. **26** (1975), 307-308.

[78] A.G. KUROSH, *"The Theory of Groups"* (in Russian), Nauka, Moskow, 1967.

[79] P. LAM-ESTRADA, F. BARRERA-MORA, and G.D. VILLA-SALVADOR, *On Kneser extensions*, J. Algebra **201** (1998), 703-717.

[80] S. LANG, *"Algebra"*, Addison-Wesley Publishing Company, Reading, Massachusetts, 1965.

[81] W. MAY, *Multiplicative groups under field extensions*, Canad. J. Math. **31** (1979), 436-440.

[82] S. MONTGOMERY, *"Hopf Algebras and Their Actions on Rings"*, CBMS, Regional Conference Series in Mathematics, Volume 82, American Mathematical Society, Providence, RI, 1993.

[83] L.J. MORDELL, *On the linear independence of algebraic numbers*, Pacific J. Math. **3** (1953), 625-630.

[84] C. NĂSTĂSESCU and F. VAN OYSTAEYEN, *"Graded Ring Theory"*, North Holland Publishing Company, Amsterdam New York Oxford, 1982.

[85] J. NEUKIRCH, *"Algebraische Zahlentheorie"*, Springer-Verlag, Berlin Heidelberg New York, 1992.

[86] M. NORRIS and W.Y. VÉLEZ, *Structure theorems for radical extensions of fields*, Acta Arith. **38** (1980), 111-115.

[87] L.S. PONTRYAGIN, *"Topological Groups"*, Gordon and Breach, New York, London, 1977.

[88] L. RÉDEI, *"Algebra"*, Akademische Verlaggesellschaft, Leipzig, 1959.

[89] P. RIBENBOIM, *"Algebraic Numbers"*, Wiley-Interscience, New York, London, Sydney, and Toronto, 1972.

[90] I. RICHARDS, *An application of Galois theory to elementary arithmetic*, Adv. Math. **13** (1974), 268-273.

[91] A. SCHINZEL, *On linear dependence of roots*, Acta Arith. **28** (1975), 161-175.

[92] A. SCHINZEL, *Abelian binomials, power residues, and exponential congruences*, Acta Arith. **32** (1977), 245-274.

[93] A. SCHINZEL, *"Selected Topics on Polynomials"*, Ann Arbor, The University of Michigan Press, 1982.

[94] A. SCHINZEL, *"Polynomials with Special Regard to Reducibility"*, Cambridge University Press, Cambridge, 2000.

[95] R. SCHMIDT, *"Subgroup Lattices of Groups"*, Walter de Gruyter, Berlin and New York, 1994.

[96] J.-P. SERRE, *"Cohomologie Galoisienne"*, Lecture Notes in Mathematics, Vol. 5, Springer-Verlag, Berlin, Heidelberg, and New York, 1964.

[97] J.-P. SERRE, *"Corps Locaux"*, Hermann, Paris, 1962.

[98] C.L. SIEGEL, *Algebraische Abhängigkeit von Wurzeln*, Acta Arith. **21** (1972), 59-64.

[99] D. ŞTEFAN, *Cogalois extensions via strongly graded fields*, Comm. Algebra **27** (1999), 5687-5702.

[100] B. STENSTRÖM, *"Rings of Quotients"*, Springer-Verlag, Berlin, Heidelberg, and New York, 1975.

[101] M. SUZUKI, *"Structure of a Group and the Structure of its Lattice of Subgroups"*, Ergebnisse der Mathematik und ihrer Grenzgebiete **10**, Springer-Verlag, Berlin, 1956.

[102] M.E. SWEEDLER, *"Hopf Algebras"*, Benjamin, New York, 1969.

[103] N. TCHEBOTARÖW, *"Grundzüge der Galoisschen Theorie"*, Noordhoff, Groningen Djakarta, 1950.

[104] M. ŢENA, *Radical, G-Kneser, and G-Cogalois extensions*, Ph.D. Thesis, Bucharest University, 2000.

[105] M. ŢENA, *Abelian G-Cogalois and cyclotomic field extensions of the rational field*, Bull. Soc. Sci. Math. Roumanie, **42 (90)** (1999), 151-158.

[106] H.D. URSELL, *The degrees of radical extensions*, Canad. Math. Bull. **17** (1974), 615-617.

[107] K.TH. VAHLEN, *Über reductible Binome*, Acta Math. **19** (1895), 195-198.

[108] W.Y. VÉLEZ, *On normal binomials*, Acta Arith. **36** (1980), 113-124.

[109] W.Y. VÉLEZ, *Correction to the paper "Structure theorems for radical extensions of fields"*, Acta Arith. **42** (1983), 427-428.

[110] W.Y. VÉLEZ, *Several results on radical extensions*, Arch. Math. (Basel) **45** (1985), 342-349.

[111] W. WATERHOUSE, *Profinite groups are Galois groups*, Proc. Amer. Math. Soc. **42** (1974), 639-640.

[112] E. WENDT, *Über die Zerlegbarkeit der Function $x^n - a$ in einem beliebigen Körper*, Math. Ann. **53** (1900), 450-456.

[113] J.-P. ZHOU, *On the degree of extensions generated by finitely many algebraic numbers*, J. Number Theory **34** (1990), 133-141.

Index